Results and Problems in Cell Differentiation

49

Series Editors
Dietmar Richter, Henri Tiedge

Holger Kalthoff (ed.)

Death Receptors and Cognate Ligands in Cancer

Springer

Editor
Holger Kalthoff
Sektion für Molekulare Onkologie
Institut für Experimentelle Tumorforschung Krebszentrum Nord - CCC
UK S-H, Campus Kiel Arnold-Heller-Strasse 3 (Haus 18)
24105 Kiel
Germany
hkalthoff@email.uni-kiel.de

Series Editors
Dietmar Richter
Center for Molecular Neurobiology
University Medical Center Hamburg-Eppendorf (UKE)
University of Hamburg
Martinistrasse 52
20246 Hamburg
Germany
richter@uke.uni-hamburg.de

Henri Tiedge
The Robert F. Furchgott Center for Neural and Behavioral Science
Department of Physiology and Pharmacology
Department of Neurology
SUNY Health Science Center at Brooklyn
Brooklyn, New York 11203
USA
htiedge@downstate.edu

ISBN 978-3-642-03044-4 e-ISBN 978-3-642-03045-1
DOI 10.1007/978-3-642-03045-1
Springer Heidelberg Dordrecht London New York

Results and Problems in Cell Differentiation ISSN 0080-1844

Library of Congress Control Number: 2009932686

Cover design: WMXDesign GmbH, Heidelberg

Printed on acid-free paper

Springer is part of Springer Science+Business Media (www.springer.com)

Contents

Introduction[1]

Following the Death TRAIL to Hunt Down Tumor Cells: Translating Programmed Cell Death Signaling Mechanisms into Clinical Practice

Since the discovery of TNFα (tumor necrosis factor alpha) in 1975 as a factor that induces the necrosis of tumors (Carswell et al., 1975), more than 100, 000 articles describing TNF functions have been published. TNF acts through binding to its receptors TNF-R1 and TNF-R2 (Wajant et al., 2003). On the basis of the sequence homology, several related ligands and receptors have been identified and classified as members of the TNF ligand- or TNF receptor-superfamily (Hehlgans and Pfeffer, 2005).

Within the TNFR-superfamily, seven receptors display a highly conserved so-called death domain and are able to induce cell death (Wajant et al., 2003). These proteins; TNF-R1 (tumor necrosis factor alpha-receptor1), CD95/Fas, TRAIL-R1 (TNF-related apoposis-inducing ligand-receptor 1), TRAIL-R2, DR6, DR3, EDAR (ectodysplasin A receptor), and p75NGFR (low affinity nerve growth factor receptor) are thus called death receptors. Within this group of receptors, the best characterized are TNF-R1, CD95, TRAIL-R1, and TRAIL-R2. For these receptors, it has been shown that ligand binding or antibody mediated aggregation leads to the activation of caspases. These proteases govern the tightly controlled course of cell destruction producing at the end of the process membrane enclosed apoptotic bodies that are removed in the tissue by phagocyting neighbouring cells.

The induction of cell death is the most studied and well-characterized function of death receptors. Recently, it has been shown that death receptors, following their stimulation with the respective death ligands, are also able to induce a number of nonapoptotic signaling pathways such as MAP-kinases (JNK, p38, ERK1/ERK2), PKC and NF-kappaB (Falschlehner et al., 2007; Wajant et al., 2003; Park et al.,

[1] Partially overlapping descriptions have not been omitted from the various chapters in which the individual authors have been asked to write their own overview in a conclusive manner. The editor feels it is worthy for the reader to assess individual discussion and opinions from different experts, particularly when the field is rapidly advancing and thus still somewhat controversial.

Major parts and groundbreaking knowledge in the field of death-ligands and their cognate receptors have been gained from the field of basic immunology. Thus in some articles, for example Richard Siegels' group (Ramaswamy et al., this volume) on Fas Ligand, the contributions of immunological data on the specific topics have been elaborated despite the major focus of the entire book on oncology.

2005; Trauzold et al., 2001; Trauzold et al., 2005; Trauzold et al., 2006; Siegmund et al., 2007). The balance between the activation of apoptotic and nonapoptotic signal transduction pathways determines the outcome of the death receptor stimulation.

Thus, depending on the cellular context, triggering of the death receptors may result in apoptosis, proliferation, differentiation, as well as in the secretion of proinflammatory and invasion-promoting proteins. Differences in the cellular systems need to be considered definitively: early strategies to explore TNF regulation and function in Jurkat cells have been very powerful, but then extending this knowledge to all other cell types has sometimes been hampered by dogmatic interpretations. An insightful illustration on the wide reaching impact of cellular context is given by the different regulation of CD95 ligand by different sets of transcription factors with respect to lymphoid and nonlymphoid cells (see Ramaswamy et al., *this volume).*

The repertoire of complex biological processes needs to be stringently kept under control: the regulation of mRNA- and protein expression is an obvious example here, but the regulation of protein processing, secondary modification, and sub-/or extracellular localization is of utmost importance as well. Post-transcriptional protein modification like phosphorylation, ubiquitination, and proteolytic cleavage are inseparably associated with both apoptotic and nonapoptotic signaling. Distinct ubiquitin modifications govern the fate of ubiquitinated substrates. Thus, ubiquitinated proteins may be directed for degradation or change their signaling properties. Since ubiquitination regulates nearly all steps of the signal transduction of death receptors, we have introduced an additional section (Fujita & Srinivasula, *this volume*) discussing the mechanisms and role of protein ubiquitination in TNF signaling.

An important and only recently discovered aspect of TNFR1 and CD95 signaling is the impact of their internalization in the apoptotic and nonapoptotic signaling. Schütze and Schneider-Brachert (*this volume*) show that immediately following ligand binding, a signaling complex that activates antiapoptotic signaling is formed at the plasma membrane. In contrast, proapopotic signal transduction starts only after the receptor internalization by endocytosis and intracellular formation of so-called death inducing signaling complex (DISC). Thus, receptor compartmentalization separates both signal transduction pathways and provides an additional regulatory platform.

In addition to the activation of their cognate receptors, for some of the TNF-family members, the capability to induce retrograde signal transduction has been reported. Moreover, CD95L has been shown to act as a positive and negative costimulatory molecule in T-cell activation. Therefore, an extension of this retrograde concept suggests that following ligand binding to the death receptor, both interacting cells; the ligand donor cell, and the acceptor cell, respond with the activation of diverse signal transduction pathways. The role of CD95L in immune regulation, its expression, the interacting proteins, as well as the issue of retrograde signaling is reviewed by Lettau et al. (*this volume*).

As death receptors can (sometimes effectively) kill tumor cells, the use of death ligands for cancer treatment represents an attractive therapeutic option.

Unfortunately, neither TNF nor CD95L are able to be safely used for treatment of patients unless used under specialized conditions like loco-regional limb infusion for example. In the case of TNF, it has been shown that it may induce tumor progression and systemic application leads to septic shock depending on the dose. Interestingly, slight stimulation of tumor growth may be a therapeutic option if combined with a powerful (proliferation-dependent) antitumor drug. This has been demonstrated by us previously, when we successfully treated pancreatic tumor patients with a combination of TNF and anti-EGF Receptor antibodies (Schmiegel et al., 1997).

More recently, we have shown in a clinically adapted mouse model that exogenous TNF strongly enhances pancreatic tumor growth and metastasis. We also demonstrated that tumor recurrence and liver metastasis after surgical resection of pancreatic ductal adenocarcinoma (PDAC) are substantially driven by tumor cell-derived TNF. The generally well-accepted context of inflammation and cancer is clearly applicable to PDAC, since other inflammatory stimuli like IL1ß, IL6, FasL, and TRAIL have also been found to activate these tumor cells in an autocrine manner. Moreover, regarding the characteristic (and sometimes massive) peri- and intratumoral infiltration of human tumors by inflammatory cells, the issue of inflammation driven tumor progression gains impact. Inhibition of endogenous (i.e. pancreatic tumor cell) TNF was recently shown by us to exhibit a very positive therapeutic effect, particularly in the adjuvant setting, and remarkably without any additional chemo- or radiation therapy (Egberts et al., 2008).

In mice, CD95L application results in liver failure caused by hepatocyte apoptosis, followed by death of the animals shortly after ligand application.

This problem fostered technological achievements by designing recombinant variants of CD95L (as well as other death ligands), which specifically target only malignant cells (see Gerspach, Wajant, and Pfizenmaier, *this volume*). As an outlook for therapeutic intervention strategies based on the enormous amount of data on death ligands and their cognate receptors in tumor targeting, a dual (context-dependent) approach may be designed by activating only those death receptors which are colocalized with other tumor-associated receptors or surface molecules.

Clearly, TRAIL has attracted considerable attention for its potential use in tumor therapy, as it induces apoptosis in tumor cells without harming normal healthy cells (Cordier, Papenfuss, and Walczak, *this volume*).

Currently, the recombinant ligand as well as agonistic anti-TRAIL-R1 and anti-TRAIL-R2 antibodies are being tested in the treatment of different malignancies in clinical studies (Newsom-Davis et al., 2009; Bellail et al., 2009). On the other hand, we have shown that TRAIL, similar to CD95L, strongly induces the expression of inflammation and invasiveness-promoting proteins such as IL-8, MCP-1, and uPA and enhances the invasion of apoptosis resistant PDAC cells in vitro. Most importantly, we demonstrated that TRAIL-treatment of pancreatic tumor-bearing mice results in dramatically increased liver metastasis and peritoneal carcinomatosis (Trauzold et al., 2006). Thus, TRAIL is able to stimulate the entire complex metastatic cascade of solid tumors under in vivo conditions. Very recently, in vitro promotion of migration and invasion was reported for apoptosis-resistant cholangiocarcinoma cells and was explained to be a consequence of TRAIL-induced activation of NF-kappaB (Ishimura et al., 2006). Moreover, a proinflammatory, growth stimulating as well as proangiogenic activity of TRAIL was also demonstrated (Beguc et al., 2006; Li et al., 2003; Morel et al., 2005; Secchiero et al., 2004). Similar to TRAIL, agonistic anti-TRAIL-R antibodies are also able to strongly induce nonapoptotic signaling at least in PDAC cells (our unpublished data). Interestingly, apoptosis induction by TRAIL as well as by agonistic, TRAIL-receptor specific antibodies can be significantly enhanced by a variety of established tumor treatments especially those that target NF-kappaB pathways. Such combinatorial therapy could not only sensitize for apoptosis induction, but in addition, block nonapoptotic TRAIL signaling pathways and thus diminish the unwanted tumor promoting therapy induced side effects. TRAIL induced signaling as well as an overview of preclinical and clinical studies with TRAIL and TRAIL-receptor agonists is described in two chapters within this book because of the great impact TRAIL-R targeting is assumed to have in the future of tumor therapy. Yet, the "legend" about TRAIL as being specific for tumor cell killing is still "open for discussion": Why should evolution provide us with a tumor-specific drug? Is it not much more likely that TRAIL fulfills multiple (still mostly unknown) physiological functions that go "wrong" (i.e. killing as a side effect) in cancer cells? Why have humans developed two TRAIL receptors with the "license to kill" and how good (representative) then are the "real" murine models?

In addition to potentially protumoral activity of TRAIL, recent work of Todaro et al. (2008) shows that most primary tumor cells are resistant to TRAIL. However, many of the resistant tumor cells can be efficiently killed by a combination of TRAIL and particularly irradiation or chemotherapy as described by Niemöller and Belka (*this volume*).

The less well-studied member of the TNF superfamily, APRIL, fulfils very diverse cellular functions and provides us with insights into the biology of death ligands that is likely to be elucidated in the future and opening a new horizon into death ligand/receptor research. Specifically, it is demonstrated that the mode of application has a great impact, since binding of APRIL to HSPG modulates activity. Moreover, humans exhibit a much higher complexity in splice forms of APRIL compared to mice. Such mechanisms need to be understood in greater detail before we will be able to specifically inhibit APRIL – but not the closely related and physiologically wanted BAFF activity. Such challenging tasks are discussed by Kimberly, Medema, and Hahne (*this volume*) and can be regarded as representative for the entire field.

Two other interesting members of the TNF/TNF-Receptor-superfamily are TWEAK and Fn14 since the respective ligand-receptor activity is frequently elevated in inflammatory as well as malignant lesions. The issues of cellular context and receptor crosstalk show up here again as it was very recently shown that targeting of Fn14 on tumor cells may preferentially eliminate malignant cells. Yet, Fn14 lacks a death domain, whereas the short cytoplasmic domain is actively recruiting TRAF adaptor proteins (Michaelson and Burkly, *this volume*).

A generally important issue is the timing of therapeutically administered drugs, and this holds true even for "conventional drugs" such as the well-known glucocorticosteroids, particularly in combination regimens with chemotherapeutics (see Herr et al., *this volume*). Furthermore, glucocorticosteroids are a clear example of the great impact that cellular differentiation, micromilieu, and context has with regard to drug activity (induction of or prevention of apoptosis). This aspect gains importance in the context of epithelial cells, which are (even in the transformed state) mostly interconnected, e.g., by gap junctions and also receive signals from extracellular membrane components. Consequently, it clearly matters whether one tries to target an epithelial cell disseminated somewhere in the body (or the tissue culture flask), being positioned ectopically in another niche or an epithelial layer, or as an epithelial 3D-complex structure. Striking examples of such different susceptibilities have been described previously by Bernstorff et al. (2002) for CD95L-mediated apoptosis, and by the group of Walczak for TRAIL susceptibility in this book (see *this volume*).

From a therapeutic point of view and regarding the perspectives for the future, a challenging task would be not only to understand and dissect the multiple steps of apoptosis and other forms of cell death, e.g., autophagy, at a molecular level, but also to be able to reassemble them in a dedicated fashion for newly designed treatment strategies against resistant cancer cells. A solid base for such a far reaching goal is currently provided by the experts focusing on the specific signaling mechanisms resulting in autophagic cell death (see: Yousefi and Simon, *this volume*). The better we design our killing strategies the more we might be able to avoid treatment side effects and particularly to overcome possible escape mechanisms. It can be hypothesized, that the early induction of multiple forms of cell death in tumors may result in better therapeutic outcome then a step-wise procedure (as mostly performed currently in clinical practice), which is likely to give the tumor cells a better chance to accumulate novel strength and resistance mechanisms.

A prototype example for such a designed killing strategy may be given by irradiation-mediated shifting of type-II to type-I apoptosis signaling, eventually circumventing mitochondrial resistance mechanisms due to increased expression of antiapoptotic members of BcL2-family, typically seen in many malignant cells (see: Niemöller and Belka, *this volume*). We still know too little about the timing and doses of "combined treatment" regimens including TRAIL and other "new (therapeutic) kids on the block." The complex combined treatment regiments are also likely to yield new combined side effects (such as elevation of endogenous CD95L by irradiation). When we start to combine "new biologicals" with "classical chemotherapeutics" plus irradiation, novel aspects of cellular behavior are likely to emerge.

The complexity and multiplicity of crosstalk signaling are as of yet poorly elaborated. This will be the great task of system biology approaches.

Acknowledgment The authors sincerely thank Dr. Sanjay Tiwari for critical reading of the manuscript. Their own work has been funded by the DFG within the SFB415 (project A3).

References

Begue B, Wajant H, Bambou JC, Dubuquoy L, Siegmund D, Beaulieu JF, Canioni D, Berrebi D, Brousse N, Desreumaux P, Schmitz J, Lentze MJ, Goulet O, Cerf-Bensussan N, Ruemmele FM (2006) Implication of TNF-related apoptosis-inducing ligand in inflammatory intestinal epithelial lesions. Gastroenterology 130:1962–1974

Bellail AC, Qi L, Mulligan P, Chhabra V, Hao C (2009) TRAIL agonists on clinical trials for cancer therapy: the promises and the challenges. Rev Recent Clin Trials 4:34–41

Bernstorff WV, Glickman JN, Odze RD, Farraye FA, Joo HG, Goedegebuure PS, Eberlein TJ (2002) Fas (CD95/APO-1) and fas ligand expression in normal pancreas and pancreatic tumors. Implications for immune privilege and immune escape. Cancer 94:2552–2560

Carswell EA, Old LJ, Kassel RL, Green S, Fiore N, Williamson B (1975) An endotoxin-induced serum factor that causes necrosis of tumors. Proc Natl Acad Sci U S A 72:3666–3670

Egberts JH, Cloosters V, Noack A, Schniewind B, Thon L, Klose S, Kettler B, von Forstner C, Kneitz C, Tepel J, Adam D, Wajant H, Kalthoff H, Trauzold A (2008) Anti-tumor necrosis factor therapy inhibits pancreatic tumor growth and metastasis. Cancer Res 68:1443–1450

Falschlehner C, Emmerich CH, Gerlach B, Walczak H (2007) TRAIL signalling: decisions between life and death. Int J Biochem Cell Biol 39:1462–1475

Hehlgans T, Pfeffer K (2005) The intriguing biology of the tumour necrosis factor/tumour necrosis factor receptor superfamily: players, rules and the games. Immunology 115:1–20

Ishimura N, Isomoto H, Bronk SF, Gores GJ (2006) Trail induces cell migration and invasion in apoptosis-resistant cholangiocarcinoma cells. Am J Physiol Gastrointest Liver Physiol 290:G129–1236

Li JH, Kirkiles-Smith NC, McNiff JM, Pober JS (2003) TRAIL induces apoptosis and inflammatory gene expression in human endothelial cells. J Immunol 171:1526–1533

Morel J, Audo R, Hahne M, Combe B (2005) Tumor necrosis factor-related apoptosis-inducing ligand (TRAIL) induces rheumatoid arthritis synovial fibroblast proliferation through mitogen-activated protein kinases and phosphatidylinositol 3-kinase/Akt. J Biol Chem 280: 15709–15718

Newsom-Davis T, Prieske S, Walczak H (2009) Is TRAIL the holy grail of cancer therapy? Apoptosis 14:607–623

Park SM, Schickel R, Peter ME. (2005) Nonapoptotic functions of FADD-binding death receptors and their signaling molecules. Curr Opin Cell Biol 17:610–616

Schmiegel W, Schmielau J, Henne-Bruns D, Juhl H, Roeder C, Buggisch P, Onur A, Kremer B, Kalthoff H, Jensen EV (1997) Cytokine-mediated enhancement of epidermal growth factor receptor expression provides an immunological approach to the therapy of pancreatic cancer. Proc Natl Acad Sci U S A 94:12622–12626

Secchiero P, Gonelli A, Carnevale E, Corallini F, Rizzardi C, Zacchigna S, Melato M, Zauli G (2004) Evidence for a proangiogenic activity of TNF-related apoptosis-inducing ligand. Neoplasia 6:364–373

Siegmund D, Klose S, Zhou D, Baumann B, Röder C, Kalthoff H, Wajant H, Trauzold A. (2007) Role of caspases in CD95L- and TRAIL-induced non-apoptotic signalling in pancreatic tumour cells. Cell Signal 19: 1172–1184

Todaro M, Lombardo Y, Francipane MG, Alea MP, Cammareri P, Iovino F, Di Stefano AB, Di Bernardo C, Agrusa A, Condorelli G, Walczak H, Stassi G (2008) Apoptosis resistance in epithelial tumors is mediated by tumor-cell-derived interleukin-4. Cell Death Differ 15:762–772

Trauzold A, Wermann H, Arlt A, Schutze S, Schafer H, Oestern S, Roder C, Ungefroren H, Lampe E, Heinrich M, Walczak H, Kalthoff H (2001) CD95 and TRAIL receptor-mediated activation of protein kinase C and NF-kappaB contributes to apoptosis resistance in ductal pancreatic adenocarcinoma cells. Oncogene 20:4258–4269

Trauzold A, Röder C, Sipos B, Karsten K, Arlt A, Jiang P, Martin-Subero J-I, Siegmund D, Müerköster S, Siebert R, Wajant H, Kalthoff H (2005) CD95 and TRAF2 promote invasiveness of pancreatic cancer cells. Faseb J 6:620–622

Trauzold A, Siegmund D, Schniewind B, Sipos B, Egberts J, Zorenkov D, Emme D, Roder C, Kalthoff H, Wajant H (2006) TRAIL promotes metastasis of human pancreatic ductal adenocarcinoma. Oncogene 25:7434–7439

Wajant H, Pfizenmaier K, Scheurich P (2003) Tumor necrosis factor signaling. Cell Death Differ 10:45–65

Kiel, June 2009

Holger Kalthoff (ed.)
Anna Trauzold

The Role of TNF in Cancer

Harald Wajant

Abstract Tumor necrosis factor (TNF) is an extraordinarily pleiotropic cytokine with a central role in immune homeostasis, inflammation, and host defense. Dependent on the cellular context, it can induce such diverse effects as apoptosis, necrosis, angiogenesis, immune cell activation, differentiation, and cell migration. These processes are of great relevance in tumor immune surveillance, and also play crucial roles in tumor development and tumor progression. It is therefore no surprise that TNF in a context-dependent manner displays pro- and antitumoral effects. Modulation of the activity of the TNF–TNF receptor system thus offers manifold possibilities for cancer therapy. In fact, TNF in combination with melphalan is already an established treatment option in the therapy of advanced soft tissue sarcoma of the extremities and many preclinical data suggest that TNF neutralization could also be exploited to fight cancer or cancer-associated complications.

Abbreviations AOM: Azoxymethane; APC: Antigen presenting cell; COX-2: Cyclo-oxygenase-2; cIAP1/2: Cellular inhibitor of apoptosis protein-1/2; DMBA: 7,12-di-methylbenz[α]-anthracene; DSS: Dextran sulfate sodium salt; EMT: Epithelial–mesenchymal transition; ERK: Extracellular-regulated kinase; FADD: Fas-associated death domain; GSK3: βGlycogen synthase kinase; ILP: Isolated limb perfusion; JNK: cJun N-terminal kinase; LMCV: Lymphocytic choriomeningitis virus; MCA: 3´-Methylcholanthrene; MCP1: Monocyte chemoattractant protein-1; MDR2: Multidrug resistance p-glycoprotein 2; MMP-9: Matrix metalloprotease-9; OA: Ocadaic acid; PDAC: Pancreatic ductal adenocarcinoma; p38 MAPK: p38 mitogen-activated protein (MAP) kinase; RANK: Receptor activator of NF-kappaB; TPA: 12-0-Tetradecanoyl-phorbol-13-acetate; TRPV1; Transient receptor potential channel vanilloid type 1; VEGFR2: Vascular endothelial growth factor (VEGF) receptor

H. Wajant
Department of Molecular Medicine,
Medical Clinic and Polyclinic II, University of Würzburg,
Röntgenring 11, 97070, Wuerzburg, Germany
e-mail: harald.wajant@mail.uni-wuerzburg.de

Results Probl Cell Differ, doi:10.1007/400_2008_26

1 Introduction

The prime sources of TNF are activated immune cells, especially macrophages and T-cells, but it can also be produced by a variety of other cell types including fibroblasts and tumor cells (Wajant et al. 2003). TNF is a trimeric type II transmembrane protein consisting of about 80 amino acid residues comprising proline-rich cytoplasmic domain involved in membrane trafficking and receptor binding-induced reverse signaling, a single transmembrane domain (TM) and an extracellular domain containing the characteristic TNF homology domain (THD), which is separated from the TM by a stalk region (Bodmer et al. 2002). The THD mediates trimerization of the molecule and is also responsible for receptor binding. The stalk region contains a processing site for the matrix metalloprotease TNFα converting enzyme (TACE)/ADAM17, allowing the release of a soluble trimeric form of TNF (Fig. 1). Transmembrane TNF, as well as soluble TNF, interacts with

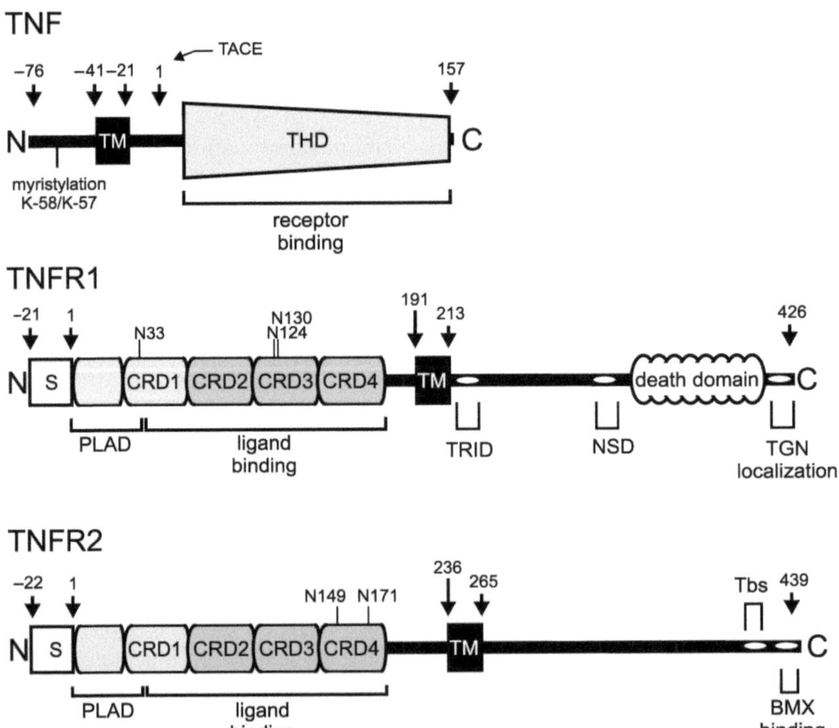

Fig. 1 Human TNF, TNFR1, and TNFR2. Amino acid numbering refers to the mature receptors and processed soluble TNF, respectively. Myristylated lysine residues in TNF and glycosylated asparagine residues of TNFR1 and TNFR2 are indicated. *S* signal peptide, *CRD* cysteine rich domain, *NSM* neutral sphingomyelinase activating domain, *PLAD* pre-ligand binding assembly domain, *Tbs* TRAF2 binding site, *TGN* trans-Golgi network, *THD* TNF homology domain, *TM* transmembrane domain, *TRID* TNFR1 internalization domain. For details see text

two distinct receptors, TNFR1 and TNFR2, both belonging to the TNF receptor superfamily (Wajant et al. 2003). Importantly, binding of transmembrane TNF results in strong activation of each of the two TNF receptors, while binding of soluble TNF triggers only TNFR1 signaling (Grell et al. 1995). TNFR1 and TNFR2 belong to different subgroups of the TNF receptor superfamily. TNFR1 contains a death domain (DD) in the cytoplasmic part and interacts by virtue of this module with cytoplasmic death domain-containing adapter proteins (Fig. 1). In contrast, TNFR2 lacks a death domain and is instead able to directly interact with adapter proteins of the TNF receptor associated factor (TRAF) family by a short TRAF binding site (Bodmer et al. 2002; Wajant et al. 2003). Notably, TNFR1 also exploits TRAF adapter proteins for signal transduction, but does not directly interact with these molecules (Fig. 2). TNF binding to TNFR1 results in the recruitment of the death domain-containing adapter protein TRADD (TNF receptor associated death domain) and the death domain-containing serine–threonine kinase RIP (receptor interacting protein). The latter together with TRAF2, which is indirectly recruited in the TNFR1 signaling complex due to its association with TRADD, recruit and activate the inhibitor of IκB kinase (IKK) complex, the crucial bottle neck for activation of the classical NFκB pathway (Perkins 2007; Wajant et al. 2003). TRAF2 and RIP also mediate activation of the cJun N-terminal kinase (JNK) and p38 mitogen-activated protein (MAP) kinase cascades. TRAF2 strongly binds to cIAP1 (cellular inhibitor of apoptosis protein-1) and cIAP2 and therefore also directs these caspase inhibitory E3 ligases into the TNFR1 signaling complex (Wang et al. 1998; Fig. 2). There is evidence that these proteins fulfill two distinct functions: first, they contribute to NFκB activation by ubiquitination of RIP and second they prevent activation of caspase-8 and thus prevent TNFR1-induced apoptosis (Wang et al. 1998).

By help of its death domain TNFR1 also induces, by less understood mechanisms, the activation of acidic sphingomyelinase and the extracellular-regulated kinase (ERK) signaling pathway (Schwandner et al. 1998). By death domain-independent mechanisms, TNFR1 is also linked to the stimulation of neutral sphingomyelinase and again the ERK signaling pathway (Adam-Klages et al. 1996). TNFR1 can also induce cell death by two distinct pathways, both emerging from its death domain. First, TNFR1 induce necrosis, which is mediated by RIP- and TRAF2-driven excessive generation of reactive oxygen species and subsequent prolonged JNK signaling (Lin et al. 2004). Second, TNFR1 induce caspase-mediated apoptosis, which involves the death domain-containing adapter proteins TRADD and FADD (Fas-associated death doamin) and also the FADD-associated initiator caspase, caspase-8 (Wajant et al. 2003). Notably, there is evidence that the pro-apoptotic interplay of TRADD, FADD, and caspase-8 takes place in a secondary cytoplasmic multiprotein complex that is formed after release from TNFR1. This complex might also contain TRAF2, RIP, and the TRAF-associated IAP proteins cIAP1 and cIAP2 and might also recruit the caspase-8 inhibitory FLIP protein. Dependent from the presence of the inhibitory FLIP and IAP proteins, the cytoplasmic complex is able to process and activate caspase-8 and apoptotic cell death (Muppidi et al. 2004). There is also evidence that TNFR1-induced apoptosis in contrast to

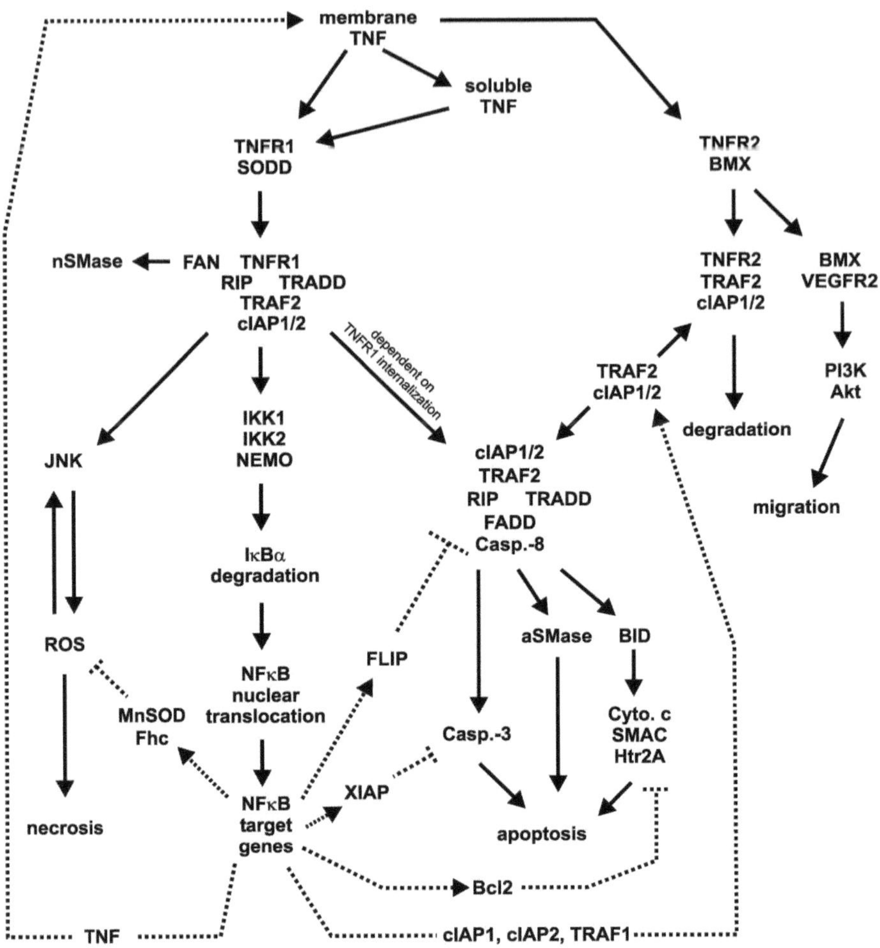

Fig. 2 The TNF signaling network. Dotted lines refer to TNF receptor-induced events requiring protein synthesis. For details see text

TNFR1-induced NFκB activation depend on internalization of the receptor signaling complex. How the later finding match together with the formation of the proapoptotic secondary complex is currently not clearly understood (Schütze et al. 2008). As caspase-8 can cleave RIP, the apoptotic pathway actively represses necrosis induction (Lin et al. 1999). Consequently, TNF-induced necrosis become especially apparent in cells treated with pharmacological inhibitors of caspases or cells that for other reasons are resistant against caspase-mediated apoptosis. TNFR2 can also activate the classical NFκB signaling pathway and the various MAP kinase cascades, but the underlying signaling mechanisms are mainly unknown (Wajant et al. 2003). Activated TNFR2 interacts with TRAF2, which secondarily recruits

TRAF1, cIAP1, and cIAP2 into the TNFR2 signaling complex (Fig. 2), but the relevance of these components of the TNFR2 complex for activation of the aforementioned pathways has not been clarified so far (Rothe et al. 1994, 1995). The TNFR2-TRAF2 interaction came along with translocation into lipid rafts and depletion of cytosolic TRAF2 pools (Fotin-Mleczek et al. 2002). The latter is enhanced with time by lipid raft-associated proteasomal degradation of TRAF2 (Fotin-Mleczek et al. 2002; Li et al. 2002; Wu et al. 2005). In accordance with the crucial role of TRAF2 and the associated IAP proteins in preventing TNFR1-induced caspase-8 activation, TNFR2-induced depletion and degradation of TRAF2 result in strong enhancement of TNFR1-induced apoptosis (Weiss et al. 1997; Chan and Lenardo 2000; Fotin-Mleczek et al. 2002). TNFR2 further constitutively associates with the tyrosine kinase BMX (Pan et al. 2002). TNFR2 stimulation results in interaction of BMX with the receptor tyrosine kinase VEGFR2 (vascular endothelial growth factor (VEGF) receptor-2) and reciprocal transphosphorylation of the two molecules (Zhang et al. 2003). Phosphorylated BMX than serves as a docking site for the p85 subunit of PI3 kinase, leading to the activation of the Akt signaling pathway. This way TNFR2 can mediate proliferation and migration of endothelial cells (Pan et al., 2002; Zhang et al., 2003).

2 Antitumoral Effects of TNF

The literature on TNF is clearly dominated by its pleiotropic proinflammatory functions and its crucial role in autoimmune pathologies such as rheumatoid arthritis, Crohn's disease, and psoriasis. As it is already evident from its name, however, TNF has been originally identified as a tumor necrosis inducing factor (Wang et al. 1985). The observation that microbial challenge or treatment with microbial compounds, such as LPS, lead to the protection of mice against experimental cancer resulted in the identification of TNF as the main responsible factor. Notably, TNF has been independently identified as cachectin, a proinflammatory factor that mediates cancer-associated fatigue and muscle wasting (Beutler et al. 1985). Early efforts to use recombinant TNF in tumor therapy, prompted by the before mentioned findings, failed, however, due to the severe inflammatory side effects associated with systemic TNF receptor activation. Nevertheless, safe local administration of TNF in combination with melphalan by isolated limb perfusion (ILP) is now an established treatment option for locally advanced soft limb sarcomas, and a variety of preclinical (van Horssen et al. 2006) and clinical studies are underway aiming antibodies to restrict TNF activity to the tumor area or to inhibit the therapy limiting side effects of TNF without affecting its antitumor properties (see chapter by Gerspach et al). Notably, it turned out that the before mentioned antitumoral effects of TNF in ILP are not caused by its capability to trigger apoptotic and necrotic signaling pathways, but is instead rather an indirect consequence of its inflammation-related capability to regulate endothelial permeability. Thus, TNF induces hyperpermeability in the tumor-associated vessels facilitating tumor entry of blood cells, and also enhances accumulation of melphalan, together yielding a strong antitumoral effect (van Horssen et al. 2006).

A more "natural" antitumoral role of TNF has been further observed in mice expressing the p53 and retinoblastoma protein inhibitory SV40 large T antigen under control of the rat insulin promoter (RIP-Tag2 mice), causing multistage carcinogenesis (Müller-Hermelink et al. 2008). Transfer of in vitro activated T antigen-specific Th1 $CD4^+$ cells in RIP-Tag2 mice enhanced their survival time and inhibited tumor growth and angiogenesis in a TNF- and IFNγ-dependent manner. Notably, this inhibitory effect was not related to apoptosis induction. When either IFNγ action was blocked by antibodies or TNFR1 activation were prevented by crossing RIP-Tag2 mice with TNFR1 knockout mice, the Tag-specific Th1 cells enhanced tumor development (Müller-Hermelink et al. 2008). In a variation of this model that is based on the use of RIP-Tag2 mice that has been crossed with mice carrying the lymphocytic choriomeningitis virus (LMCV) glycoprotein (GP) under control of the RIP promoter, there was further evidence for a role of $CD8^+$ T-cells in tumor control (Calzascia et al. 2007). Adoptively transferred GP-specific $CD8^+$ T-cells showed significant reduced proliferation after recovering from pancreatic draining lymphe nodes when they lack TNF or TNFR2 expression, but behaved normal when TNFR1 was absent. Thus, TNFR2-mediated costimulation seems to be important in this model to ensure proper activation of tumor antigen-specific $CD8^+$ T cells. On the other site, TNFR1 expression in the host cell was also necessary in this study to reach optimal activation of the adoptively transferred GP-specific $CD8^+$ T-cells pointing to an additional role of TNF in this model related to TNFR1 signaling in antigen presenting cells (APC; Calzascia et al., 2007).

3 Pro-Tumoral Functions of TNF

3.1 *TNF and Skin Carcinogenesis*

First genetic evidence pointing to a role of TNF in malignancy was obtained by analysis of experimental skin carcinogenesis. Mice of different genetic background that were sequentially treated with a single dose of the carcinogen 7,12-di-methylbenz[α]-anthracene (DMBA) and 2–3 months with 12-0-tetradecanoyl-phorbol-13-acetate (TPA) or ocadaic acid (OA) develop in up to 100% of challenged mice papillomas. In this two step model of skin carcinogenesis, DMBA acts as a tumor initiator causing genetic alterations, while TPA or OA promote tumor development by facilitating the dominant growth of cells bearing appropriate mutations. In TNF-deficient mice or mice treated with a TNF-neutralizing monoclonal antibody, the DMBA/TPA model of skin carcinogenesis, however, results in reduced papilloma development and diminished papilloma numbers per mouse (Moore et al. 1999; Suganuma et al. 1999). While analysis of DNA-adduct formation delivered no evidence for a role of TNF in DMBA-induced tumor initiation, TPA-induced activation of PKCα and AP1 are reduced in TNF-deficient mice, pointing to role in tumor promotion (Arnott et al. 2002). Analysis of TNFR1 and TNFR2 knockout mice further revealed that both TNF receptors contribute to DMBA/TPA-induced carcinogenesis with a more important role of TNFR1 (Arnott et al. 2004). TNF is upregulated in the epidermis within hours after TPA treatment, and TNF-deficient mice show much lower infiltration of neutrophils and eosinophils after TPA treatment than wild-type mice. TPA-induced TNF production, epidermal hyperplasia, and leukocyte infiltration are blocked in skin derived from transgenic mice expressing the NFκB inhibitory Smad7 protein (Hong et al. 2007). In accordance with evidence suggesting that clonal expansion of mutated follicular stem cells underlies carcinogenesis in the DMBA/TPA model (Binder et al. 1997), transcription of matrix metalloprotease-9 (MMP-9) is induced in follicular epithelial cells early in TPA promotion and after repeated TPA treatments also in interfollicular keratinocytes (Scott et al. 2004). Notably, migration of keratinocytes in vitro is dependent on endogenous TNF and MMP-9 (Scott et al. 2004). Thus, the tumor promoting activity of TPA considerably relies on induction of TNF and TNF-dependent pro-inflammatory events. Transformation and anchorage-independent growth of the epidermal cell line CI41 by benzo[α]pyrene-7,8-diol-9,10-epoxide, a carcinogenic metabolite of benzo[α]pyrene, which has been implicated in smoking-related cancer, also base significantly on TNF induction (Ouyang et al. 2007). However, TNF not necessarily has an obligate role in carcinogenesis. In fact, 3′-methylcholanthrene (MCA)-triggered formation of fibrosarcoma has been found to be significantly enhanced in TNF-deficient mice (Swann et al. 2007). The qualitatively contrasting effects of TNF in the DMBA/TPA and MCA model of carcinogenesis might be related to the strength of the associated inflammatory processes. While the DMBA/TPA model is strongly associated with inflammation, development of MCA-related tumors has only a minor inflammatory component.

3.2 *TNF and Hepatic Carcinogenesis*

Compensatory hepatocyte proliferation together with suppression of apoptosis occurs regularly after liver injury by non-genotoxic compounds and can induce preneoplastic alterations that might ultimately result in tumor formation. The injury-induced regenerative response includes expansion of a progenitor cell compartment called oval cells that become either hepatocytes or biliary epithelial cells. Upon liver damage induced with a choline-deficient ethionine-supplemented (CDE) diet, TNF can be found in oval cells and infiltrated leukocytes. More relevant, TNFR1-deficient mice show reduced oval cell number after 2–4 weeks of CDE diet and less liver tumors after long time diet, while TNFR2-deficient mice showed no changes compared to wt mice (Knight et al. 2000). Thus, these data suggest that TNF contributes to liver carcinogenesis early in the preneoplastic phase by driving oval cell proliferation (Fig. 3). TNF and NFκB activation in hepatocytes have also been identified as crucial factors accelerating tumor progression in hepatocarcinogenesis, which spontaneously occur in multidrug resistance p-glycoprotein 2 (Mdr2)-knockout mice. In this model hepatocytes display enhanced proliferation associated with increased hyperploidy, parenchymal infiltration of TNF expressing inflammatory cells, and dysplasia, which over an adenoma-like intermediate step progress into hepatocellular carcinoma at an age of about 7 months (Pikarsky et al. 2004). Experiments with transgenic Mdr2-knockout mice allowing tetracyclin-regulated expression of a nondegradable deletion mutant of IκBα in hepatocytes showed that NFκB signaling in the latter is dispensable for the occurrence of the early neoplastic events while it is required during tumor progression to prevent apoptosis (Pikarsky et al. 2004). As treatment with TNF-blocking antibodies displayed a similar effect, it is tempting to speculate that TNF facilitates tumor progression in this model by inducing NFκB-regulated anti-apoptotic proteins in hepatocytes (Fig. 3).

3.3 TNF and Gastrointestinal Carcinogenesis

The importance of chronic inflammation for gastrointestinal carcinogenesis is particularly evident for gastric cancer and colitis-associated cancer. While gastric cancer is associated with chronic inflammation related to infections with *Helicobacter pylori*, the latter primarily develops in patients suffering from ulcerative colitis and is responsible for up to 5% of colorectal cancers (Correa 2003). In a mouse model of ulcerative colitis-associated carcinogenesis, the incidence of tumors induced by a single challenge with the pro-carcinogen azoxymethane (AOM) is enhanced by triggering colitis-like ongoing colonic inflammation by repeated cycles of dextran sulfate sodium salt (DSS) administration. Based on a former study showing a crucial role of NFκB activity in the AOM/DSS model of colon carcinogenesis (Greten et al. 2004), it has been recently demonstrated that tumor incidence and tumor-associated symptoms such as body weight loss and diarrhea are reduced in TNFR1-deficient mice (Fig. 3; Popivanova et al. 2008). TNF is not or poorly expressed in untreated or AOM challenged mice, but become readily detectable after DSS-induced inflammation in infiltrating mononuclear cells (Popivanova et al. 2008). TNF was also observed in colon biopsies from patients with active ulcerative colitis or colorectal cancer (Popivanova et al. 2008). While apoptosis in AOM/DSS-induced cancer was not affected in TNFR1 knockout mice, expression of cyclo-oxygenase-2 (COX-2) and the chemokines CXCL-1/KC and monocyte chemoattractant protein-1 (MCP1) as well as tumor infiltration by their target cells (neutrophils and macrophages) were strongly reduced. The relevance of this inflammatory defect was further substantiated by the observation that chimeric mice with transplanted TNFR1-deficient bone marrow cells showed a strongly reduced tumor incidence in the AOM/DSS model compared with chimeric mice that have

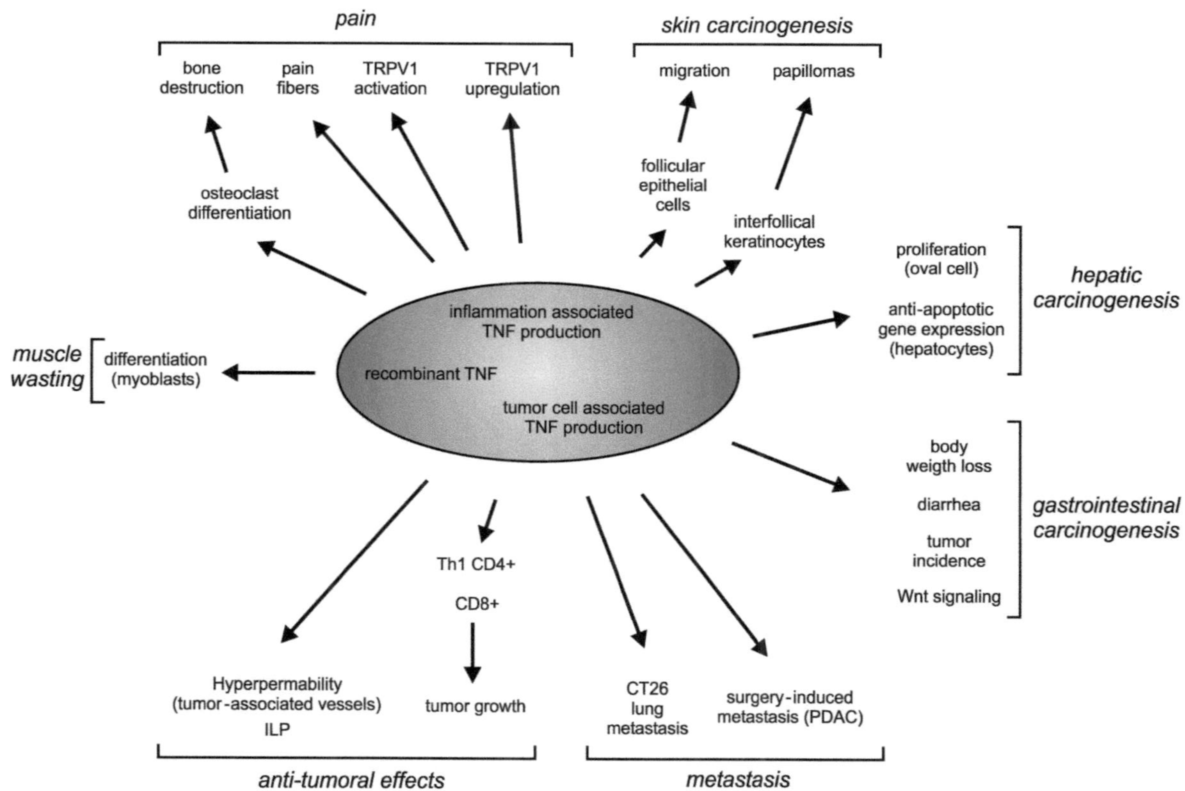

Fig. 3 Cancer-related effects of TNF. In accordance with its immunostimulatory functions, TNF can contribute to tumor suppression by the immune system. However, these capabilities can be hijacked by cancer cells in later stages of tumorigenesis to drive, for example, migration and angiogenesis or to evade apoptosis induction. TNF is also crucially involved in cancer-associated pain. The main sources of TNF in cancer-related processes are tumor infiltrating leukocytes, tumor-associated fibroblasts, and the tumor cells itself. TNF might also be delivered exogenously during isolated limb perfusion

received bone marrow cells from wild-type mice. Last but not least, tumor development in AOM/DSS-treated wild-type mice could be significantly inhibited by therapeutic intervention with the TNF antagonist etanercept even if the treatment is not started until the finalization of the AOM/DSS treatment scheme (Popivanova et al. 2008). In accordance with the special role of the Wnt/β-catenin pathway in the development of gastrointestinal cancers, a lower number of adenocarcinomatous lesions and reduced accumulation of β-catenin in the nucleus have also been observed in tumor cells of TNFR1-knockout mice (Popivanova et al. 2008). Furthermore, in transgenic mice that develop dysplastic lesions in the stomach due to expression of Wnt1 in gastric epithelial cells (K19 promoter), β-catenin accumulating cells associate with infiltrated macrophages, which in turn in vitro activate the Wnt/β-catenin pathway in gastric cancer cell lines via TNFR1 stimulation and inhibitory phosphorylation of glycogen synthase kinase (GSK)-3β, the major negative regulator of the Wnt/β-catenin pathway (Oguma et al. 2008).

A crucial role of hematopoietic cell-derived TNF has also been found in a murine metastasis model with the colon cell line CT26. In this model, LPS-induced NFκB-mediated metastasis into the lung was reduced upon transplantation of irradiated mice with TNF-deficient bone marrow cells (Luo et al. 2004). Likewise, pretreatment of CT26 cells with TNF caused increased metastasis in lung and liver after transplantation (Choo et al. 2005). Metastasis driving TNF-activity must not be necessarily initiated by immune cells. In a murine orthotopic xenotransplantation model with pancreatic ductal adenocarcinoma (PDAC) cell lines mimicking tumor recurrence and metastasis after surgical tumor resection, we identified by use of human TNF-specific blocking antibodies a crucial role of tumor cell-derived TNF on recurrent tumor growth and metastasis (Egberts et al. 2008). A pro-metastatic role of TNF was also shown in the B16-F10 melanoma model by the help of a TNF-neutralizing soluble TNFR1 fusion protein, TNF-blocking antibodies, or TNF autoavaccination-induced self anti-TNF antibodies (Waterston et al. 2004; Cubillos et al. 1997). The principle capacity of TNF to promote metastasis has also been shown using treatment with exogenous TNF (Orosz et al. 1993) or tumor cells genetically engineered to express TNF (Qin et al. 1993). In further accordance with a metastasis promoting function of TNF there is evidence from in vitro studies that TNF can induce epithelial–mesenchymal transition (EMT), the hallmark in the progression of benign carcinomas into more aggressive invasive tumors (Bates and Mercurio 2003; Chuang et al. 2008).

4 TNF and Cancer-Associated Pain

TNF not only has crucial roles in tumor development and tumor progression, but is also of relevance in several tumor-associated complications such as bone destruction, cancer-related pain, fatigue, and muscle wasting. Bone destruction is a major source of tumor-associated pain and is characterized by the imbalanced differentiation and activation of osteoblasts. RANK (receptor activator of NF-kappaB) ligand (RANKL), which is expressed on stroma cells and osteoblasts, and its corresponding receptor RANK, which is expressed on osteoclasts, are the main regulators of osteoclast differentiation and activation (Boyce and Xing 2008). However, there is good evidence from in vitro studies that TNF via its two receptors TNFR1 and TNFR2 can regulate differentiation and activation of osteoclasts independent from the RANKL-RANK system (Kobayashi et al. 2000). Moreover, there is evidence that the latter and TNF can cooperate in osteoclastogenesis (Zhang et al. 2001). However, the TNF–TNF receptor system is not only involved in cancer-related pain due to its osteoclastogenesis promoting function, but might also be directly involved in pain sensing. It has been observed that TNF increases the number of pain signaling neurons (pain fibers or nociceptors) in the tumor area and contributes this way to mechanical and heat hyperalgesia (Constantin et al. 2008). It has been further demonstrated that TNF via TNFR2 sensitizes for signaling via the pain-related TRP channel, transient receptor potential channel vanilloid type 1 (TRPV1) within second via a p38- and PKC-dependent signaling pathway (Constantin et al. 2008). Moreover, TNF again via TNFR2 is involved in the upregulation of TRPV1. The fatigue and muscle wasting effect of TNF is evident from its NFκB-mediated inhibitory effect on MyoD-mediated differentiation of myocytes, which has been shown in vitro and in vivo (Guttridge et al. 2000).

TNF has been implicated in renal cell carcinoma and TNF blockage by Infliximab has been recently investigated in two phase II trials, with patients showing disease progression after cytokine therapy with IFNα and/or IL2 (Harrison et al. 2007). In these studies, 6 of 19 and 11 of 18 patients showed clinical response (mainly stable disease) while the others showed ongoing disease progression (Harrison et al. 2007). Besides mild side effects, the major caution note in these trials was one patient who died due to a non-neutropenic sepsis, a complication that could be related or exacerbated by TNF blockade. Fatigue and muscle wasting can be exacerbated by chemotherapeutic drugs and thus also limit the maintenance of dose-intensity during conventional chemotherapy. This opens the possibility that TNF neutralization improves the tolerability of chemotherapy and allows prolonged treatment with appropriate chemotherapeutic drugs. In fact, it has been recently published that patients with advanced malignancies belonging to a cohort that has been treated with docetaxel and etanercept report an improved fatigue symptom inventory (FSI) score compared to patients of a cohort receiving only docetaxel. Moreover, in 12 of 18 patients treated with doxetaxel and etanercept a clinical benefit was reported (Monk et al. 2006). Because of the absence of a placebo control, however, this pilot study has to be considered with caution, but inspire corresponding clinical studies. Likewise, recent pilot studies with Infliximab showed disease stabilization and mainly good tolerability, but in one study there was also occurrence of severe infections (Tookman et al. 2008; Brown et al. 2008). In a multicenter, phase II, placebo-controlled study, however, where cachexia in patients suffering on advanced pancreatic cancer was treated with gemcitabine or gemcitabine and infliximab, no statistically significant differences in safety or efficacy were observed compared with the placebo group (Widdenmann et al. 2008).

References

Adam-Klages S, Adam D, Wiegmann K, Struve S, Kolanus W, Schneider-Mergener J, Krönke M (1996) FAN, a novel WD-repeat protein, couples the p55 TNF-receptor to neutral sphingomyelinase. *Cell* 86:937–947

Arnott CH, Scott KA, Moore RJ, Hewer A, Phillips DH, Parker P, Balkwill FR, Owens DM (2002) Tumour necrosis factor-alpha mediates tumour promotion via a PKC alpha- and AP-1-dependent pathway. *Oncogene* 21:4728–4738

Arnott CH, Scott KA, Moore RJ, Robinson SC, Thompson RG, Balkwill FR (2004) Expression of both TNF-alpha receptor subtypes is essential for optimal skin tumour development *Oncogene* 23:1902–1910

Bates RC, Mercurio AM (2003) Tumor necrosis factor-alpha stimulates the epithelial-to-mesenchymal transition of human colonic organoids. *Mol Biol Cell* 14:1790–1800

Beutler B, Greenwald D, Hulmes JD, Chang M, Pan YC, Mathison J, Ulevitch R, Cerami A (1985) Identity of tumour necrosis factor and the macrophage-secreted factor cachectin. *Nature* 316:552–554

Binder RL, Gallagher PM, Johnson GR, Stockman SL, Smith BJ, Sundberg JP, Conti CJ (1997) Evidence that initiated keratinocytes clonally expand into multiple existing hair follicles during papilloma histogenesis in SENCAR mouse skin. *Mol Carcinog* 20:151–158

Bodmer JL, Schneider P, Tschopp J (2002) The molecular architecture of the TNF superfamily. *Trends Biochem Sci* 27:19–26

Boyce BF, Xing L (2008) Functions of RANKL/RANK/OPG in bone modeling and remodeling *Arch Biochem Biophys* 473:139–146

Brown ER, Charles KA, Hoare SA, Rye RL, Jodrell DI, Aird RE, Vora R, Prabhakar U, Nakada M, Corringham RE, DeWitte M, Sturgeon C, Propper D, Balkwill FR, Smyth JF (2008) A clinical study assessing the tolerability and biological effects of infliximab, a TNF-alpha inhibitor, in patients with advanced cancer. *Ann Oncol* 19:1340–1346

Calzascia T, Pellegrini M, Hall H, Sabbagh L, Ono N, Elford AR, Mak TW, Ohashi PS (2007) TNF-alpha is critical for antitumor but not antiviral T cell immunity in mice. *J Clin Invest* 117:3833–3845

Chan FK, Lenardo MJ (2000) A crucial role for p80 TNF-R2 in amplifying p60 TNF-R1 apoptosis signals in T lymphocytes. *Eur J Immunol* 30:652–660

Choo MK, Sakurai H, Koizumi K, Saiki I (2005) Stimulation of cultured colon 26 cells with TNF-alpha promotes lung metastasis through the extracellular signal-regulated kinase pathway. *Cancer Lett* 230:47–56

Chuang MJ, Sun KH, Tang SJ, Deng MW, Wu YH, Sung JS, Cha TL, Sun GH (2008) Tumor-derived tumor necrosis factor-alpha promotes progression and epithelial-mesenchymal transition in renal cell carcinoma cells. *Cancer Sci* 99:905–913

Constantin CE, Mair N, Sailer CA, Andratsch M, Xu ZZ, Blumer MJ, Scherbakov N, Davis JB, Bluethmann H, Ji RR, Kress M (2008) Endogenous tumor necrosis factor alpha (TNFalpha) requires TNF receptor type 2 to generate heat hyperalgesia in a mouse cancer model *J Neurosci* 28:5072–5081

Correa P (2003) Helicobacter pylori infection and gastric cancer. *Cancer Epidemiol Biomarkers Prev* 12:238s–241s

Cubillos S, Scallon B, Feldmann M, Taylor P (1997) Effect of blocking TNF on IL-6 levels and metastasis in a B16-BL6 melanoma/mouse model. *Anticancer Res* 17:2207–2211

Egberts JH, Cloosters V, Noack A, Schniewind B, Thon L, Klose S, Kettler B, von Forstner C, Kneitz C, Tepel J, Adam D, Wajant H, Kalthoff H, Trauzold A (2008) Anti-tumor necrosis factor therapy inhibits pancreatic tumor growth and metastasis. *Cancer Res Mar* 68:1443–1450

Fotin-Mleczek M, Henkler F, Samel D, Reichwein M, Hausser A, Parmryd I, Scheurich P, Schmid JA, Wajant H (2002) Apoptotic crosstalk of TNF receptors: TNF-R2-induces depletion of TRAF2 and IAP proteins and accelerates TNF-R1-dependent activation of caspase-8. *J Cell Sci* 115:2757–2770

Grell M, Douni E, Wajant H, Löhden M, Clauss M, Maxeiner B, Georgopoulos S,Lesslauer W, Kollias G, Pfizenmaier K, Scheurich P (1995) The transmembrane form of tumor necrosis factor is the prime activating ligand of the 80 kDa tumor necrosis factor receptor. *Cell* 83:793–802

Greten FR, Eckmann L, Greten TF, Park JM, Li ZW, Egan LJ, Kagnoff MF, Karin M (2004) IKKbeta links inflammation and tumorigenesis in a mouse model of colitis-associated cancer *Cell* 118:285–296

Guttridge DC, Mayo MW, Madrid LV, Wang CY, Baldwin AS Jr (2000) NF-kappaB-induced loss of MyoD messenger RNA: possible role in muscle decay and cachexia. *Science* 289:2363–2366

Harrison ML, Obermueller E, Maisey NR, Hoare S, Edmonds K, Li NF, Chao D, Hall K, Lee C, Timotheadou E, Charles K, Ahern R, King DM, Eisen T, Corringham R, DeWitte M, Balkwill F, Gore M (2007) Tumor necrosis factor alpha as a new target for renal cell carcinoma: two sequential phase II trials of infliximab at standard and high dose. *J Clin Oncol* 25:4542–4549

Hong S, Lim S, Li AG, Lee C, Lee YS, Lee EK, Park SH, Wang XJ, Kim SJ (2007) Smad7 binds to the adaptors TAB2 and TAB3 to block recruitment of the kinase TAK1 to the adaptor TRAF2. *Nat Immunol* 8:504–513

van Horssen R, Ten Hagen TL, Eggermont AM (2006) TNF-alpha in cancer treatment: molecular insights, antitumor effects, and clinical utility. *Oncologist* 11:397–408

Knight B, Yeoh GC, Husk KL, Ly T, Abraham LJ, Yu C, Rhim JA, Fausto N (2000) Impaired preneoplastic changes and liver tumor formation in tumor necrosis factor receptor type 1 knockout mice. *J Exp Med* 192:1809–1818

Kobayashi K, Takahashi N, Jimi E, Udagawa N, Takami M, Kotake S, Nakagawa N, Kinosaki M, Yamaguchi K, Shima N, Yasuda H, Morinaga T, Higashio K, Martin TJ, Suda T (2000) Tumor necrosis factor alpha stimulates osteoclast differentiation by a mechanism independent of the ODF/RANKL-RANK interaction. *J Exp Med* 191:275–286

Li X, Yang Y, Ashwell JD (2002) TNF-RII and c-IAP1 mediate ubiquitination and degradation of TRAF2. *Nature* 416:345–347

Lin Y, Devin A, Rodriguez Y, Liu ZG (1999) Cleavage of the death domain kinase RIP by caspase-8 prompts TNF-induced apoptosis. *Genes Dev* 13:2514–2526

Lin Y, Choksi S, Shen HM, Yang QF, Hur GM, Kim YS, Tran JH, Nedospasov SA, Liu ZG (2004) Tumor necrosis factor-induced nonapoptotic cell death requires receptor-interacting protein-mediated cellular reactive oxygen species accumulation. *J Biol Chem* 279:10822–10828

Luo JL, Maeda S, Hsu LC, Yagita H, Karin M (2004) Inhibition of NF-kappaB in cancer cells converts inflammation- induced tumor growth mediated by TNFalpha to TRAIL-mediated tumor regression. *Cancer Cell* 6:297–305

Monk JP, Phillips G, Waite R, Kuhn J, Schaaf LJ, Otterson GA, Guttridge D, Rhoades C, Shah M, Criswell T, Caligiuri MA, Villalona-Calero MA (2006) Assessment of tumor necrosis factor alpha blockade as an intervention to improve tolerability of dose-intensive chemotherapy in cancer patients. *J Clin Oncol* 24:1852–1859

Moore RJ, Owens DM, Stamp G, Arnott C, Burke F, East N, Holdsworth H, Turner L, Rollins B, Pasparakis M, Kollias G, Balkwill F (1999) Mice deficient in tumor necrosis factor-alpha are resistant to skin carcinogenesis. *Nat Med* 5:828–831

Muppidi JR, Tschopp J, Siegel RM (2004) Life and death decisions: secondary complexes and lipid rafts in TNF receptor family signal transduction. *Immunity* 21:461–465

Müller-Hermelink N, Braumüller H, Pichler B, Wieder T, Mailhammer R, Schaak K, Ghoreschi K, Yazdi A, Haubner R, Sander CA, Mocikat R, Schwaiger M, Förster I, Huss R, Weber WA, Kneilling M, Röcken M (2008) TNFR1 signaling and IFN-gamma signaling determine whether T cells induce tumor dormancy or promote multistage carcinogenesis. *Cancer Cell* 13:507–518

Oguma K, Oshima H, Aoki M, Uchio R, Naka K, Nakamura S, Hirao A, Saya H, Taketo MM, Oshima M (2008) Activated macrophages promote Wnt signalling through tumour necrosis factor-alpha in gastric tumour cells. *EMBO J* 27:1671–1681

Orosz P, Echtenacher B, Falk W, Rüschoff J, Weber D, Männel DN (1993) Enhancement of experimental metastasis by tumor necrosis factor. *J Exp Med* 177:1391–1398

Ouyang W, Hu Y, Li J, Ding M, Lu Y, Zhang D, Yan Y, Song L, Qu Q, Desai D, Amin S, Huang C (2007) Direct evidence for the critical role of NFAT3 in benzo[a]pyrene diol-epoxide-induced cell transformation through mediation of inflammatory cytokine TNF induction in mouse epidermal Cl41 cells. *Carcinogenesis* 28:2218–2226

Pan S, An P, Zhang R, He X, Yin G, Min W (2002) Etk/Bmx as a tumor necrosis factor receptor type 2-specific kinase: role in endothelial cell migration and angiogenesis. *Mol Cell Biol* 22:7512–7523

Perkins ND (2007) Integrating cell-signalling pathways with NF-kappaB and IKK function. *Nat Rev Mol Cell Biol* 8:49–62

Pikarsky E, Porat RM, Stein I, Abramovitch R, Amit S, Kasem S, Gutkovich-Pyest E, Urieli-Shoval S, Galun E, Ben-Neriah Y (2004) NF-kappaB functions as a tumour promoter in inflammation-associated cancer. *Nature* 431:461–466

Popivanova BK, Kitamura K, Wu Y, Kondo T, Kagaya T, Kaneko S, Oshima M, Fujii C, Mukaida N (2008). Blocking TNF-alpha in mice reduces colorectal carcinogenesis associated with chronic colitis. *J Clin Invest* 118:560–570

Qin Z, Krüger-Krasagakes S, Kunzendorf U, Hock H, Diamantstein T, Blankenstein T (1993) Expression of tumor necrosis factor by different tumor cell lines results either in tumor suppression or augmented metastasis. *J Exp Med* 178:355–360

Rothe M, Wong SC, Henzel WJ, Goeddel DV (1994) A novel family of putative signal transducers associated with the cytoplasmic domain of the 75 kDa tumor necrosis factor receptor. *Cell* 78:681–692

Rothe M, Pan MG, Henzel WJ, Ayres TM, Goeddel DV (1995) The TNFR2-TRAF signaling complex contains two novel proteins related to baculoviral inhibitor of apoptosis proteins. *Cell* 83:1243–1252

Schütze S, Tchikov V, Schneider-Brachert W (2008) Regulation of TNFR1 and CD95 signalling by receptor compartmentalization. *Nat Rev Mol Cell Biol* 9(8):655–662

Schwandner R, Wiegmann K, Bernardo K, Kreder D, Kronke M (1998) TNF receptor death domain-associated proteins TRADD and FADD signal activation of acid sphingomyelinase. *J Biol Chem* 273:5916–5922

Scott KA, Arnott CH, Robinson SC, Moore RJ, Thompson RG, Marshall JF, Balkwill FR (2004) TNF-alpha regulates epithelial expression of MMP-9 and integrin alphavbeta6 during tumour promotion. A role for TNF-alpha in keratinocyte migration? *Oncogene* 23:6954–6966

Suganuma M, Okabe S, Marino MW, Sakai A, Sueoka E, Fujiki H (1999) Essential role of tumor necrosis factor alpha (TNF-alpha) in tumor promotion as revealed by TNF-alpha-deficient mice. *Cancer Res* 59:4516–4518

Swann JB, Vesely MD, Silva A, Sharkey J, Akira S, Schreiber RD, Smyth MJ (2007) Demonstration of inflammation-induced cancer and cancer immunoediting during primary tumorigenesis. *Proc Natl Acad Sci USA* 105:652–656

Tookman AJ, Jones CL, Dewitte M, Lodge PJ (2008) Fatigue in patients with advanced cancer: a pilot study of an intervention with infliximab. *Support Care Cancer* 16(10):1131–1140

Wajant H, Pfizenmaier K, Scheurich P (2003) Tumor necrosis factor signaling. *Cell Death Differ* 10:45–65

Wang AM, Creasey AA, Ladner MB, Lin LS, Strickler J, Van Arsdell JN, Yamamoto R, Mark DF (1985) Molecular cloning of the complementary DNA for human tumor necrosis factor. *Science* 228:149–154

Wang CY, Mayo MW, Korneluk RG, Goeddel DV, Baldwin AS Jr (1998) NF-kappaB antiapoptosis: induction of TRAF1 and TRAF2 and c-IAP1 and c-IAP2 to suppress caspase-8 activation. *Science* 281:1680–1683

Waterston AM, Salway F, Andreakos E, Butler DM, Feldmann M, Coombes RC (2004) TNF autovaccination induces self anti-TNF antibodies and inhibits metastasis in a murine melanoma model. *Br J Cancer* 90:1279–1284

Weiss T, Grell M, Hessabi B, Bourteele S, Müller G, Scheurich P, Wajant H (1997) Enhancement of TNF receptor p60-mediated cytotoxicity by TNF receptor p80: requirement of the TNF receptor-associated factor-2 binding site. *J Immunol* 158:2398–2404

Wiedenmann B, Malfertheiner P, Friess H, Ritch P, Arseneau J, Mantovani G, Caprioni F, Van Cutsem E, Richel D, DeWitte M, Qi M, Robinson D Jr, Zhong B, De Boer C, Lu JD, Prabhakar U, Corringham R, Von Hoff D (2008) A multicenter, phase II study of infliximab plus gemcitabine in pancreatic cancer cachexia. *J Support Oncol* 6(1):18–25

Wu CJ, Conze DB, Li X, Ying SX, Hanover JA, Ashwell JD (2005) TNF-alpha induced c-IAP1/TRAF2 complex translocation to a Ubc6-containing compartment and TRAF2 ubiquitination. *EMBO J* 24:1886–1898

Zhang YH, Heulsmann A, Tondravi MM, Mukherjee A, Abu-Amer Y (2001) Tumor necrosis factor-alpha (TNF) stimulates RANKL-induced osteoclastogenesis via coupling of TNF type 1 receptor and RANK signaling pathways. *J Biol Chem* 276:563–568

Zhang R, Xu Y, Ekman N, Wu Z, Wu J, Alitalo K, Min W (2003) Etk/Bmx transactivates vascular endothelial growth factor 2 and recruits phosphatidylinositol 3-kinase to mediate the tumor necrosis factor-induced angiogenic pathway. *J Biol Chem* 278:51267–51276

Many Checkpoints on the Road to Cell Death: Regulation of Fas–FasL Interactions and Fas Signaling in Peripheral Immune Responses

Madhu Ramaswamy, Sophia Y. Cleland, Anthony C. Cruz, and Richard M. Siegel

Abstract Interactions between the TNF-family receptor Fas (CD95) and Fas Ligand (FasL, CD178) can efficiently induce apoptosis and are critical for the maintenance of immunological self-tolerance. FasL is kept under strict control by transcriptional and posttranslational regulation. Surface FasL can be cleaved by metalloproteases, resulting in shed extracellular domains, and FasL can also traffic to secretory lysosomes. Each form of FasL has distinct biological functions. Fas is more ubiquitously expressed, but its apoptosis-inducing function is regulated by a number of mechanisms including submembrane localization, efficiency of receptor signaling complex assembly and activation, and bcl-2 family members in some circumstances. When apoptosis is not induced, Fas–FasL interactions can also trigger a number of activating and proinflammatory signals. Harnessing the apoptosis-inducing potential of Fas for therapy of cancer and autoimmune disease has been actively pursued, and despite a number of unexpected side-effects that result from manipulating Fas–FasL interactions, this remains a worthy goal.

1 Introduction: Fas–Fas Ligand Interactions in Immune Responses

The discovery in the early 1990s that antibodies to the cell surface TNF-family member receptor Fas (CD95) could mediate rapid protein-synthesis independent apoptosis of a number of transformed and nontransformed cell types set the stage for the investigation of engaging Fas and related 'death receptors' as possible targets for intervention in cancer therapy. Fas also plays a critical role in immunological self-tolerance through the deletion of a number of cell types that contribute to autoimmunity. Mutations in Fas and its TNF family ligand Fas Ligand (CD178,

R.M. Siegel(✉), M. Ramaswamy, S.Y. Cleland, and A.C. Cruz
Immunoregulation Unit, Autoimmunity Branch, NIAMS,
National Institutes of Health, Bethesda MD, USA, 20892
e-mail: rsiegel@nih.gov

Results Probl Cell Differ, doi:10.1007/400_2008_24

FasL) are responsible for the single gene autoimmune *lpr* and *gld* phenotypes in mice (Ramsdell et al. 1994b; Watanabe-Fukunaga et al. 1992a) and most cases of the strikingly similar autoimmune lymphoproliferative syndrome (ALPS) in humans, which is associated in a majority of patients (Type IA ALPS) with dominant-interfering Fas mutations (Straus et al. 1999). Not surprisingly for an interaction that can permanently eliminate cells through apoptosis, it has become clear that there are many levels of regulation of Fas–FasL interactions. Both FasL synthesis and trafficking are subject to strict control, which limit the production of biologically active ligands to a few cell types. Although most activated lymphocytes express Fas, there are many levels of regulation that control the efficiency of Fas-induced apoptosis, both at the level of assembly and activation of the Fas signaling complex, and at the level of signal integration at the mitochondria. These mechanisms cooperate to create a situation where Fas–FasL interactions can efficiently eliminate autoreactive T and B cells, while having little impact on most immune responses to pathogens.

Fas–FasL interactions have been shown to be responsible for much of the apoptosis that occurs when activated CD4$^+$ T cells are restimulated through the T-cell receptor (TCR) (Dhein et al. 1995; Ju et al. 1995). Since this process is molecularly distinct from much of the T cell death that occurs during initial T cell activation, we refer to this process as Restimulation Induced Cell Death, or RICD. Most of the death that restimulated CD4$^+$ T cells undergo is through RICD by FasL, while FasL appears to play a subsidiary role in CD8$^+$ T cells to other proteins contained in cytotoxic T cell granules such as perforin and granzymes (Davidson et al. 2002). As we will discuss in this chapter, although most activated and memory lymphocytes express cell surface Fas, RICD only kills activated T cells under conditions of chronic T-cell restimulation, due to controls on FasL expression and processing and Fas signaling that render this pathway inactive under other circumstances. Different functional subsets of CD4$^+$ T cells may also use the Fas–FasL pathway of apoptosis to greater or lesser extents. The majority of cell death that occurs after T cell activation appears to be apoptosis caused by inadequate supply of cytokines such as IL7 and IL15 that signal through the gamma-chain containing cytokine receptors and Jak/STAT proteins to increase expression and function of Bcl-2 family proteins. We term this type of cell death postactivation cell death (PACD). Experiments in which activated lymphocytes are infused into IL-7 and IL-15 deficient mice have shown that these two cytokines cooperate to allow survival of T cells after activation, and genetic or pharmacological delivery of these cytokines can prolong T cell survival (Sprent and Surh 2002; Tan et al. 2002).

In most circumstances, the massive proliferation of activated T cells during immune responses outgrows the cytokine supply and results in a balance of pro and antiapoptotic Bcl-2 family members that favor apoptosis. This is dramatically illustrated by mice that lack the BH-3-only proapoptotic family member Bim. There is accumulation of excess lymphocytes in these mice and antigen-specific T cells are impaired in their ability to undergo cell death after acute antigen stimulation, while the RICD of activated T cells is not affected. Conversely, Fas deficient animals have nearly normal cell death of T cells after acute activation. Mice in which Bim and Fas have both been genetically ablated have greatly enhanced pathology and autoimmune disease compared to each mutant alone, providing genetic evidence that these two pathways are distinct (Hughes et al. 2008; Hutcheson et al. 2008; Weant et al. 2008). RICD helps to maintain peripheral tolerance by eliminating reactive T cells and reducing the chance for reactive T cells to act on target cells beyond their specific effector function. Fas–FasL interactions can alter the response of activated T cells during infections, and impairment of any of the mechanisms that promote FasL mediated apoptosis discussed below may predispose towards autoimmunity.

In this chapter, we discuss the basic biology of Fas and Fas Ligand with emphasis on the role of Fas–FasL interactions in the immune system, pointing out a number of steps that strictly regulate the scope of cells that die via Fas–FasL interactions. In particular, we discuss findings by our group and others showing that much of the regulation of Fas signaling lies in the early steps of assembly and activation of the Fas signaling complex. We will also discuss the status of attempts to harness Fas-induced apoptosis for therapeutic use in autoimmunity, transplantation and cancer therapy.

2 Fas Ligand, a Highly Regulated TNF Family Member

Fas Ligand (FasL or CD178), the sole known TNF-family ligand for Fas, is synthesized as a 281 amino acid type II transmembrane protein. Compared with other TNF family members, FasL has a relatively long N-terminal cytoplasmic domain that has been found to contain multiple sorting motifs governing the trafficking of FasL. Some groups have reported that the cytoplasmic domain of FasL can also mediate ‘reverse signaling’ upon binding to Fas. The extracellular TNF-homology domain of FasL can also be cleaved from the membrane by metalloproteinases to become a secreted trimer. However, trimeric FasL is highly unstable and appears to be largely inert as an apoptosis-inducing ligand. These aspects of the biology of FasL make control of trafficking and cleavage of this protein as important as the regulation of FasL expression.

2.1 Regulation of FasL Gene Expression

While FasL expression on activated T cells is transient, FasL expression on non-immune cells is generally constitutive. The signaling pathways and transcription factors mediating inducible FasL expression have best been studied in T cells. The calcium-sensitive NFAT (Nuclear factor of activated T-cells) transcription factor family mediates a major part of the signal by which the TCR induces FasL. Calcineurin activation results in NFAT activation and translocation to the nucleus, and inhibitors of calcineurin such as cyclosporine A block FasL expression in activated T cells (Anel et al. 1994; Brunner et al. 1996; Dhein et al. 1995; Latinis et al. 1997a). The FasL promoter has two NFAT sites, with the distal NFAT site on the FasL promoter being more important for TCR-mediated FasL expression in

CD4⁺ T cells (Latinis et al. 1997b). The Egr family of transcription factors is induced by NFAT and may act synergistically with NFAT in inducing FasL expression in some cell types (Dzialo-Hatton et al. 2001; Mittelstadt and Ashwell 1999; Rengarajan et al. 2000). C-myc binds to a separate site on the FasL promoter and has been shown to be required for TCR-induced FasL expression (Brunner et al. 2000; Wang et al. 1998). TCR-induced FasL expression can also be negatively regulated by a number of mechanisms. The CIITA (MHC class II transactivator) transcription factor as well as retinoic acid can block NFAT function and inhibit FasL transcription (Gourley and Chang 2001; Lee et al. 2002). TGF-β (Transforming Growth Factor-beta) can also inhibit TCR-induced FasL expression through downmodulating c-myc expression (Genestier et al. 1999).

In parallel with NFAT, the NF–κB transcription factors, which have separate binding sites in the FasL promoter, can also induce FasL expression. Through the action of protein kinase C theta (PKC-θ), the TCR activates NF–κB and can synergize with calcineurin-dependent signaling to induce FasL expression. (Kasibhatla et al. 1999; Villalba et al. 1999; Villunger et al. 1999). The AP-1 transcription factor complex, which is activated by the TCR through MAP kinase signaling, also activates FasL expression (Matsui et al. 2000). Interestingly, inducers of nitric oxide (NO) inhibit FasL expression through blocking AP-1 activity (Melino et al. 2000). Interferon Regulatory Factors (IRFs), a family of transcription factors that induce the transcription of interferons in response to viral infection, may also cooperate with other transcription factors to maximally induce FasL in response to TCR and other stimuli (Chow et al. 2000). Viral IRF homologs from human herpesvirus 8 (HHV8), the cause of Kaposi's Sarcoma in immunocompromised patients, interferes with IRF-1 binding and downregulates FasL transcription, which may aid in the escape of infected T cells from FasL mediated apoptosis (Kirchhoff et al. 2002). A separate set of transcription factors govern basal and constitutive FasL expression in both lymphoid and nonlymphoid cells. The transcription factor Sp1 regulates the basal FasL expression in Jurkat T cells and constitutive expression of FasL in Sertoli cells (McClure et al. 1999). Sp1 regulates FasL expression on smooth muscle cells (SMCs) by cooperating with the transcription factor Ets-1 (Kavurma et al. 2002, 2001). Transcriptional regulation of FasL results in constitutive expression in tissues such as the eye and testis that may contribute to immune tolerance through inducing apoptosis in infiltrating lymphocytes, whereas FasL dynamically expressed on T cells can eliminate Fas-sensitive cells in tissues where FasL may not be expressed (Bellgrau et al. 1995; Bonfoco et al. 1998).

2.2 *Three Forms of FasL Controlled by Posttranslational Modification and Subcellular Trafficking*

Post-translational modification of FasL results in dramatically different trafficking of FasL in cells and in its ability to induce apoptosis. FasL is initially synthesized as a type II transmembrane protein containing a TNF homology domain at the

C-terminal that traffics to the plasma membrane through the golgi. FasL expressed ectopically on the plasma membrane can be a strong stimulator of apoptosis (Jodo et al. 2001; Suda et al. 1997). A portion of FasL has also been reported to partition into glycosphingolipid-enriched membrane 'rafts', which may also enhance its death-inducing function (Cahuzac et al. 2006). Extracellular FasL can be cleaved by the metalloproteinase ADAM10, resulting in shedding of a 20–26 KDa free extracellular domain, and the intracellular domain can be cleaved and released into the cytosol by the signal peptidase-like protease SPPL2a (Kirkin et al. 2007; Schulte et al. 2007). Soluble FasL released in this way is generally thought to be inactive or even inhibitory for FasL-mediated apoptosis, (Jodo et al. 2001; Suda et al. 1997) so metalloproteinase-dependent cleavage could be an inactivating event for FasL function as a membrane-bound ligand. The intracellular fragment traffics to the nucleus, but the function of this fragment is not clear. In addition, surface FasL can be internalized and sorted into multivesicular bodies, which when fused with the plasma membrane allow secretion of membrane-bound FasL into secretory microvesicles, also known as secretory lysosomes. FasL secreted in microvesicles can be highly biologically active, so understanding of the mechanisms that regulate FasL trafficking into this compartment is important.

Sorting of FasL into secretory lysosomes occurs only in cells that have this specialized trafficking pathway, such as lymphocytes and myeloid cell lines. In fibroblasts and epithelial cell lines, FasL predominantly traffics to the plasma membrane (Blott and Griffiths 2002). Sorting of FasL into the secretory lysosome pathway requires post-translational modification and association of specific domains in the intracellular portion of FasL with several proteins involved in protein and organelle trafficking, cytoskeletal reorganization and formation of the immunological synapse. The proline rich domain (PRD) of FasL binds to Fgr, a Src family tyrosine kinase, and deletions of the PRD as well as mutations in Fgr result in more surface FasL and less FasL in secretory lysosomes. Interestingly, similar mutations also appear to inhibit 'reverse signaling' through FasL that has been reported to occur after Fas binding and function to costimulate $CD8^+$ T cell activation (Sun et al. 2007). Ubiquitination and tyrosine phosphorylation at specific residues in the N-terminal portion of FasL also control trafficking of FasL into secretory lysosomes and deletion or mutation of residues important in these processes redirect FasL to the cell surface (Jodo et al. 2005; Zuccato et al. 2007). Kinases associated with actin remodeling, such as Nck, have also been shown to colocalize with FasL (Lettau et al. 2006) and direct it to secretory lysosomes, providing another level of regulation of FasL trafficking (Fig.1).

These mechanisms cooperate to generate two different waves of FasL produced by T cells acutely stimulated through the TCR. The first phase of cell surface FasL occurs within 10 min of T cell stimulation, and is thought to derive from the fusion of FasL stored in secretory vesicles with the plasma membrane. The continued stimulation of T cells results in a second wave of surface expression of FasL derived from newly synthesized protein peaking 2–4 h after stimulation (He and Ostergaard 2007; Lettau et al. 2004). FasL secreted in exosomal vesicles may derive from either of these pools and several nonlymphoid cells and tissues such as

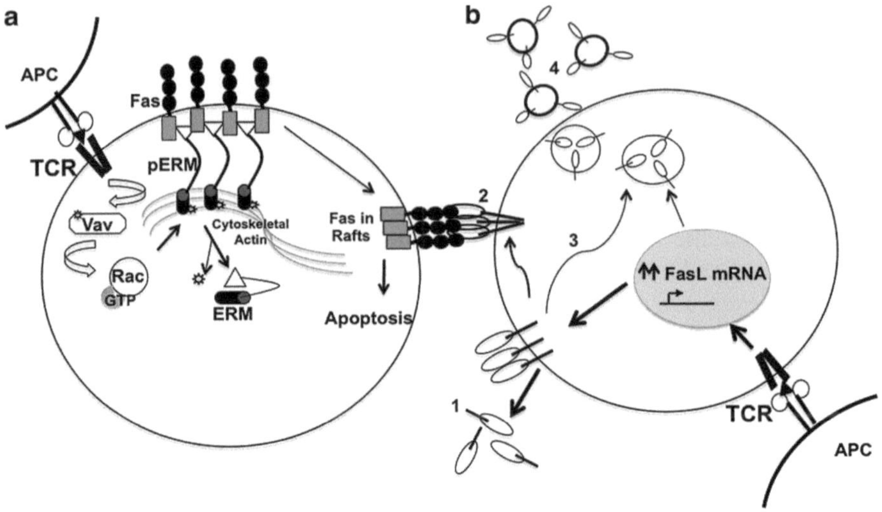

Fig. 1 Schematic of regulatory checkpoints and trafficking of Fas and FasL. (**a**) Fas Receptor- In most activated T cells, surface Fas is not associated with lipid rafts or preassociated, but is linked in monomeric form to the cytoskletal actin by phosphorylated Ezrin–Radixin–Moesin linker proteins (pERM). TCR stimulation by antigen/MHC complexes on APC activates Vav GEF and Rac GTPases, which in turn activate phosphatases. Phosphatase activity may inactivate pERM, releasing Fas to translocate to lipid rafts, where apoptosis signaling is more efficient. (**b**) FasL regulation- TCR stimulation in T cells also induces transcription of FasL in the nucleus, resulting in upregulation of the protein on the surface. Surface expressed FasL is either cleaved by the ADAM10 proteases to release non-apoptosis inducing soluble FasL (1) or forms a membrane bound stable trimer (2) that binds to Fas expressing cells to induce effective signaling. Some can also be released from the cells in microvesicles (4), after endocytosis (3) from the plasma membrane or from newly synthesized FasL

dendritic cells (DC) and other myeloid cells can express FasL in these various forms. Certain DC subsets have been reported to be able to kill Fas-expressing $CD4^+$ T cells (Suss and Shortman 1996). FasL on lymph node dendritic cells have been recently shown to regulate the magnitude of $CD8^+$ effector T cell responses in the lung in the context of influenza infection through induction of T cell apoptosis via FasL (Legge and Braciale 2005). FasL expression on macrophages results in both macrophage and T cell apoptosis (Kiener et al. 1997; Ma et al. 2004; Monari et al. 2005; Villena et al. 2008). Epithelial cells such as those present in tissues, most likely express constitutive surface FasL.

3 Fas: An Apoptosis-Inducing TNF-Family Receptor

Unlike its ligand, Fas is expressed in many diverse cell types, tissues and organs. The initial discovery of Fas was made by two different groups screening for apoptosis-inducing antibodies against cell surface antigens. Two such antibodies, anti-Fas (for FS-7 associated surface antigen) (Yonehara et al. 1989) and anti-Apo1 (Trauth et al. 1989) bound the same 35–52 KDa protein that was termed Fas/Apo-1 (Trauth et al. 1989). Cloning and characterization studies classified Fas as a prototype of the TNF-receptor superfamily, and further designated Fas as CD95/TNFRSF6. Early work on the human Fas antigen was done mostly in lymphoma cell lines, indicating that it was highly expressed on both T and B cell lymphomas (Oehm et al. 1992). In the mouse, Fas expression was seen in the heart, liver, ovary and the thymus (Watanabe-Fukunaga et al. 1992b). This fits with the expression patterns of the other members of the TNF/TNFR superfamily, where expression of ligands have a propensity to be more restricted and dynamically regulated than the receptors, whose expression is regulated between cell lineages but tends to be more constant over time.

3.1 Role of Fas in Lymphocyte Biology

The observation that Fas-deficient *lpr* mice and *gld* mice, which carry recessive disabling mutations in the Fas and FasL genes respectively, produce autoantibodies and have excessive accumulation of $CD4^{-}CD8^{-}$ (double negative) T cells initially pointed towards a role for Fas in thymic negative selection (Watanabe-Fukunaga et al. 1992a). However, despite expression of Fas on thymocytes and the susceptibility of most thymocytes to Fas-induced apoptosis, negative selection does not appear to depend on Fas–FasL interactions, since self-reactive T cells are deleted effectively in the thymus of both *lpr* and *gld* mice (Singer and Abbas 1994). Rather, Fas participates in the elimination of self-reactive T cells by a process known as restimulation induced cell death (RICD), an important 'safety net' for maintaining self-tolerance in T cells that have escaped central thymic tolerance. Despite being constitutively expressed on most T lymphoma cell lines, naive T cells do not have surface Fas expression and therefore are highly refractory to Fas-mediated apoptosis, whereas, memory T cells have high Fas levels. Activation of the resting naive cells via TCR stimulation upregulates surface Fas within 24 h after activation, with highest surface levels occurring with in 6 days of stimulation (Klas et al. 1993; Miyawaki et al. 1992). However, the regulation of Fas-induced apoptosis is a multilayered process and receptor expression alone does not render cells sensitive to Fas-induced apoptosis. It has been observed that recently activated Fas positive cells are refractory to cell death unless cultured in IL-2 for at least 48 additional hours (Peter et al. 1997). This "propriocidal death" is due to IL-2 induced cell cycle progression, which is necessary to make T cells sensitive to TCR and Fas-induced apoptosis (Lenardo et al. 1999). The T-cell receptor provides a critical and physiologically significant signal that also sensitizes T cells to Fas-mediated apoptosis, and is the basis for antigen-specific deletion of activated T cells. As discussed above, TCR engagement of activated T cells results in FasL gene upregulation and secretion (Dhein et al. 1995; Ju et al. 1995). However, mixing experiments with T cells of different specificities showed that the FasL produced by the antigen specific T cells mediates apoptosis of only the restimulated clonotype and not other Fas-expressing bystander cells. This feature of restimulation-induced cell death ensures that only chronically stimulated T cells undergo Fas-mediated apoptosis, and is probably responsible for the restricted role of Fas in elimination of T cells specific for autoantigens and chronic pathogens. The signal mediated via the TCR that sensitizes these cells to Fas-induced apoptosis is termed the "competency to die" signal. TCR and Fas engagement synergize to induce apoptosis in a manner unaffected by protein synthesis inhibitors (Combadiere et al. 1998; Hornung et al. 1997; Wong et al. 1997). Recent work has shown that part of the TCR-induced 'competency to die' signal induces translocation of Fas to lipid rafts, also known as glycosphingolipids enriched membrane microdomains (Muppidi and Siegel 2004). The role of lipid raft microdomains in Fas signaling is discussed in detail below.

The role of Fas in elimination of chronically restimulated T cells is largely confined to $CD4^+$ T cells, which are thought to be the key T cells that can provide help to autoreactive B cells to allow autoantibody secretion. $CD8^+$ T cells, like their $CD4^+$ counterparts, upregulate Fas upon activation, and can be induced to undergo apoptosis through Fas (Miyawaki et al. 1992). However, restimulation of $CD8^+$ T cells from Fas-deficient mice or patients with ALPS induces normal levels of cell death, indicating that Fas-independent mechanisms contribute to RICD of $CD8^+$ T cells. The granzyme/perforin cytotoxic serine protease pathway is a major player in attrition of antigen specific cytotoxic T lymphocyte (CTL) response. Regulation of granzyme B is critical, since the cytoplasmic granzyme B can cause self-directed injury to the CTL producing it. Recent work identifies the presence of a serine protease inhibitor (SPI6) that binds to cytoplasmic granzyme B to form a stable complex, thereby ensuring normal antigenic response by preventing CTL loss by suicide (Zhang et al. 2006).

3.2 *Function of Fas in B Cells and Dendritic Cells*

Fas also regulates B cell autoantibody production and antigen presenting cell function. As in T cells, Fas is dispensable for B cell development, but important in mediating peripheral B cell tolerance. Fas is not expressed on resting B cells, but is upregulated on activated B cells and highly expressed on germinal center B cells, some of which acquire autoreactive specificities and are eliminated through BCR ligation. Since B cells are not thought to upregulate FasL upon BCR stimulation, Fas-mediated B cell apoptosis probably depends on FasL produced by other cells. In this way T cells may indirectly regulate autoreactive B cells in the periphery. Study of antigen-specific T and B cell interactions showed that $CD4^+$ T cells can specifically eliminate autoreactive B cells in a Fas-dependent manner (Rathmell et al. 1995). In autoimmune Fas-deficient mice, nephritis could still occur to some degree in animals engineered to prevent antibody secretion, showing that the antigen presenting function of B cells is important in the pathogenesis of nephritis in this model system (Shlomchik et al. 1994). *Lpr* mice lacking B cells did not develop nephritis, and interestingly, this was accompanied by a concomitant reduction in the accumulation of memory phenotype $CD4^+$ T cells normally present in *lpr* mice (Chan and Shlomchik 1998). Thus, autoreactive B cells that fail to be eliminated through Fas may sustain autoimmunity through acting as antigen presenting cells for autoantigens and further activating autoreactive T cells. An essential role for B cell expression of Fas in maintaining self-tolerance was also shown in mice in which Fas was specifically eliminated in the B cell compartment. These mice developed characteristic lymphadenopathy, splenomegaly, high autoantibody titers and also accumulation of T cells, reemphasizing the role of Fas in the maintenance of peripheral B cell tolerance (Stranges et al. 2007).

Like B cells, DC may be eliminated through Fas–FasL interactions, and this may serve to downmodulate antigen presentation. Antigen-pulsed DC injected into mice were observed to disappear after 2–3 days from the draining lymph node if antigen-specific T cells are present, suggesting that T–DC interactions may be responsible for the elimination of DC (Ingulli et al. 1997). Further, accumulation of DC occurs in autoimmune diseases and animal models where apoptosis pathways are disrupted, such as in human patients harboring caspase-10 mutations and *lpr* mice (Fields et al. 2001; Wang et al. 1999). Transgenic mice in which the caspase inhibitor p35 was overexpressed in DC resulted in accumulation of DC in lymph nodes, T cell hyperplasia and development of antinuclear antibodies in older mice. These death-resistant DC also increased the rapidity of autoimmune manifestations in an autoimmune-prone mouse strain (Chen et al. 2006). Mice in which Fas was specifically deleted in DC also developed autoantibodies (Stranges et al. 2007). In both of these models the development and titer of autoantibodies were lower than in mice with universally disrupted Fas function. Taken together, these studies indicate that DC apoptosis is likely to occur physiologically during antigen presentation to T cells, and that this mechanism also contributes to peripheral self-tolerance. Remarkably, elimination of Fas expression in T cells, B cells and DC all contribute to this function. This pleiotropic role of Fas probably explains why mutations affecting Fas and FasL confer such potent susceptibility to autoimmune diseases. Although additional susceptibility genes govern the nature and severity of autoimmune disease pathology, development of autoantibody production is remarkably high in mice and humans with mutations disabling this pathway.

3.3 Fas on Nonimmune Cells

It is important to note that other cell types can also express Fas. One of the major nonimmune sites of Fas expression is in hepatocytes, which are quite sensitive to Fas-induced-apoptosis (Rouquet et al. 1996). Although Fas-deficient *lpr* mice do not develop liver hyperplasia, a small amount of Fas protein may still be produced by the *lpr* mutant Fas allele, and mice engineered to completely lack Fas protein did exhibit liver hyperplasia (Adachi et al. 1995). Though Fas is expressed at high levels, FasL is not expressed by hepatocytes. However, liver sinusoids do contain T cells which may express FasL. Administration of anti-mouse Fas antibodies results in lethal acute hepatic necrosis that is dependent on hepatocyte Fas expression and Fc-mediated crosslinking of these antibodies (Adachi et al. 1995; Ogasawara et al. 1993; Xu et al. 2003). During viral and other forms of hepatitis, FasL expressed on activated T cells may also play a role in causing hepatocyte damage (Kondo et al. 1997; Seino et al. 1997). Fas expression by target tissues of T-cell mediated autoimmune disease, such as the thyroid and pancreatic islets, may also play a role in tissue destruction in these conditions (Signore et al. 1998; Stassi and De Maria 2002).

4 Fas Receptor Signaling for Apoptosis: Ordered Assembly of Oligomeric Protein Complexes to Activate Caspase-8

Activation of the Fas signaling pathway begins with the binding of FasL or other receptor agonists, resulting in recruitment of the adaptor protein FADD (Fas associated death domain) and the cysteinyl aspartic proteases, caspase-8 (and caspase-10 in humans) to form a proximal signaling platform called the Death Inducing Signaling Complex (DISC) (Kischkel et al. 1995). The DISC can be detected within seconds of receptor engagement and functions to activate caspase-8/10, an essential step in initiation of programmed cell death. The recruitment and signaling specificities are maintained by alpha-helical modular domains that interact with each other and in some cases, also self-associate. The death domains (DD) in the intracytoplasmic region of Fas and the C-terminal of FADD interact to recruit FADD to the receptor. FADD also contains an amino-terminal death-effector domain (DED), structurally related to the DD but with affinity for other DED modules. FADD DED binds to DEDs in the prodomain of caspase-8/10, bringing them into the DISC. Aggregation and complex formation are necessary in the DISC to catalyze caspase-8 cleavage and downstream cleavage and to activate the effector caspase-3, which then culminates in apoptosis. The apoptotic machinery is irrevocable once effector caspases are activated, and many mechanisms have evolved to safeguard against wanton activation of cell death via Fas. Although posttranslational modifications such as phosphorylation or ubiquitination are not required for assembly and activation of the DISC, these steps are regulated at different stages, beginning with surface receptor clustering, ligand binding and efficiency of cytoplasmic complex formation at stages downstream of the DISC which will be discussed in turn below. Fas-induced apoptosis in cell lines was originally divided into 'Type I' or 'Type II' pathways depending on the ability of Fas-induced apoptosis to be blocked by overexpression of antiapoptotic bcl-2 family members (Scaffidi et al. 1998). More recent data has shown that these two signaling pathways also reflect differences between more proximal events in receptor signaling, such as the localization of receptors to lipid raft microdomains which in turn regulates the preassociation of receptors prior to ligand binding. In primary T cells, as we will discuss below, there is a spectrum of sensitivity to Fas-induced apoptosis that can be regulated by cytoskeletal remodeling through the Rac family of small GTPases and probably though other mechanisms as well.

Fas receptor, a 45 kDa type-I transmembrane glycoprotein, is a prototypic TNFRSF death receptor with a cytoplasmic 80 amino acid DD and three cysteine rich domains (CRD) in the extracytoplasmic region. Mutational analysis studies indicated that ligand binding for optimal signaling required the presence of all three CRDs (Orlinick et al. 1997). From the crystal structures of TNF ligands bound to their receptors, it was found that TNF ligands exist in trimers. The stoichiometry of ligand-receptor complexes was 3:3, and trimeric structures have also been found for adaptor molecules downstream of the receptor (Bodmer et al. 2002). This supported a model of signaling where there is cooperativity and dissemination of signaling via formation of ligand:receptor:adaptor heterocomplexes. It was therefore not surprising that in patients with Type1A ALPS, where heterozygous mutations of the Fas receptor are localized mostly to the DD, lymphocytes were highly resistant to Fas-mediated apoptosis, even though they have an equal gene dosage of wild-type and mutant Fas genes. Expression of Fas constructs containing ALPS-associated mutations dominantly interfered with Fas-induced apoptosis even in the presence of wild type Fas, supporting the theory of cooperativity in signaling where inclusion of mutant Fas disrupts formation of large complexes of ligand-receptor molecules. This results in inefficient DISC formation and caspase-8 activation (Fisher et al. 1995; Martin et al. 1998, 1999; Vaishnaw et al. 1999).

The above model posits that FasL binding initiates receptor trimerization, and downstream cytoplasmic events. However, in some patients with type 1A ALPS, Fas mutations occurred in the extracellular domain that disrupted FasL binding, but had intact intracellular signaling domains. Surprisingly, it was found that these mutants could also interfere with Fas-induced apoptosis. This finding gave rise to a new hypothesis that receptors could form preassociated complexes, without necessarily requiring ligand crosslinking. A 'preligand receptor association domain' (PLAD) in the N-terminal CRD1 portion of the receptor, distinct from the ligand binding site was subsequently identified in Fas, TNFR1, CD40, TACI, and TRAIL-receptors, and is probably a general feature of TNF receptors. (Chan et al. 2000; Clancy et al. 2005; Garibyan et al. 2007; Siegel et al. 2000). Dominant interference with receptor signaling by mutations in Fas were found to be dependent on the PLAD, favoring the hypothesis that receptor preassociation is required for and precedes ligand binding. An intact PLAD is also required for ligand binding, since ligand binding Fas mutants preassociate with wildtype receptors, but do not signal, indicating that receptor preassociation also aligns the receptors to maximally bind ligand. Interestingly, all the dominant interfering mutations identified so far in ALPS patients have an intact CRD1/PLAD region, even though some of them have mutations in the ligand binding region of CRD2, again emphasizing the role of receptor preassociation for signaling (Siegel et al. 2000). These findings make the PLAD a plausible therapeutic target to regulate TNFSF responses, and bacterially synthesized or synthetic peptides comprising of the PLAD were effective in blocking TNFR1 signaling in vitro and in a TNF-dependent model of inflammatory arthritis (Deng et al. 2005).

Engagement of Fas by FasL or agonistic anti Fas antibodies, induces a series of events; most of which begin with receptor clustering, formation of surface microaggregates and microclusters, receptor capping and internalization. The final outcome of these receptor-aggregated superclusters at the surface is efficient DISC assembly, thereby ensuring caspase-8 activation. Formation of SDS-stable Fas microaggregates occurs simultaneously with the formation of the DISC, and microscopically, surface receptor aggregates can be visualized in the same time frame. These structures were termed SPOTS, for Signaling Protein Oligomerization Transduction Structures. FADD recruitment was seen to be a necessary step in the formation of these large Fas surface clusters, since the SPOTS formation was drastically reduced in cells deficient for FADD or expressing Fas mutations unable to bind FADD. However, activation of caspase-8 was not necessary for the formation of SPOTS (Siegel et al. 2004). The formation of receptor microclusters is further stabilized and maintained via FADD homotypic associations. In addition to its function as an adaptor protein bridging Fas and caspase -8, FADD has an innate ability to oligomerize and form lateral interactions with itself, resulting in a characteristic filamentous structure termed 'death effector filaments'. This self-association resides in the DED and is independent of Fas and caspase-8 interactions. Mutations that disrupt FADD self-association resulted in dominant interfering mutants, indicating that caspase-8 recruitment is dependent on FADD self-association (Muppidi et al. 2006; Sandu et al. 2006).

As outlined above, early events in Fas signaling proceed in a stepwise manner, beginning with receptor preassociation, formation of microaggregates upon receptor binding and finally formation of large lateral signaling platforms, SPOTS. Until this step, there is no feedback regulation of downstream molecules in the DISC on the receptor aggregation. However, after the formation of SPOTS, the Fas signaling complex forms polar aggregates on one side of the cell, referred to as "capping," which are then internalized (Cremesti et al. 2001). Internalization itself may amplify Fas-induced apoptosis through further concentration of DISC components and activation of caspase-8. Receptor internalization is both actin and caspase dependent, indicating a feed-forward mechanism by which caspase-8 activity can enhance its own cleavage. The internalized receptor is then targeted to an endosomal pathway, since it colocalizes with the transferrin receptor (Algeciras-Schimnich et al. 2002). Internalization and capping occur more rapidly and prominently in Type I cells which correlates with more rapid DISC activation (Algeciras-Schimnich and Peter 2003; Eramo et al. 2004; Siegel et al. 2004). Receptor endocytosis can occur either through the clathrin or caveolin mediated pathway. Fas endocytosis is exclusively dependent on the clathrin coated pits, specialized membrane vesicles formed with the help of AP-2 (adaptor proteins) and surface dynamin resulting in cytoplasmic clathrin coated vesicles. These are targeted to the endosomes and give rise to the early endosome. RNAi mediated-knockdown of endogenous clathrin heavy chain or adaptor protein complex molecules resulted in accumulation of receptor clusters on the surface, and inhibited recruitment of FADD and caspase-8 cleavage in the DISC, resulting in abrogation of apoptosis by crosslinked soluble Fas ligand. Similar observations were seen also in primary human activated T cells, indicating that endocytosis via clathrin is required for Fas-mediated apoptosis. Interestingly, blocking endocytosis and DISC formation resulted in activation of the MAP kinase (mitogen activated protein kinase) as well as NF–κB pathway, indicating that endocytosis mediates either a proapoptotic or proliferative outcome downstream of Fas signaling (Lee et al. 2006). A feed-forward loop in which caspase-8 activates endocytosis, which in turn promotes further receptor aggregation and caspase-8 cleavage can explain how caspases appear to be both 'upstream' and 'downstream' of receptor endocytosis.

4.1 Lipid Raft Microdomains as Platforms for Efficient Fas Signaling

Lipid rafts are highly dynamic microdomains rich in sphingolipids and cholesterol, which facilitate the formation of many membrane-bound receptor signaling complexes. Lipid rafts are characterized by their insolubility in low ionic detergents, such as Triton X-100 or Brij98 (Munro 2003; Simons and Toomre 2000). They are less fluid than the traditional plasma membrane bilayer due to their lipid composition, allowing for isolation by low-density gradient centrifugation. Lipid raft distribution across the plasma membrane varies with cell type. For instance, polarized epithelial cells preferentially have lipid rafts localized to their apical surface. In lymphocytes, rafts tend to distribute over the cell surface without any defined polarity (Garcia et al. 2003). Although the size of individual rafts also varies, the consensus estimates range from 20 to 100 nm in diameter, making them essentially submicroscopic in nonpolarized cells when visualized by conventional microscopy. However, lipid rafts can coalesce and form much larger structures during signaling, as seen in the immunological synapse (Patra 2008; Simons and Toomre 2000).

Many recent studies have shown that lipid rafts play a critical role in immune cell signaling through the organization of signaling proteins, adaptor molecules and surface receptors at focal points on the cell membrane. It has been shown that lipid rafts act as signaling platforms for FcεRI (IgE) receptors, as well as the T cell and B cell receptor complexes, allowing the receptor to localize within close proximity to adaptor signaling components constitutively found in lipid microdomains (Cherukuri et al. 2001; Dykstra et al. 2003). Rafts are dynamic rather than static structures, allowing membrane proteins to flow in and out of them, thus changing the properties of the local protein milieu. This is the case with the FcεRI receptor on mast cells and basophils, where crosslinking of the FcεRI leads to translocation into rafts, where Lyn is a constituitive resident, and subsequent recruitment of Syk and PLCγ1 (Dykstra et al. 2003). In the case of T cell receptor (TCR) activation, rafts concentrate the coreceptors CD4 and CD8 (Resh 2006), as well as Src-family kinase Lck and many of the adaptor components needed for signaling, such as LAT (Kabouridis 2006). Upon TCR engagement, many additional components of the signaling cascade are recruited to lipid rafts, such as PKCθ and ZAP-70, among others (Bi and Altman 2001; Bi et al. 2001; Viola et al. 1999). CD4 partitions to lipid rafts via its interaction with Lck as well as its preferential S-palmitoylation (Resh 2006). CD4 stimulation enhances signaling by the TCR by inducing aggregation of lipid rafts and formation of molecular assemblies at the site of the immunological synapse. During TCR activation, many of the cytoplasmic signaling proteins become detergent-insoluble, probably because of the association with lipid rafts (Kabouridis 2006; Viola et al. 1999). Studies have shown that monomeric TCR complexes have weak raft affinity compared to receptor crosslinking, which increases raft-associated TCR molecules and the amount of TCR found in detergent-insoluble raft domains. Treatment of cells with methyl-β-cyclodextrin (MβCD) can dissociate these proteins from rafts and inactivate the signaling cascade (Janes et al. 1999; Montixi et al. 1998; Simons and Toomre 2000). The cascade of interactions occurring at the site of TCR stimulation builds up the immunological synapse, with the lipid raft microdomains critical to the stability and function of this complex and dynamic signaling assembly.

Membrane-anchored signaling kinases do not participate in TNF receptor family signal transduction, but the local membrane microenvironment can be just as important for efficiency of signaling. Work performed by many labs has shown very distinct functional outcomes for both TNFR1 and Fas signaling with regard to lipid rafts. TNFR1 translocated to lipid rafts very quickly after TNF treatment in HT1080 cells. Subsequent recruitment of TNFR signaling molecules RIP, TRADD and TRAF2 occurs very quickly: within 2 min of treatment, the TNF-induced signaling complex can be identified in lipid rafts, initiating NF–κB signaling through phosphorylation of IκBα (Legler et al. 2003). Cholesterol chelation (and subsequent disruption of lipid rafts) via cyclodextrin treatment inhibited IκB phosphorylation and induced apoptosis. Similarly, blockade of signaling molecule recruitment to the lipid rafts via dipalmitoyl–phosphatidylethanolamine (DPPE) (Legler et al. 2001) also impaired IκBα phosphorylation and increased apoptosis (Legler et al. 2003).

Lipid rafts have recently emerged as important regulators of Fas-induced apoptosis through regulating the efficiency of early events in Fas signaling. We have found that in type I cells, which make a stronger DISC, a fraction of Fas resides in lipid raft constitutively, while in Type II cells the receptor seems to be excluded from rafts during the early signaling events (Muppidi and Siegel 2004). This preassociation of Fas with lipid rafts in Type I cells allows them to undergo apoptosis even in the presence of low-valency Fas stimuli, while Type II cells cannot. Disruption of the lipid rafts through cholesterol chelation restores a requirement for Fas-crosslinking in Type I cells, while having no effect on Type II cells (Muppidi and Siegel 2004). In mouse thymocytes, Fas recruitment and localization in the lipid rafts was critical for efficient DISC formation and subsequent cell death (Hueber et al. 2002). Formation of crosslinkable preassociated receptor complexes through the N-terminal preligand assembly domain (PLAD), is more efficient in cells in which Fas partitions into lipid rafts (Muppidi and Siegel 2004; Siegel et al. 2000). Taken together, these studies suggest that lipid raft microdomains are necessary for efficient Fas signaling. Primary human $CD4^+$ T cell cultures generally respond to Fas stimuli in a Type II manner. Upon TCR engagement, however, Fas redistributes to lipid rafts and renders these cells sensitive to noncrosslinked anti Fas antibodies or natively synthesized FasL (Muppidi and Siegel 2004). As we will discuss in other sections, Rac-1 dependent cytoskeletal remodeling is required for this to occur.

Posttranslational modification of Fas may also play a critical role in its function in lipid rafts. Modification of proteins with saturated acyl groups can result in lipid raft localization, such as Srk-family kinases (Resh 2006). Also, proteins linked to saturated acyl chains, such as those directly acylated with two or more palmitate or a palmitate and myristate chain, can also be targeted to rafts. In fact, many membrane proteins localized to rafts carry posttranslational acyl modifications, such as N-myristoylation and/or S-palmitoylation. S-palmitoylation is a reversible modification involving addition of a 16-carbon palmitate moiety to a cysteine residue via a thioester linkage, and can be readily cleaved by palmitoyl thioesterases. An interesting feature of S-palmitoylation is its dynamic nature: cycles of palmitoylation and depalmitoylation occur in a regulated fashion for many proteins, allowing for translocation in and out of lipid raft microdomains (Resh 2006). Fas is palmitoylated at cysteine 199, just proximal to the cytoplasmic juxamembrane region (Chakrabandhu et al. 2007; Feig et al. 2007). Disruption of palmitoylation using competitive inhibitors or cleaving the thioester bond between palmitate and Fas receptor blocked Fas translocation to lipid rafts and inhibited formation of SDS-stable $CD95^{hi}$ aggregates associated with DISC formation (Feig et al. 2007). Mutation of cysteine 199 to prevent palmitoylation impaired DISC formation and inhibited Fas-induced cell death (Chakrabandhu et al. 2007; Feig et al. 2007). Palmitoylation of Fas is essential for raft association, and the apparent difference in raft-associated Fas observed in Type I-like vs. Type II-like cells could be due to the differential ability of the receptor to be palmitoylated. Interestingly, the death domain of TNFR1 is required for targeting to rafts, as deletion of the domain prevented the receptor from targeting to lipid rafts and resulted in more uniform distribution across the plasma membrane (Cottin et al. 2002). It is quite evident that posttranslational palmitoylation of Fas, allowing translocation to the lipid raft microdomains, is important for effective signaling. Whether there is a difference in palmitoylation states in Type I-like cells compared to Type II-like cells remains to be seen. The restricted lateral diffusion of membrane proteins found in the lipid microdomains, compared to the fluid plasma membrane, would favor oligimerization and formation of supra-clustering signaling complexes. Therefore, receptor preclustering via lipid raft targeting would act as a presignaling complex, whereby a Type I-like cell would respond to a low-level Fas stimulus more efficiently that a Type II-like cell. Fas is sequestered away from the rafts in Type II-like cells and is less likely to oligimerize into larger signaling complexes from a low-level signal. They require a stronger stimulus, probably involving crosslinking of multiple Fas receptors, to bring them into closer proximity and subsequent DISC formation and signaling.

The divergent outcomes of receptor signaling of both Fas and TNFR1 reflect the role of lipid rafts in receptor signaling efficiency in the primary signaling complexes. The apparent contradiction in the outcomes of lipid raft dissociation between Fas and TNFR1 become clearer when you take their respective signaling into account. TNFR1 signals through NF–κB via receptor-associated complex I, while the death signal is transduced through complex II, which is not associated with the receptor (Micheau and Tschopp 2003; Muppidi et al. 2004). In contrast, Fas signals through the receptor-associated DISC to trigger cell death. In one instance, cell death is promoted through dissolution of receptor-associated signaling (TNFR1), while in another apoptosis is abrogated due to inefficient formation of a signaling complex (Fas). As a result, lipid raft associated signaling has two very distinct outcomes in TNFR1 and Fas. Fas signaling should be seen as more of a dynamic process that involves more than just ligand binding to receptor, and that which also takes into account local microenvironment, subcellular localization, secondary modifications of Fas and global cell signaling pathways.

4.2 Cytoskeletal Reorganization as a Regulator of Fas-Induced Apoptosis

TCR induced lipid raft translocation of Fas is essential for the occurrence of Fas-mediated RICD in T cells. Our recent finding that Rac GTPases are critical for the death-inducing function of the TCR (Ramaswamy et al. 2007) implicates actin-based cytoskeletal remodeling in this process. Cytoskeletal remodeling was already known to play an important role in mediating sustained signaling as well as in strengthening and stabilizing T cell-APC contacts during initial T cell activation in the 'immune synapse' (IS). The IS is an area of dynamic contact between a T cell and its activating APC and its formation requires cytoskeletal rearrangements that occur as a consequence of initial signaling events. The IS results in sequestration of many signaling complexes with in lipid raft moieties of the membrane with rearrangement of the actin cytoskeleton which is critical for this event. Actin cytoskeletal rearrangement is dependent on Arp2/3 proteins, which are responsible for nucleating the F-actin and mediating T cell shape change to facilitate IS formation. Cytoskeletal remodeling also plays a role in negatively regulating T cell activation by terminating IS formation by the end of 6–8 h (Dustin and Chan 2000; Huang and Wange 2004).

Rac GTPases are one among a number of subfamilies of small G proteins that function in cytoskeletal remodeling, and act in parallel with Rho and CDC42 GTPases to alter the dynamics in T cells. We have found a unique role for Rac GTPases in TCR signaling to sensitize cells to die via Fas (Ramaswamy et al. 2007). Upon TCR ligation, the guanine-nucleotide exchange factor (GEF) Vav-1, activates Rac to mediate actin-based cytoskeletal changes (Tybulewicz 2005). Vav1 mediated Rac activation also leads to sustained lipid raft clustering in T cells (Villalba et al. 2001). In cell lines as well as in primary human cells, Rac protein knockdown blocks TCR induced Fas apoptosis. Unlike the more widespread effects of Vav1, Rac1-mediated sensitization to Fas is independent of early TCR signaling events, but appears linked to Rac1 dependent cytoskeletal reorganization, likely through dephosphorylation of The Ezrin Radixin Moesin (ERM) proteins (Ramaswamy et al. 2007). The ERM family of proteins link membrane proteins to the underlying cytoskeleton and modulate various biological functions such as adhesion, motility, polarization and signaling in lymphocytes. Dephosphorylation of ERM proteins induces a conformational change that dissociates the N-terminal ERM domain from the cytoplasmic tails of membrane proteins with which they interact (Bretscher et al. 2002). Receptor triggered ERM dephosphorylation, in many instances, is a Rac1 dependent process (Nijhara et al. 2004) and during T cell activation by antigen presenting cells, Rac1 mediated ERM dephosphorylation results in a generalized increase in T cell deformability. This allows for more efficient T cell–APC conjugation and potentiates T cell activation (Faure et al. 2004). Since Fas has been found linked to the cytoskeleton through ezrin (Parlato et al. 2000), ERM dephosphorylation may allow Fas to dissociate from the cytoskeleton and migrate to lipid raft microdomains where Fas-induced apoptosis signaling is potentiated (Muppidi and Siegel 2004). Rac1 has also been shown to promote repositioning of the centrosome toward the area of contact with target cells, which is important for efficient CTL and NK-cell mediated cytotoxicity (Billadeau et al. 1998; Gomez et al. 2007; Stinchcombe et al. 2006). Whether or not this form of cytoskeletal rearrangement could also be important in Fas–FasL mediated apoptosis is not known.

Since Rac GTPases mediate signaling by many receptors on activated T cells, other receptors alter the threshold for Fas-induced apoptosis through this mechanism. CD44 is one such surface molecule that also activates Rac1, and enhances Fas-induced apoptosis. T cells from CD44-deficient mice are resistant to RICD and have increased pathology in a ConA-induced hepatitis model, known to be controlled by Fas elimination of activated T cells (Chen et al. 2001). Thus, CD44 acts as an endogenous sensitizer of Fas. Deliberate activation of receptors such as CD44 that activate Rac1 or low-level TCR stimulation may form a basis for curtailing autoimmune effector T cells by sensitizing them to Fas-induced apoptosis (Ramaswamy et al. 2007).

4.3 Intracellular Regulation of the Fas-Induced Death Signal

In Type II cells, DISC formation is inefficient and activation of caspase-8 is delayed, resulting in insufficient initiator caspases to trigger apoptosis without amplification through the mitochondrial death pathway. Activation of the mitochondrial pathway is induced by intrinsic DNA damage signals or growth factor deprivation that promotes release of cytochrome C present in the intermembrane space of the outer mitochondrial membrane. The loss of mitochondrial outer membrane potential (MOMP) and cytochrome C release potentiates formation of the apoptosome complex, which is equivalent to the DISC and consists of cytochrome C complexed to APAF-1 (apoptotic protease activating factor-1). SMAC/Diablo, another molecule released from mitochondria, inactivates inhibitory proteins bound to caspases (Verhagen and Vaux 2002). Activated APAF-1, cytochrome-c and caspase -9 form a complex (the 'apoptosome') that very efficiently activates the effector caspases, caspase 3 and 7 (Riedl and Salvesen 2007). Apoptosis-inducing receptors such as Fas can activate the intrinsic cell death pathway via a member of the Bcl-2 protein family called Bid. Bid, a BH3 (Bcl-2 homology domain 3) only proapoptotic BCL-2 family member, is a substrate of active caspase-8, which cleaves it into a truncated form, designated tBid (Li et al. 1998; Luo et al. 1998). tBid translocates to mitochondria and allows multimerization of proapoptotic Bcl2 family proteins, BAK/BAX that induce loss of MOMP and cytochrome C release (Youle and Strasser 2008). The limited amounts of caspase-8 generated in the Type II DISC is sufficient to cleave Bid, but not effector caspase-3. Hence, Type II cells are very dependent on the mitochondrial amplification via Bid and are susceptible to Bcl-2 inhibition when stimulated with Fas. Bid knockout mice have no developmental phenotype, but exhibit a dramatic resistance to the hepatic necrosis induced by anti Fas antibody treatment, most likely due to hepatocyte resistance to Fas-induced apoptosis. However, resistance to Fas-induced apoptosis was not seen in lymphocytes, possibly because they are mostly Type I and do not require a mitochondrial amplification loop to undergo cell death (Yin et al. 1999).

In T lymphocytes the intrinsic pathway of apoptosis acts as a negative regulator in cells undergoing various forms of stress, including lack of growth factors or cytokines. Bcl-2 blocks oligomerization of Bak/Bax, thus, preventing loss of MOMP. Mice lacking Bcl-2 have severe immune defects, especially in T cell activation, and also polycystic kidney disease, whereas mice with transgenic over expression of Bcl-2 have lymphadenopathy and abnormal survival of peripheral T cells. Bim, a BH3 only Bcl-2 family member is a cytosolic protein that acts on other Bcl-2 family members to induce MOMP. Bim knock-out mice show severe lymphadenopathy and kidney disease due to autoimmune attack (Bouillet et al. 1999). In Bcl-2 deficient mice, parallel knock out of even one allele of Bim rescues many of the proliferation defects as well as the severe kidney disease, indicating that most of the Bim proapoptotic function is counteracted by Bcl-2. Bim knockout mice also have severe defects in thymic negative selection. T cells expressing a Bcl-2 transgene or lacking Bim, but not Fas-deficient T cells, had severe defects in undergoing apoptosis when exposed to cytokine withdrawal and acute antigenic stimulation. However, in a more chronic stimulation model, Fas-deficient T cells were apoptosis resistant, indicating that repeated TCR stimulation as occurs in chronic infections as well as with endogenous antigens elicits the Fas pathway (Hildeman et al. 2002; Van Parijs et al. 1998). This notion of separate contributions of the intrinsic and extrinsic cell death pathways to immune homeostasis and self tolerance was borne out by studies of mice doubly deficient in Bim and Fas(Hughes et al. 2008; Hutcheson et al. 2008; Weant et al. 2008). These animals develop massive splenomegaly, and lymphadenopathy due to increases in not only T cells, but also B cells and DC numbers. Acute viral infection with HSV-1 resulted in accumulation of virus specific T cells only in Bim deficient mice, with no synergistic increase in the Bim/Fas double deficient ones. Interestingly, in a model of chronic infection, there is cooperation between Fas and Bim, with synergistic accumulation of virus-specific T cells (Hughes et al. 2008; Hutcheson et al. 2008; Weant et al. 2008). These findings indicate that the Fas death receptor signaling intersects with the intrinsic cell death machinery at two levels, one at the membrane proximal signaling through Bid and secondly in a more indirect manner with Bim. Different models of chronic infection and autoimmunity likely involve more or less cross-talk between these two pathways.

5 Nonapoptotic Signaling Through Fas

Members of the TNF-receptor superfamily induce apoptosis, inflammation and proliferation. Unlike TNFR1, which induces either apoptosis or inflammation, Fas is considered a prototypic death receptor, as signaling induces apoptosis in most circumstances. However, Fas can also induce nonapoptotic signaling resulting in cell proliferation. Anti-Fas antibodies are known to costimulate T cell proliferation and secretion of cytokines such as IL-2 and IFN–γ, in the presence of minimal TCR stimulation (Alderson et al. 1993). This effect, blocked by Fas antagonists and synergistically increased by addition of exogenous FasL, appears to be caspase dependent, since caspase inhibitors block T cell proliferation co-stimulated through Fas (Kennedy et al. 1999). Fas co-stimulation has most often been elicited in vitro with anti-Fas antibodies or recombinant FasL. In vivo, however, T cell activation defects or lymphopenia was not observed in Fas or FasL deficient mice or in ALPS patients, suggesting that physiologically, Fas–FasL functions are mostly inducers of apoptosis.

Another nonapoptotic axis of Fas signaling is mediated via the NF–κB pathway. This pathway is dependent on FADD/caspase-8 and independent of mitochondrial activation, since caspase-8 inhibitor, cFLIP, but not Bcl-2 overexpression in Type II cells, inhibits NF–κB activation (Imamura et al. 2004; Miwa et al. 1998). Interestingly, varying signaling thresholds for the apoptotic pathway appear to determine activation of alternate Fas signaling pathways. In ALPS patients with DD mutations, NF–κB signaling is activated in spite of the dominant interference caused by mutant receptors (Legembre et al. 2004). Biochemically, internalization of the Fas receptor induces the apoptotic pathway; however noninternalized receptors result in activation of NF–κB (Lee et al. 2006). Furthermore, in Fas resistant tumor cells, induction of NF–κB as well as activation of MAPK pathways increases motility and invasiveness, specifically in response to Fas crosslinking, but not to TNF-α or TRAIL treatment (Barnhart et al. 2004). Therefore, Fas, similar to the dual signaler TNFR1, induces inflammatory outcomes resulting in tumor formation, dependent on different physiological conditions as well as receptor and ligand signaling thresholds. However, these alternate functions of Fas, dependent on the presence of intact downstream DISC components, may reflect nonapoptotic functions of FADD or caspase-8, two important mediators of developmental regulation as well as lymphoproliferation.

Remarkably, FADD, caspase-8 and c-FLIP, essential components of the Fas apoptosis-inducing complex, all appear to be required for embryonic development and early events in T cell signaling that lead to NF–κB activation. FADD is essential for embryonic development, since its deletion results in embryonic lethality. This is in contrast to a lack of lethality in mice lacking receptors such as Fas or TRAIL-R that use FADD as an adapter. In order to study the role of FADD in immune system development, $FADD^{-/-}$ embryonic stem cells were reconstituted into $RAG1^{-/-}$ mice. B cell development and T cell activation required FADD function, since these mice had no B cells and very few peripheral T cells. The peripheral $FADD^{-/-}$ T cells however were totally resistant to Fas-induced apoptosis, indicating that the proliferative function of FADD in the thymus is separate from the apoptosis/adaptor function in peripheral T cells (Zhang et al. 1998). However, conditional knockout studies indicate that FADD is not essential for thymic development, but is important in peripheral T cell activation and proliferation as well as in Fas-induced apoptosis (Zhang et al. 2005). The proliferative function of FADD depends on the phosphorylation of a serine residue, which is essential for cell cycle progression but not apoptosis induction (Park et al. 2005). Deficiency of caspase-8 as well as another DISC component, cFLIP, results in embryonic lethality, and similar to the FADD deficient mice, T cell-specific ablation of c-FLIP or caspase-8 result in similar defects in T cell activation, proliferation and postactivation survival (Siegel 2006). More recently, caspase-8 was found to activate TCR induced NF–κB activation mediated through the BCM complex (Bcl-10/CARMA1/MALT1) by targeting IKK to the complex and stabilizing NF–κB activation (Su et al. 2005). A Fas-mediated caspase-8 independent pathway of apoptosis or necrosis also occurs in some primary human as well as murine T cells, where, presence of caspase inhibitor, zVAD did not prevent cell death. However, this caspase independent death was found to be dependent on FADD. It was also found that the necrotic cell death was dependent on the receptor interacting protein (RIP), which interacted with Fas in a FADD dependent manner (Kataoka et al. 2000).

6 Differential Function of Fas in CD4$^+$ T Cell Subsets

Upon encountering an antigen, a naïve CD4 T cell undergoes a series of events beginning with activation by the antigen presenting APC followed by an enormous clonal expansion and then attrition of the response by cytokine withdrawal and finally, generation of memory cells. Among the different checkpoints that exist in controlling these events, Fas-mediated apoptosis occurs usually in cells that overcome the clonal contraction phase. However, not all T cell subset homeostasis is governed by Fas-induced apoptosis and recent studies indicate that memory T cells have intrinsic sensitivity to Fas apoptosis. Memory T cells, formed from the repertoire of antigen specific cells, are long-lived and characterized by quick induction and secretion of effector cytokines. Differential expression of the chemokine receptor CCR7 and CD27, a TNFRSF member, delineates functional memory subtypes into central and effector memory cells (Sallusto et al. 2000), with the central memory cells retaining the ability to recirculate to the lymph node, and the effector cells residing predominantly in target tissues producing cytokines more efficiently. There have been a number of studies recently indicating that deregulated memory T cells may be important in driving autoimmune pathology. In Fas-deficient *lpr* mice, memory T cells accumulate abnormally and a similar phenomenon has been observed in patients with systemic lupus erythematosus (SLE), a polygenic autoimmune disease (Fritsch et al. 2006). Furthemore, T cells from SLE patients were found to be resistant to TCR and Fas-mediated apoptosis (Kovacs et al. 1996). In normal donors, the effector, but not the central memory cells are intrinsically sensitive to apoptosis and have upregulated levels of FasL and Bim. One mechanism of enhanced survival of central memory cells was ascribed to activation effects of the AKT kinase signaling pathway, which leads to inactivation of FOXO3a, a fork-head family transcription factor that upregulates both FasL and Bim expression (Riou et al. 2007). Since effector memory cells are potent effector cytokine producers with tissue homing abilities and a high turnover rate, any dysregulation in their numbers could lead to pathological consequences. Regulation of Fas-induced apoptosis, thereby, serves to counterbalance autoreactive T cell escape of these terminally differentiated cells.

Cytokine secretion patterns define lineages of peripheral $CD4^+$ T cells. Classically, $CD4^+$ T cells were divided into two lineages, Th1 or Th2, depending on upregulation of specific transcription factors, which induced lineage specific cytokines. Due to the diverse cytokine profiles, the lineages differ in their functionality, with Th1 effecting cell-mediated immunity and Th2 cells providing humoral defense (Abbas et al. 1996). Differential Fas sensitivity was seen in human T cell clones, where, Th1, but not Th2 clones were susceptible to TCR induced RICD, mostly regulated at the transcriptional level of FasL (Ramsdell et al. 1994a).

Interferon-γ (IFN-γ), the principal cytokine secreted by Th1 T cells, also regulates the sensitivity to Fas-mediated apoptosis, since both inhibition of IFN-γ secretion by blocking antibodies or $IFN\gamma^{-/-}$ T cells are resistant to RICD (Liu and Janeway 1990; Refaeli et al. 2002). Another important CD4 T cell subset that is regulated by the Fas pathway is the regulatory T cell lineage (Treg). This subset is necessary for maintaining peripheral tolerance through immune suppression of other effector cell types. It has been found that Tregs exhibit an activated effector phenotype, being CD25 high and CD45RO positive. Tregs are susceptible to Fas-induced apotposis exvivo (Fritzsching et al. 2005). Recently, however, a small subset of Fas resistant $CD45RA^+$ 'naive' Treg were identified in cord blood (Fritzsching et al. 2006). The susceptibility of Treg to Fas-induced apoptosis may have more to do with their memory-like phenotype when commonly isolated from adults. Recent research has identified new $CD4^+$ subsets secreting IL-17, IL-21 and IL-22 (Laurence and O'Shea 2007; Laurence et al. 2008; Tato et al. 2006). The susceptibility of these subsets to Fas- mediated apoptosis is currently being investigated.

7 Therapeutic Use of Fas–FasL Interactions in Cancer and Immunological Diseases

The potential of Fas to permanently alter cell fate and eliminate cells through apoptosis has sparked efforts to harness Fas to eliminate pathogenic T and B cells or Fas-positive tumor cells in a number of clinical settings. Although anti-Fas antibodies induced acute hepatic necrosis, systemic dosing of FasL or antibodies not cross linked by Fc receptors results in reduced hepatotoxicity, and it may be possible to use this approach to selectively eliminate Fas-sensitive tumors or autoreactive lymphocytes. This approach is limited by the fact that in many cell types, both from tumors and normal tissues, Fas expression does not correlate well with sensitivity to Fas-induced apoptosis. The well-known potential of tumor cell lines to become resistant to Fas-induced apoptosis after prolonged culture, due to loss of Fas or components in the Fas signaling pathway, has been helpful in isolating mutants lacking Fas signaling components, but also raises concerns about the development of resistance to anti-Fas therapy if it were attempted in cancers. Stimulation of Fas on tumor cell lines resistant to Fas-induced apoptosis induces characteristics of motility and invasiveness (Barnhart et al. 2004), raising further concerns about unexpected effects of Fas stimulation in cancer.

Since endogenous FasL expression on selected tissues can induce immune tolerance and elimination of reactive T cells (Stuart et al. 1997), genetic delivery of FasL to grafts in a transplantation setting was also hoped to prolong graft survival through elimination of alloreactive lymphocytes. Despite some early reports of success with this approach (Lau et al. 1996), the vast majority of experiments in which FasL is ectopically expressed on graft tissue by retroviral transduction or transgenes under strong promoters, results in hyperacute rejection with a predominantly neutrophilic infiltrate in the graft (Kang et al. 1997; Takeuchi et al. 1999). A direct chemotactic effect of soluble FasL on neutrophils has been reported (Ottonello et al. 1999), but in vivo, membrane-bound or vesicular FasL induces rapid apoptosis of macrophages, which then secrete proinflammatory cytokines and chemotactic factors, of which IL-1 and IL-17 appear to be important in neutrophil recruitment (Hohlbaum et al. 2001; Umemura et al. 2004). Whether this proinflammatory effect of ectopic expression of FasL can be overcome to deliver a purely apoptotic signal to infiltrating lymphocytes in the setting of graft rejection is not known. More promising are studies in which FasL has been implicated as being more potent in mediating graft-vs. host disease than beneficial graft-vs. leukemia effects (Jiang et al. 2001; Schmaltz et al. 2001). These studies suggest that blocking Fas–FasL interactions could potentially ameliorate graft-vs. host disease in bone marrow transplants while leaving intact the beneficial graft-vs. leukemia activity by donor cells. Interestingly, as discussed above, effector memory T cells, which are very susceptible to TCR and Fas-induced apoptosis, have the ability in vivo to selectively mediate graft-vs. leukemia effects (Zheng et al. 2008). This raises the interesting possibility that alloantigen-induced Fas-mediated apoptosis in effector memory T cells may cull the immune repertoire of pathogenic graft-vs. host disease causing T cells while leaving intact graft-vs. leukemia effects, which are predominantly mediated by mechanisms other than Fas–FasL interactions.

Although the barriers to using direct ligation of Fas in immunotherapy in autoimmune disease remain high, interventions to eliminate autoreactive lymphocytes by enhancing apoptosis via Fas–FasL interactions may still prove effective. In diseases in which autoantigens are known, chronic stimulation of the TCR with antigen has been shown to mediate Fas-induced apoptosis and ameliorate animal models of antigen-specific autoimmune diseases (Critchfield et al. 1994). As the understanding of the pathways that regulate Fas-induced apoptosis grows, enhancing sensitivity to Fas-mediated apoptosis through stimulation of receptors that activate the 'competancy to die' signal or otherwise enhance FasL sensitivity remains a worthy therapeutic goal.

References

Abbas AK, Murphy KM, Sher A (1996) Functional diversity of helper T lymphocytes. Nature 383:787–793

Adachi M, Suematsu S, Kondo T, Ogasawara J, Tanaka T, Yoshida N, Nagata S (1995) Targeted mutation in the Fas gene causes hyperplasia in peripheral lymphoid organs and liver. Nat Genet 11:294–300

Alderson MR, Armitage RJ, Maraskovsky E, Tough TW, Roux E, Schooley K, Ramsdell F, Lynch DH (1993) Fas transduces activation signals in normal human T lymphocytes. J Exp Med 178:2231–2235

Algeciras-Schimnich A, Peter ME (2003) Actin dependent CD95 internalization is specific for Type I cells. FEBS Lett 546:185–188

Algeciras-Schimnich A, Shen L, Barnhart BC, Murmann AE, Burkhardt JK, Peter ME (2002) Molecular ordering of the initial signaling events of CD95. Mol Cell Biol 22:207–220

Anel A, Buferne M, Boyer C, Schmitt-Verhulst AM, Golstein P (1994) T cell receptor-induced Fas ligand expression in cytotoxic T lymphocyte clones is blocked by protein tyrosine kinase inhibitors and cyclosporin A. Eur J Immunol 24:2469–2476

Barnhart BC, Legembre P, Pietras E, Bubici C, Franzoso G, Peter ME (2004) CD95 ligand induces motility and invasiveness of apoptosis-resistant tumor cells. EMBO J 23:3175–3185

Bellgrau D, Gold D, Selawry H, Moore J, Franzusoff A, Duke RC (1995) A role for CD95 ligand in preventing graft rejection. Nature 377:630–632

Bi K, Altman A (2001) Membrane lipid microdomains and the role of PKCtheta in T cell activation. Semin Immunol 13:139–146

Bi K, Tanaka Y, Coudronniere N, Sugie K, Hong S, van Stipdonk MJ, Altman A (2001) Antigen-induced translocation of PKC-theta to membrane rafts is required for T cell activation. Nat Immunol 2:556–563

Billadeau DD, Brumbaugh KM, Dick CJ, Schoon RA, Bustelo XR, Leibson PJ (1998) The Vav-Rac1 pathway in cytotoxic lymphocytes regulates the generation of cell-mediated killing. J Exp Med 188:549–559

Blott E, Griffiths G (2002) Secretory lysosomes. Nat Rev Mol Cell Biol 3:122–131

Bodmer JL, Schneider P, Tschopp J (2002) The molecular architecture of the TNF superfamily. Trends Biochem Sci 27:19–26

Bonfoco E, Stuart PM, Brunner T, Lin T, Griffith TS, Gao Y, Nakajima H, Henkart PA, Ferguson TA, Green DR (1998) Inducible nonlymphoid expression of Fas ligand is responsible for superantigen-induced peripheral deletion of T cells. Immunity 9:711–720

Bouillet P, Metcalf D, Huang DC, Tarlinton DM, Kay TW, Kontgen F, Adams JM, Strasser A (1999) Proapoptotic Bcl-2 relative Bim required for certain apoptotic responses, leukocyte homeostasis, and to preclude autoimmunity. Science 286:1735–1738

Bretscher A, Edwards K, and Fehon RG (2002) ERM proteins and merlin: integrators at the cell cortex. Nat Rev Mol Cell Biol 3:586–599

Brunner T, Yoo NJ, LaFace D, Ware CF, Green DR (1996) Activation-induced cell death in murine T cell hybridomas. Differential regulation of Fas (CD95) versus Fas ligand expression by cyclosporin A and FK506. Int Immunol 8:1017–1026

Brunner T, Kasibhatla S, Pinkoski MJ, Frutschi C, Yoo NJ, Echeverri F, Mahboubi A, Green DR (2000) Expression of Fas ligand in activated T cells is regulated by c-Myc. J Biol Chem 275:9767–9772

Cahuzac N, Baum W, Kirkin V, Conchonaud F, Wawrezinieck L, Marguet D, Janssen O, Zornig M, Hueber AO (2006) Fas ligand is localized to membrane rafts, where it displays increased cell death-inducing activity. Blood 107:2384–2391

Chakrabandhu K, Herincs Z, Huault S, Dost B, Peng L, Conchonaud F, Marguet D, He HT, Hueber AO (2007) Palmitoylation is required for efficient Fas cell death signaling. EMBO J 26:209–220

Chan O, Shlomchik MJ (1998) A new role for B cells in systemic autoimmunity: B cells promote spontaneous T cell activation in MRL-lpr/lpr mice. J Immunol 160:51–59

Chan FK, Chun HJ, Zheng L, Siegel RM, Bui KL, Lenardo MJ (2000) A domain in TNF receptors that mediates ligand-independent receptor assembly and signaling. Science 288:2351–2354

Chen D, McKallip RJ, Zeytun A, Do Y, Lombard C, Robertson JL, Mak TW, Nagarkatti PS, Nagarkatti M (2001) CD44–deficient mice exhibit enhanced hepatitis after concanavalin A injection: evidence for involvement of CD44 in activation-induced cell death. J Immunol 166:5889–5897

Chen M, Wang Y-H., Wang Y, Huang L, Sandoval H, Liu Y-J., Wang J (2006) Dendritic Cell Apoptosis in the Maintenance of Immune Tolerance. Science 311:1160–1164

Cherukuri A, Dykstra M, Pierce SK (2001) Floating the raft hypothesis: lipid rafts play a role in immune cell activation. Immunity 14:657–660

Chow WA, Fang JJ, Yee JK (2000) The IFN regulatory factor family participates in regulation of Fas ligand gene expression in T cells. J Immunol 164:3512–3518

Clancy L, Mruk K, Archer K, Woelfel M, Mongkolsapaya J, Screaton G, Lenardo MJ, Chan FK (2005) Preligand assembly domain-mediated ligand-independent association between TRAIL receptor 4 (TR4) and TR2 regulates TRAIL-induced apoptosis. Proc Nat Acad Sci U S A 102:18099–18104

Combadiere B, Reis e Sousa C, Trageser C, Zheng LX, Kim CR, Lenardo MJ (1998) Differential TCR signaling regulates apoptosis and immunopathology during antigen responses in vivo. Immunity 9:305–313

Cottin V, Doan JE, Riches DW (2002) Restricted localization of the TNF receptor CD120a to lipid rafts: a novel role for the death domain. J Immunol 168:4095–4102

Cremesti A, Paris F, Grassme H, Holler N, Tschopp J, Fuks Z, Gulbins E, Kolesnick R (2001) Ceramide enables fas to cap and kill. J Biol Chem 276:23954–23961

Critchfield JM, Racke MK, Zuniga-Pflucker JC, Cannella B, Raine CS, Goverman J, Lenardo MJ (1994) T cell deletion in high antigen dose therapy of autoimmune encephalomyelitis. Science 263:1139–1143

Davidson WF, Haudenschild C, Kwon J, Williams MS (2002) T cell receptor ligation triggers novel nonapoptotic cell death pathways that are Fas-independent or Fas-dependent. J Immunol 169:6218–6230

Deng GM, Zheng L, Chan FK, Lenardo M (2005) Amelioration of inflammatory arthritis by targeting the pre-ligand assembly domain of tumor necrosis factor receptors. Nat Med 11:1066–1072

Dhein J, Walczak H, Baumler C, Debatin KM, Krammer PH (1995) Autocrine T-cell suicide mediated by APO-1/(Fas/CD95) Nature 373:438–441

Dustin ML, Chan AC (2000) Signaling takes shape in the immune system. Cell 103:283–294

Dykstra M, Cherukuri A, Sohn HW, Tzeng SJ, Pierce SK (2003) Location is everything: lipid rafts and immune cell signaling. Annu Rev Immunol 21:457–481

Dzialo-Hatton R, Milbrandt J, Hockett RD, Jr., Weaver CT (2001) Differential expression of Fas ligand in Th1 and Th2 cells is regulated by early growth response gene and NF-AT family members. J Immunol 166:4534–4542

Eramo A, Sargiacomo M, Ricci-Vitiani L, Todaro M, Stassi G, Messina CG, Parolini I, Lotti F, Sette G, Peschle C, De Maria R (2004) CD95 death-inducing signaling complex formation and internalization occur in lipid rafts of type I and type II cells. Eur J Immunol 34:1930–1940

Faure S, Salazar-Fontana LI, Semichon M, Tybulewicz VL, Bismuth G, Trautmann A, Germain RN, Delon J (2004) ERM proteins regulate cytoskeleton relaxation promoting T cell-APC conjugation. Nat Immunol 5:272–279

Feig C, Tchikov V, Schutze S, Peter ME (2007) Palmitoylation of CD95 facilitates formation of SDS-stable receptor aggregates that initiate apoptosis signaling. EMBO J 26:221–231

Fields ML, Sokol CL, Eaton-Bassiri A, Seo S, Madaio MP, Erikson J (2001) Fas/Fas ligand deficiency results in altered localization of anti-double-stranded DNA B cells and dendritic cells. J Immunol 167:2370–2378

Fisher GH, Rosenberg FJ, Straus SE, Dale JK, Middleton LA, Lin AY, Strober W, Lenardo MJ, Puck JM (1995) Dominant interfering Fas gene mutations impair apoptosis in a human autoimmune lymphoproliferative syndrome. Cell 81:935–946

Fritsch RD, Shen X, Illei GG, Yarboro CH, Prussin C, Hathcock KS, Hodes RJ, Lipsky PE (2006) Abnormal differentiation of memory T cells in systemic lupus erythematosus. Arthritis Rheum 54:2184–2197

Fritzsching B, Oberle N, Eberhardt N, Quick S, Haas J, Wildemann B, Krammer PH, Suri-Payer E (2005) In contrast to effector T cells, CD4+CD25+FoxP3+ regulatory T cells are highly susceptible to CD95 ligand- but not to TCR-mediated cell death. J Immunol 175:32–36

Fritzsching B, Oberle N, Pauly E, Geffers R, Buer J, Poschl J, Krammer P, Linderkamp O, Suri-Payer E (2006) Naive regulatory T cells: a novel subpopulation defined by resistance toward CD95L-mediated cell death. Blood 108:3371–3378

Garcia A, Cayla X, Fleischer A, Guergnon J, Alvarez-Franco Canas F, Rebollo MP, Roncal F, Rebollo A (2003) Rafts: a simple way to control apoptosis by subcellular redistribution. Biochimie 85:727–731

Garibyan L, Lobito AA, Siegel RM, Call ME, Wucherpfennig KW, Geha RS (2007) Dominant-negative effect of the heterozygous C104R TACI mutation in common variable immunodeficiency (CVID) J Clin Invest 117:1550–1557

Genestier L, Kasibhatla S, Brunner T, Green DR (1999) Transforming growth factor beta1 inhibits Fas ligand expression and subsequent activation-induced cell death in T cells via downregulation of c-Myc. J Exp Med 189:231–239

Gomez TS, Kumar K, Medeiros RB, Shimizu Y, Leibson PJ, Billadeau DD (2007) Formins regulate the actin-related protein 2/3 complex-independent polarization of the centrosome to the immunological synapse. Immunity 26:177–190

Gourley TS, Chang CH (2001) Cutting edge: the class II transactivator prevents activation-induced cell death by inhibiting Fas ligand gene expression. J Immunol 166:2917–2921

He JS, Ostergaard HL (2007) CTLs contain and use intracellular stores of FasL distinct from cytolytic granules. J Immunol 179:2339–2348

Hildeman DA, Zhu Y, Mitchell TC, Bouillet P, Strasser A, Kappler J, Marrack P (2002) Activated T cell death in vivo mediated by proapoptotic bcl-2 family member bim. Immunity 16:759–767

Hohlbaum AM, Gregory MS, Ju ST, Marshak-Rothstein A (2001) Fas ligand engagement of resident peritoneal macrophages in vivo induces apoptosis and the production of neutrophil chemotactic factors. J Immunol 167:6217–6224

Hornung F, Zheng L, Lenardo MJ (1997) Maintenance of clonotype specificity in CD95/Apo-1/Fas-mediated apoptosis of mature T lymphocytes. J Immunol 159:3816–3822

Huang Y, Wange RL (2004) T cell receptor signaling: beyond complex complexes. J Biol Chem 279:28827–28830

Hueber AO, Bernard AM, Herincs Z, Couzinet A, He HT (2002) An essential role for membrane rafts in the initiation of Fas/CD95-triggered cell death in mouse thymocytes. EMBO Rep 3:190–196

Hughes PD, Belz GT, Fortner KA, Budd RC, Strasser A, Bouillet P (2008) Apoptosis regulators Fas and Bim cooperate in shutdown of chronic immune responses and prevention of autoimmunity. Immunity 28:197–205

Hutcheson J, Scatizzi JC, Siddiqui AM, Haines GK, 3rd, Wu T, Li QZ, Davis LS, Mohan C, Perlman H (2008) Combined deficiency of proapoptotic regulators Bim and Fas results in the early onset of systemic autoimmunity. Immunity 28:206–217

Imamura R, Konaka K, Matsumoto N, Hasegawa M, Fukui M, Mukaida N, Kinoshita T, Suda T (2004) Fas ligand induces cell-autonomous NF-kappaB activation and interleukin-8 production by a mechanism distinct from that of tumor necrosis factor-alpha. J Biol Chem 279:46415–46423

Ingulli E, Mondino A, Khoruts A, Jenkins MK (1997) In vivo detection of dendritic cell antigen presentation to CD4(+) T cells. J Exp Med 185:2133–2141

Janes PW, Ley SC, Magee AI (1999) Aggregation of lipid rafts accompanies signaling via the T cell antigen receptor. J Cell Biol 147:447–461

Jiang Z, Podack E, Levy RB (2001) Major histocompatibility complex-mismatched allogeneic bone marrow transplantation using perforin and/or Fas ligand double-defective CD4(+) donor T cells: involvement of cytotoxic function by donor lymphocytes prior to graft-versus-host disease pathogenesis. Blood 98:390–397

Jodo S, Xiao S, Hohlbaum A, Strehlow D, Marshak-Rothstein A, Ju ST (2001) Apoptosis-inducing membrane vesicles. A novel agent with unique properties. J Biol Chem 276:39938–39944

Jodo S, Pidiyar VJ, Xiao S, Furusaki A, Sharma R (2005) Cutting edge: Fas Ligand (CD178) Cytoplasmic tail is a positive regulator of Fas ligand-mediated cytotoxicity . J Immunol 174:4470–4474

Ju ST, Panka DJ, Cui H, Ettinger R, el-Khatib M, Sherr DH, Stanger BZ, Marshak-Rothstein A (1995) Fas(CD95)/FasL interactions required for programmed cell death after T-cell activation. Nature 373:444–448

Kabouridis PS (2006) Lipid rafts in T cell receptor signalling. Mol Membr Biol 23:49–57

Kang SM, Schneider DB, Lin Z, Hanahan D, Dichek DA, Stock PG, Baekkeskov S (1997) Fas ligand expression in islets of Langerhans does not confer immune privilege and instead targets them for rapid destruction. Nat Med 3:738–743

Kasibhatla S, Genestier L, Green DR (1999) Regulation of fas-ligand expression during activation-induced cell death in T lymphocytes via nuclear factor kappaB. J Biol Chem 274:987–992

Kataoka T, Budd RC, Holler N, Thome M, Martinon F (2000) The caspase-8 inhibitor FLIP promotes activation of NF-C:\Documents and Settings\KGunasekaran\Desktop\WOBLB and Erk signaling pathways. Curr Biol 10:640–648

Kavurma MM, Santiago FS, Bonfoco E, Khachigian LM (2001) Sp1 phosphorylation regulates apoptosis via extracellular FasL-Fas engagement. J Biol Chem 276:4964–4971

Kavurma MM, Bobryshev Y, Khachigian LM (2002) Ets-1 positively regulates Fas ligand transcription via cooperative interactions with Sp1. J Biol Chem 277:36244–36252

Kennedy NJ, Kataoka T, Tschopp J, Budd RC (1999) Caspase activation is required for T cell proliferation. J Exp Med 190:1891–1896

Kiener PA, Davis PM, Starling GC, Mehlin C, Klebanoff SJ, Ledbetter JA, Liles WC (1997) Differential induction of apoptosis by Fas-Fas ligand interactions in human monocytes and macrophages. J Exp Med 185:1511–1516

Kirchhoff S, Sebens T, Baumann S, Krueger A, Zawatzky R, Li-Weber M, Meinl E, Neipel F, Fleckenstein B, Krammer PH (2002) Viral IFN-regulatory factors inhibit activation-induced cell death via two positive regulatory IFN-regulatory factor 1-dependent domains in the CD95 ligand promoter. J Immunol 168:1226–1234

Kirkin V, Cahuzac N, Guardiola-Serrano F, Huault S, Lückerath K, Friedmann E, Novac N, Wels W, Martoglio B, Hueber A, Zörnig M (2007) The Fas ligand intracellular domain is released by ADAM10 and SPPL2a cleavage in T-cells. Cell Death Differ 14:1678–1687

Kischkel FC, Hellbardt S, Behrmann I, Germer M, Pawlita M, Krammer PH, Peter ME (1995) Cytotoxicity-dependent APO-1 (Fas/CD95)-associated proteins form a death-inducing signaling complex (DISC) with the receptor. EMBO J 14:5579–5588

Klas C, Debatin KM, Jonker RR, Krammer PH (1993) Activation interferes with the APO-1 pathway in mature human T cells. Int Immunol 5:625–630

Kondo T, Suda T, Fukuyama H, Adachi M, Nagata S (1997) Essential roles of the Fas ligand in the development of hepatitis. Nat Med 3:409–413

Kovacs B, Vassilopoulos D, Vogelgesang SA, Tsokos GC (1996) Defective CD3-mediated cell death in activated T cells from patients with systemic lupus erythematosus: role of decreased intracellular TNF-alpha. Clin Immunol Immunopathol 81:293–302

Latinis KM, Carr LL, Peterson EJ, Norian LA, Eliason SL, Koretzky GA (1997a) Regulation of CD95 (Fas) ligand expression by TCR-mediated signaling events. J Immunol 158: 4602–4611

Latinis KM, Norian LA, Eliason SL, Koretzky GA (1997b) Two NFAT transcription factor binding sites participate in the regulation of CD95 (Fas) ligand expression in activated human T cells. J Biol Chem 272:31427–31434

Lau HT, Yu M, Fontana A, Stoeckert CJ, Jr. (1996) Prevention of islet allograft rejection with engineered myoblasts expressing FasL in mice. Science 273:109–112

Laurence A, O'Shea JJ (2007) T(H)-17 differentiation: of mice and men. Nat Immunol 8:903–905

Laurence A, O'Shea JJ, Watford WT (2008) Interleukin-22: a sheep in wolf's clothing. Nat Med 14:247–249

Lee K, Feig C, Tchikov V, Schickel R, Hallas C, Schütze S, Peter M, Chan A (2006) The role of receptor internalization in CD95 signaling. EMBO J 25:1009–1023
Lee MO, Kang HJ, Kim YM, Oum JH, Park J (2002) Repression of FasL expression by retinoic acid involves a novel mechanism of inhibition of transactivation function of the nuclear factors of activated T-cells. Eur J Biochem 269:1162–1170
Legembre P, Barnhart BC, Zheng L, Vijayan S, Straus SE, Puck J, Dale JK, Lenardo M, Peter ME (2004) Induction of apoptosis and activation of NF-kappaB by CD95 require different signalling thresholds. EMBO Rep 5:1084–1089
Legge KL, Braciale TJ (2005) Lymph node dendritic cells control CD8+ T cell responses through regulated FasL expression. Immunity 23:649–659
Legler DF, Doucey MA, Cerottini JC, Bron C, Luescher IF (2001) Selective inhibition of CTL activation by a dipalmitoyl-phospholipid that prevents the recruitment of signaling molecules to lipid rafts. FASEB J 15:1601–1603
Legler DF, Micheau O, Doucey MA, Tschopp J, Bron C (2003) Recruitment of TNF receptor 1 to lipid rafts is essential for TNFalpha-mediated NF-kappaB activation. Immunity 18:655–664
Lenardo M, Chan KM, Hornung F, McFarland H, Siegel R, Wang J, Zheng L (1999) Mature T lymphocyte apoptosis – immune regulation in a dynamic and unpredictable antigenic environment. Annu Rev Immunol 17:221–253
Lettau M, Qian J, Kabelitz D, Janssen O (2004) Activation-dependent FasL expression in T lymphocytes and Natural Killer cells. Signal Transduction 4:206–211
Lettau M, Qian J, Linkermann A, Latreille M, Larose L, Kabelitz D, Janssen O (2006) The adaptor protein Nck interacts with Fas ligand: guiding the death factor to the cytotoxic immunological synapse. Proc Natl Acad Sci U S A 103:5911–5916
Li H, Zhu H, Xu CJ, Yuan J (1998) Cleavage of BID by caspase 8 mediates the mitochondrial damage in the Fas pathway of apoptosis. Cell 94:491–501
Liu Y, Janeway CA, Jr. (1990) Interferon gamma plays a critical role in induced cell death of effector T cell: a possible third mechanism of self-tolerance. J Exp Med 172:1735–1739
Luo X, Budihardjo I, Zou H, Slaughter C, Wang X (1998) Bid, a Bcl2 interacting protein, mediates cytochrome c release from mitochondria in response to activation of cell surface death receptors. Cell 94:481–490
Ma Y, Liu H, Tu-Rapp H, Thiesen HJ, Ibrahim SM, Cole SM, Pope RM (2004) Fas ligation on macrophages enhances IL-1R1-Toll-like receptor 4 signaling and promotes chronic inflammation. Nat Immunol 5:380–387
Martin DA, Siegel RM, Zheng L, Lenardo MJ (1998) Membrane oligomerization and cleavage activates the caspase-8 (FLICE/MACHalpha1) death signal. J Biol Chem 273:4345–4349
Martin DA, Zheng L, Siegel RM, Huang B, Fisher GH, Wang J, Jackson CE, Puck JM, Dale J, Straus SE, et al (1999) Defective CD95/APO-1/Fas signal complex formation in the human autoimmune lymphoproliferative syndrome, type Ia. Proc Nat Acad Sci U S A 96:4552–4557
Matsui K, Xiao S, Fine A, Ju ST (2000) Role of activator protein-1 in TCR-mediated regulation of the murine fasl promoter. J Immunol 164:3002–3008
McClure RF, Heppelmann CJ, Paya CV (1999) Constitutive Fas ligand gene transcription in Sertoli cells is regulated by Sp1. J Biol Chem 274:7756–7762
Melino G, Bernassola F, Catani MV, Rossi A, Corazzari M, Sabatini S, Vilbois F, Green DR (2000) Nitric oxide inhibits apoptosis via AP-1-dependent CD95L transactivation. Cancer Res 60:2377–2383
Micheau O, Tschopp J (2003) Induction of TNF receptor I-mediated apoptosis via two sequential signaling complexes. Cell 114:181–190
Mittelstadt PR, Ashwell JD (1999) Role of Egr-2 in up-regulation of Fas ligand in normal T cells and aberrant double-negative lpr and gld T cells. J Biol Chem 274:3222–3227
Miwa K, Asano M, Horai R, Iwakura Y, Nagata S, Suda T (1998) Caspase 1-independent IL-1beta release and inflammation induced by the apoptosis inducer Fas ligand. Nat Med 4:1287–1292
Miyawaki T, Uehara T, Nibu R, Tsuji T, Yachie A, Yonehara S, Taniguchi N (1992) Differential expression of apoptosis-related Fas antigen on lymphocyte subpopulations in human peripheral blood. J Immunol 149:3753–3758

Monari C, Bistoni F, Casadevall A, Pericolini E, Pietrella D, Kozel TR, Vecchiarelli A (2005) Glucuronoxylomannan, a microbial compound, regulates expression of costimulatory molecules and production of cytokines in macrophages. J Infect Dis 191:127–137

Montixi C, Langlet C, Bernard AM, Thimonier J, Dubois C, Wurbel MA, Chauvin JP, Pierres M, He HT (1998) Engagement of T cell receptor triggers its recruitment to low-density detergent-insoluble membrane domains. EMBO J 17:5334–5348

Munro S (2003) Lipid rafts: elusive or illusive? Cell 115:377–388

Muppidi JR, Siegel RM (2004) Ligand-independent redistribution of Fas (CD95) into lipid rafts mediates clonotypic T cell death. Nat Immunol 5:182–189

Muppidi JR, Tschopp J, Siegel RM (2004) Life and death decisions: secondary complexes and lipid rafts in TNF receptor family signal transduction. Immunity 21:461–465

Muppidi JR, Lobito AA, Ramaswamy M, Yang JK, Wang L, Wu H, Siegel RM (2006) Homotypic FADD interactions through a conserved RXDLL motif are required for death receptor-induced apoptosis. Cell Death Differ 13:1641–1650

Nijhara R, van Hennik PB, Gignac ML, Kruhlak MJ, Hordijk PL, Delon J, Shaw S (2004) Rac1 mediates collapse of microvilli on chemokine-activated T lymphocytes. J Immunol 173:4985–4993

Oehm A, Behrmann I, Falk W, Pawlita M, Maier G, Klas C, Li-Weber M, Richards S, Dhein J, Trauth BC, et al (1992) Purification and molecular cloning of the APO-1 cell surface antigen, a member of the tumor necrosis factor/nerve growth factor receptor superfamily. Sequence identity with the Fas antigen. J Biol Chem 267:10709–10715

Ogasawara J, Watanabe-Fukunaga R, Adachi M, Matsuzawa A, Kasugai T, Kitamura Y, Itoh N, Suda T, Nagata S (1993) Lethal effect of the anti-Fas antibody in mice. Nature 364:806–809

Orlinick JR, Vaishnaw A, Elkon KB, Chao MV (1997) Requirement of cysteine-rich repeats of the Fas receptor for binding by the Fas ligand. J Biol Chem 272:28889–28894

Ottonello L, Tortolina G, Amelotti M, Dallegri F (1999) Soluble Fas ligand is chemotactic for human neutrophilic polymorphonuclear leukocytes. J Immunol 162:3601–3606

Park SM, Schickel R, Peter ME (2005) Nonapoptotic functions of FADD-binding death receptors and their signaling molecules. Curr Opin Cell Biol 17:610–616

Parlato S, Giammarioli AM, Logozzi M, Lozupone F, Matarrese P, Luciani F, Falchi M, Malorni W, Fais S (2000) CD95 (APO-1/Fas) linkage to the actin cytoskeleton through ezrin in human T lymphocytes: a novel regulatory mechanism of the CD95 apoptotic pathway. EMBO J 19: 5123–5134

Patra SK (2008) Dissecting lipid raft facilitated cell signaling pathways in cancer. Biochim Biophys Acta 1785:182–206

Peter ME, Kischkel FC, Scheuerpflug CG, Medema JP, Debatin KM, Krammer PH (1997) Resistance of cultured peripheral T cells towards activation-induced cell death involves a lack of recruitment of FLICE (MACH/caspase 8) to the CD95 death-inducing signaling complex. Eur J Immunol 27:1207–1212

Ramaswamy M, Dumont C, Cruz AC, Muppidi JR, Gomez TS, Billadeau DD, Tybulewicz VL, Siegel RM (2007) Cutting edge: Rac GTPases sensitize activated T cells to die via Fas. J Immunol 179:6384–6388

Ramsdell F, Seaman MS, Miller RE, Picha KS, Kennedy MK, Lynch DH (1994a) Differential ability of Th1 and Th2 T cells to express Fas ligand and to undergo activation-induced cell death. Int Immunol 6:1545–1553

Ramsdell F, Seaman MS, Miller RE, Tough TW, Alderson MR, Lynch DH (1994b) gld/gld mice are unable to express a functional ligand for Fas. Eur J Immunol 24:928–933

Rathmell JC, Cooke MP, Ho WY, Grein J, Townsend SE, Davis MM, Goodnow CC (1995) CD95 (Fas)-dependent elimination of self-reactive B cells upon interaction with CD4+ T cells. Nature 376:181–184

Refaeli Y, Van Parijs L, Alexander SI, Abbas AK (2002) Interferon gamma is required for activation-induced death of T lymphocytes. J Exp Med 196:999–1005

Rengarajan J, Mittelstadt PR, Mages HW, Gerth AJ, Kroczek RA, Ashwell JD, Glimcher LH (2000) Sequential involvement of NFAT and Egr transcription factors in FasL regulation. Immunity 12:293–300

Resh MD (2006) Palmitoylation of ligands, receptors, and intracellular signaling molecules. Sci STKE 2006:re14
Riedl S, Salvesen G (2007) The apoptosome: signalling platform of cell death. Nat Rev Mol Cell Biol 8:405–413
Riou C, Yassine-Diab B, Van grevenynghe J, Somogyi R, Greller LD, Gagnon D, Gimmig S, Wilkinson P, Shi Y, Cameron MJ, et al (2007) Convergence of TCR and cytokine signaling leads to FOXO3a phosphorylation and drives the survival of CD4+ central memory T cells. J Exp Med 204:79–91
Rouquet N, Carlier K, Briand P, Wiels J, Joulin V (1996) Multiple pathways of Fas-induced apoptosis in primary culture of hepatocytes. Biochem Biophys Res Commun 229:27–35
Sallusto F, Langenkamp A, Geginat J, Lanzavecchia A (2000) Functional subsets of memory T cells identified by CCR7 expression. Curr Top Microbiol Immunol 251:167–171
Sandu C, Morisawa G, Wegorzewska I, Huang T, Arechiga AF, Hill JM, Kim T, Walsh CM, Werner MH (2006) FADD self-association is required for stable interaction with an activated death receptor. Cell Death Differ 13:2052–2061
Scaffidi C, Fulda S, Srinivasan A, Friesen C, Li F, Tomaselli KJ, Debatin KM, Krammer PH, Peter ME (1998) Two CD95 (APO-1/Fas) signaling pathways. EMBO J 17:1675–1687
Schmaltz C, Alpdogan O, Horndasch KJ, Muriglan SJ, Kappel BJ, Teshima T, Ferrara JL, Burakoff SJ, van den Brink MR (2001) Differential use of Fas ligand and perforin cytotoxic pathways by donor T cells in graft-versus-host disease and graft-versus-leukemia effect. Blood 97:2886–2895
Schulte M, Reiss K, Lettau M, Maretzky T, Ludwig A, Hartmann D, de Strooper B, Janssen O, Saftig P (2007) ADAM10 regulates FasL cell surface expression and modulates FasL-induced cytotoxicity and activation-induced cell death. Cell Death Differ 10:1040–1049
Seino K, Kayagaki N, Takeda K, Fukao K, Okumura K, Yagita H (1997) Contribution of Fas ligand to T cell-mediated hepatic injury in mice. Gastroenterology 113:1315–1322
Shlomchik MJ, Madaio MP, Ni D, Trounstein M, Huszar D (1994) The role of B cells in lpr/lpr-induced autoimmunity. J Exp Med 180:1295–1306
Siegel RM (2006) Caspases at the crossroads of immune-cell life and death. Nat Rev Immunol 6:308–317
Siegel RM, Frederiksen JK, Zacharias DA, Chan FK, Johnson M, Lynch D, Tsien RY, Lenardo MJ (2000) Fas preassociation required for apoptosis signaling and dominant inhibition by pathogenic mutations. Science 288:2354–2357
Siegel RM, Muppidi JR, Sarker M, Lobito A, Jen M, Martin D, Straus SE, Lenardo MJ (2004) SPOTS: signaling protein oligomeric transduction structures are early mediators of death receptor-induced apoptosis at the plasma membrane. J Cell Biol 167:735–744
Signore A, Annovazzi A, Gradini R, Liddi R, Ruberti G (1998) Fas and Fas ligand-mediated apoptosis and its role in autoimmune diabetes. Diabet Metab Rev 14:197–206
Simons K, Toomre D (2000) Lipid rafts and signal transduction. Nat Rev Mol Cell Biol 1:31–39
Singer GG, Abbas AK (1994) The fas antigen is involved in peripheral but not thymic deletion of T lymphocytes in T cell receptor transgenic mice. Immunity 1:365–371
Sprent J, Surh CD (2002) T cell memory. Annu Rev Immunol 20:551–579
Stassi G, De Maria R (2002) Autoimmune thyroid disease: new models of cell death in autoimmunity. Nat Rev Immunol 2:195–204
Stinchcombe JC, Majorovits E, Bossi G, Fuller S, Griffiths GM (2006) Centrosome polarization delivers secretory granules to the immunological synapse. Nature 443:462–465
Stranges PB, Watson J, Cooper CJ, Choisy-Rossi CM, Stonebraker AC, Beighton RA, Hartig H, Sundberg JP, Servick S, Kaufmann G, et al (2007) Elimination of antigen-presenting cells and autoreactive T cells by fas contributes to prevention of autoimmunity. Immunity 26:629–641
Straus SE, Sneller M, Lenardo MJ, Puck JM, Strober W (1999) An inherited disorder of lymphocyte apoptosis: the autoimmune lymphoproliferative syndrome. Ann Intern Med 130:591–601
Stuart PM, Griffith TS, Usui N, Pepose J, Yu X, Ferguson TA (1997) CD95 ligand (FasL)-induced apoptosis is necessary for corneal allograft survival. J Clin Invest 99:396–402

Su H, Bidere N, Zheng L, Cubre A, Sakai K, Dale J, Salmena L, Hakem R, Straus S, Lenardo M (2005) Requirement for caspase-8 in NF-kappaB activation by antigen receptor. Science 307:1465–1468
Suda T, Hashimoto H, Tanaka M, Ochi T, Nagata S (1997) Membrane fas ligand kills human peripheral blood T lymphocytes, and soluble Fas ligand blocks the killing. J Exp Med 186:2045–2050
Sun M, Lee S, Karray S, Levi-Strauss M, Ames KT, Fink PJ (2007) Cutting edge: two distinct motifs within the Fas ligand tail regulate Fas ligand-mediated costimulation. J Immunol 179:5639–5643
Suss G, Shortman K (1996) A subclass of dendritic cells kills CD4 T cells via Fas/Fas-ligand- induced apoptosis. J Exp Med 183:1789–1796
Takeuchi T, Ueki T, Nishimatsu H, Kajiwara T, Ishida T, Jishage K, Ueda O, Suzuki H, Li B, Moriyama N, Kitamura T (1999) Accelerated rejection of Fas ligand-expressing heart grafts. J Immunol 162:518–522
Tan JT, Ernst B, Kieper WC, LeRoy E, Sprent J, Surh CD (2002) Interleukin (IL)-15 and IL-7 jointly regulate homeostatic proliferation of memory phenotype CD8+ cells but are not required for memory phenotype CD4+ cells. J Exp Med 195:1523–1532
Tato CM, Laurence A, O'Shea JJ (2006) Helper T cell differentiation enters a new era: le roi est mort; vive le roi! J Exp Med 203:809–812
Trauth BC, Klas C, Peters AM, Matzku S, Moller P, Falk W, Debatin KM, Krammer PH (1989) Monoclonal antibody-mediated tumor regression by induction of apoptosis. Science 245:301–305
Tybulewicz VL (2005) Vav-family proteins in T-cell signalling. Curr Opin Immunol 17:267–274
Umemura M, Kawabe T, Shudo K, Kidoya H, Fukui M, Asano M, Iwakura Y, Matsuzaki G, Imamura R, Suda T (2004) Involvement of IL-17 in Fas ligand-induced inflammation. Int Immunol 16:1099–1108
Vaishnaw AK, Orlinick JR, Chu JL, Krammer PH, Chao MV, Elkon KB (1999) The molecular basis for apoptotic defects in patients with CD95 (Fas/Apo-1) mutations [published erratum appears in J Clin Invest 1999 Apr;103(7):1099]. J Clin Invest 103:355–363
Van Parijs L, Peterson DA, Abbas AK (1998) The Fas/Fas ligand pathway and Bcl-2 regulate T cell responses to model self and foreign antigens. Immunity 8:265–274
Verhagen AM, Vaux DL (2002) Cell death regulation by the mammalian IAP antagonist Diablo/Smac. Apoptosis 7:163–166
Villalba M, Bi K, Rodriguez F, Tanaka Y, Schoenberger S, Altman A (2001) Vav1/Rac-dependent actin cytoskeleton reorganization is required for lipid raft clustering in T cells. J Cell Biol 155:331–338
Villalba M, Kasibhatla S, Genestier L, Mahboubi A, Green DR, Altman A (1999) Protein kinase ctheta cooperates with calcineurin to induce Fas ligand expression during activation-induced T cell death. J Immunol 163:5813–5819
Villena SN, Pinheiro RO, Pinheiro CS, Nunes MP, Takiya CM, DosReis GA, Previato JO, Mendonca-Previato L, Freire-de-Lima CG (2008) Capsular polysaccharides galactoxylomannan and glucuronoxylomannan from Cryptococcus neoformans induce macrophage apoptosis mediated by Fas ligand. Cell Microbiol 10:1274–1285
Villunger A, Ghaffari-Tabrizi N, Tinhofer I, Krumbock N, Bauer B, Schneider T, Kasibhatla S, Greil R, Baier-Bitterlich G, Uberall F, et al (1999) Synergistic action of protein kinase C theta and calcineurin is sufficient for Fas ligand expression and induction of a crmA-sensitive apoptosis pathway in Jurkat T cells. Eur J Immunol 29:3549–3561
Viola A, Schroeder S, Sakakibara Y, Lanzavecchia A (1999) T lymphocyte costimulation mediated by reorganization of membrane microdomains. Science 283:680–682
Wang J, Zheng L, Lobito A, Chan FK, Dale J, Sneller M, Yao X, Puck JM, Straus SE, Lenardo MJ (1999) Inherited human Caspase 10 mutations underlie defective lymphocyte and dendritic cell apoptosis in autoimmune lymphoproliferative syndrome type II. Cell 98:47–58
Wang R, Brunner T, Zhang L, Shi Y (1998) Fungal metabolite FR901228 inhibits c-Myc and Fas ligand expression. Oncogene 17:1503–1508
Watanabe-Fukunaga R, Brannan CI, Copeland NG, Jenkins NA, Nagata S (1992a) Lymphoproliferation disorder in mice explained by defects in Fas antigen that mediates apoptosis. Nature 356:314–317

Watanabe-Fukunaga R, Brannan CI, Itoh N, Yonehara S, Copeland NG, Jenkins NA, Nagata S (1992b) The cDNA structure, expression, and chromosomal assignment of the mouse Fas antigen. J Immunol 148:1274–1279
Weant AE, Michalek RD, Khan IU, Holbrook BC, Willingham MC, Grayson JM (2008) Apoptosis regulators Bim and Fas function concurrently to control autoimmunity and CD8+ T cell contraction. Immunity 28:218–230
Wong B, Arron J, Choi Y (1997) T cell receptor signals enhance susceptibility to Fas-mediated apoptosis. J Exp Med 186:1939–1944
Xu Y, Szalai AJ, Zhou T, Zinn KR, Chaudhuri TR, Li X, Koopman WJ, Kimberly RP (2003) Fc gamma Rs modulate cytotoxicity of anti-Fas antibodies: implications for agonistic antibody-based therapeutics. J Immunol 171:562–568
Yin XM, Wang K, Gross A, Zhao Y, Zinkel S, Klocke B, Roth KA, Korsmeyer SJ (1999) Bid-deficient mice are resistant to Fas-induced hepatocellular apoptosis. Nature 400:886–891
Yonehara S, Ishii A, Yonehara M (1989) A cell-killing monoclonal antibody (anti-Fas) to a cell surface antigen co-downregulated with the receptor of tumor necrosis factor. J Exp Med 169:1747–1756
Youle RJ, Strasser A (2008) The BCL-2 protein family: opposing activities that mediate cell death. Nat Rev Mol Cell Biol 9:47–59
Zhang J, Cado D, Chen A, Kabra NH, Winoto A (1998) Fas-mediated apoptosis and activation-induced T-cell proliferation are defective in mice lacking FADD/Mort1. Nature 392:296–300
Zhang M, Park SM, Wang Y, Shah R, Liu N, Murmann AE, Wang CR, Peter ME, Ashton-Rickardt PG (2006) Serine protease inhibitor 6 protects cytotoxic T cells from self-inflicted injury by ensuring the integrity of cytotoxic granules. Immunity 24:451–461
Zhang Y, Rosenberg S, Wang H, Imtiyaz HZ, Hou YJ, Zhang J (2005) Conditional Fas-Associated Death Domain Protein (FADD):GFP Knockout Mice Reveal FADD Is Dispensable in Thymic Development but Essential in Peripheral T Cell Homeostasis. J Immunol 175:3033–3044
Zheng H, Matte-Martone C, Li H, Anderson BE, Venketesan S, Sheng Tan H, Jain D, McNiff J, Shlomchik WD (2008) Effector memory CD4+ T cells mediate graft-versus-leukemia without inducing graft-versus-host disease. Blood 111:2476–2484
Zuccato E, Blott E, Holt O, Sigismund S, Shaw M, Bossi G, Griffiths G (2007) Sorting of Fas ligand to secretory lysosomes is regulated by mono-ubiquitylation and phosphorylation. J Cell Sci 120:191–199

FasL Expression and Reverse Signalling

M. Lettau[1], M. Paulsen[1], D. Kabelitz and O. Janssen

Abstract FasL plays a central role in the induction of apoptosis within the immune system. It mediates activation-induced cell death (AICD) of T lymphocytes and contributes to the cytotoxic effector function of T and NK cells. Moreover, FasL is discussed as direct effector molecule for the establishment of immune privilege and tumour survival. Besides its death-promoting activity, FasL has been implicated in reverse signalling and might thus also play a role in T cell development and selection and the modulation of T cell activation. Considering these diverse functions, the overall FasL expression has to be tightly controlled to avoid unwanted damage. Based on an activation-associated transcriptional control, several post-transcriptional processes ensure a safe storage, a rapid mobilisation, a target-directed activity and a subsequent inactivation. Over the past years, the identification and characterisation of FasL-interacting proteins provided novel insight into the mechanisms of FasL transport, processing and reverse signalling, which might be exemplary also for the other members of the TNF family.

1 Structural Composition of FasL

As indicated by the name-giving TNF homology domain (THD), the highest degree of homology between individual members of the TNF family is seen in the extracellular part of the type II transmembrane proteins, although the receptor-specific binding sites are located at the very C-terminus. Moreover, the extracellular parts of several TNF family members contain putative cleavage sites for metalloproteases to deliberate a soluble cytokine. In case of FasL, the putative cleavage sites are located N-terminal of a region that is required for self-aggregation and

O. Janssen (✉), M. Lettau, M. Paulsen, and D. Kabelitz
Institute of Immunology, University Hospital Schleswig-Holstein Campus Kiel, Michaelisstr. 5, D-24105 Kiel, Germany
e-mail: ojanssen@email.uni-kiel.de

[1]Equally contributing first Authors

Results Probl Cell Differ, doi:10.1007/400_2008_21

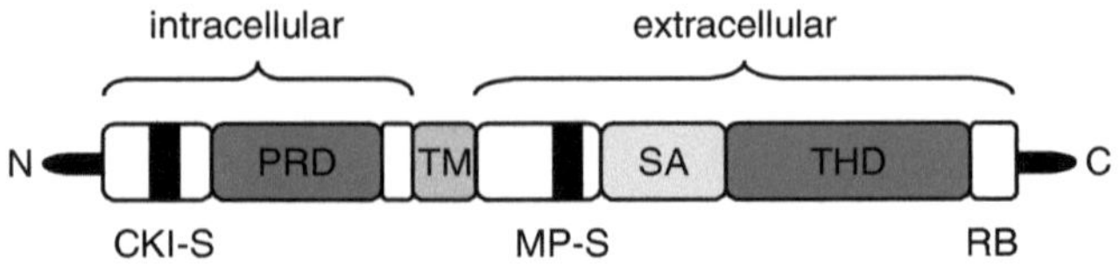

Fig. 1 Structural composition of FasL. The N-terminal cytoplasmic part (aa 1–81) of FasL exhibits a casein kinase substrate motive (CKI-S, aa 17–21) and a proline-rich domain (aa 37–70). The transmembrane region spans aa 81–102. Within the extracellular part, putative metalloprotease cleavage sites are located outside a region required for self-assembly and trimerisation (SA). The TNF homology domain (THD) contains several putative N-glycosylation sites. The receptor-binding site (RB) is located at the very C-terminal part of the molecule.

trimerisation (aa 137–183). The transmembrane area (aa 81–102) is followed by a particularly interesting intracellular region (aa 1–81 in humans), where FasL significantly differs from all other members of the TNF family (Janssen et al. 2003). Briefly, with 80 amino acids, the N-terminal region of (human) FasL is much longer than those of the functionally most related TNF family members (TNF (35 aa), LTβ (18 aa) and TRAIL (17 aa)), and only FasL contains the unique conserved proline-rich domain (PRD) (aa 45–71, with 22 prolins and 5 leucins in humans) that allows interactions with cytosolic proteins containing SH3 or WW domains. Moreover, FasL harbours a conserved casein kinase (CKI) substrate region (aa 17–21 in humans) and FasL, but none of the other TNF family members contains potential tyrosine phosphorylation sites (aa 7,9,13 in humans and aa 7 in mice) (Fig. 1). These features strongly suggest a distinct and important biological function for the N-terminal intracellular part of FasL.

2 Expression of FasL

In contrast to Fas, which is constitutively or inducibly expressed in many different cell types, FasL expression is tightly regulated and restricted to only few cell types. In $CD4^+$ T helper 1 (TH1) cells, FasL surface expression is induced upon activation (e.g. by antigen, anti-T cell receptor (TCR)/CD3 antibodies, mitogenic lectins, bacterial superantigens, phorbol ester and calcium ionophore). In contrast, some $CD4^+$ TH2 cells do not show a significant expression of FasL even after appropriate stimulation, while activated $CD8^+$ T cells usually express higher amounts of FasL than $CD4^+$ TH1 cells (Janssen et al. 2000; Suda et al. 1995). FasL is also expressed in unstimulated natural killer (NK) cells, and its expression can be further enhanced by appropriate stimulation (CD16 ligation or cytokines like IL-2 or IL-12). In cytotoxic T lymphocytes (CTLs) and NK cells, FasL is stored within the cell in cytotoxic granules/secretory lysosomes (see Sect. 4) (Blott et al. 2001). In monocytes, FasL can also be detected intracellularly and is released after treatment with immune complexes, phytohemagglutinin (PHA) or superantigen (Janssen et al. 2003). Outside the immune system, a constitutive expression of FasL on cells or tissues of so-called immune privileged sites (i.e., eyes, testis, trophoblasts) (Niederkorn 2006) and also in some tumours (Igney and Krammer 2005) has been reported. Neurons and astrocytes also express FasL, where it might also contribute to the immune privilege of the central nervous system (Niederkorn 2006). However, the precise molecular basis for this differential expression (intracellular vs. cellular surface) remains unclear to date.

3 FasL Functions in the Immune System

As a death factor, the main function of FasL is the induction of apoptosis in Fas-expressing cells. In this context, FasL is the key death factor of the immune system.

It contributes to the cytotoxicity of T and NK cells, where FasL is stored in association with cytotoxic granules/secretory lysosomes. After target cell recognition, these storage granules are transported to the site of intercellular contact, where they fuse with the plasma membrane and release cytotoxic effector molecules (i.e. granzymes and perforin) into the immunological synapse (Russell and Ley 2002). Besides these classical components of cell-mediated cytotoxicity, FasL also constitutes part of the effector machinery of cytotoxic immune cells and supplements the granzyme/perforin-mediated pathway of target cell lysis. However, depending on the expression of Fas and the sensitivity towards Fas/FasL-induced apoptosis of a given target cell, one or the other mechanism might predominate (Brunner et al. 2003).

FasL also mediates AICD of T lymphocytes. While activation of naive T cells results in clonal expansion and differentiation, the repeated stimulation of activated cells with antigen in vitro induces AICD. Thus, AICD was suggested as one mechanism to eliminate reactive T cells and to terminate immune responses. Besides FasL, other members of the TNF receptor family (TNF-R1, TRAIL-R) and also receptor-independent mechanisms contribute to AICD (Krammer et al. 2007).

FasL has also been implicated in the establishment of immune-privileged sites like the brain, the eyes or testis. The constitutive expression of FasL on certain cell types in these tissues sites is supposed to minimise local immune responses by eliminating infiltrating Fas-positive lymphocytes (Niederkorn 2006). In this context, some tumours might employ immunomodulatory factors to generate their own "immune-privilege-like" microenvironment. The constitutive expression of FasL on the cell surface or the secretion of soluble exosomal vesicle-associated FasL by tumour cells may thus contribute to immune escape and protect the tumour from the attack of activated (=Fas-positive) tumour-infiltrating lymphocytes. However, the biological significance of a so-called tumour counterattack is a still discussed controversial (Igney and Krammer 2005).

Interestingly, FasL is meanwhile also regarded as an accessory or costimulatory molecule for T cell activation. In this context, FasL ligation was described to inhibit the proliferation of murine $CD4^+$ T cells. On the other hand, FasL ligation has been shown to positively modulate the proliferation of murine $CD8^+$ rather than $CD4^+$ T cells both in vitro and in vivo. Based on its reverse signalling capacity, a participation of FasL as an accessory molecule for the positive selection of thymocytes has been claimed (Newell and Desbarats 1999; Sun and Fink 2007).

4 Regulation of FasL Expression

FasL expression is controlled by many classical transcription factors. NF-AT, NF-kB, c-myc, IRF-1, stimulating protein 1 (SP-1), cyclin B1/Cdk1 and Egr-3 are engaged in the induction of FasL expression after T cell activation, for example in the course of AICD. Transcription factors of the forkhead family play a role in the induction of FasL expression after withdrawal of growth factors, whereas AP-1 is involved in stress-induced expression of FasL. Transcriptional regulation also involves negative regulators, including, for example c-Fos. Taken together, FasL expression is controlled by a wide range of cis-acting promotor elements that enable a precise adjustment of gene activity in response to a certain stimulus (Brunner et al. 2003; Li-Weber and Krammer 2003).

Besides transcriptional control, several post-transcriptional processes modulate FasL surface expression and its death-promoting activity (Fig. 2) (Lettau et al. 2008). It is quite well-established that FasL is stored in cytolytic granules/secretory

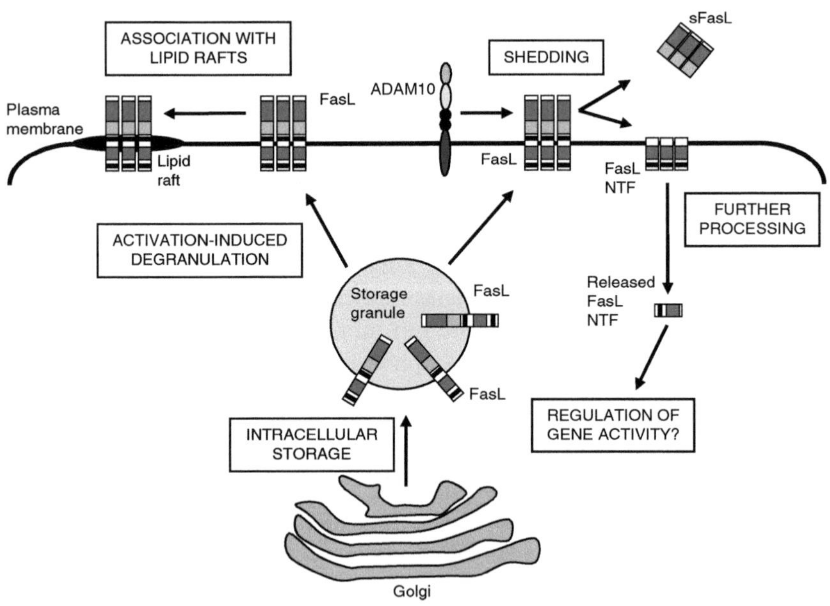

Fig. 2 Post-transcriptional regulation of FasL expression. In cytotoxic T and NK cells. FasL is stored in secretory granules/lysosomes. Upon target cell recognition, these granules are directed to the site of intercellular contact, where they fuse with the plasma membrane, thus locally exposing FasL within the immunological synapse. On the cell surface, the selective association with lipid rafts increases the death-promoting activity of FasL. Moreover, FasL surface expression is modulated by

lysosomes of cytotoxic T and NK cells. These highly specialised dual-functional organelles serve as a degradative as well as a secretory compartment. They contain the classical lysosomal machinery for protein degradation but additionally store proteins designated for secretion and thus enable their regulated exocytosis in response to appropriate stimuli. These organelles are almost exclusively found in cells belonging to the haematopoietic lineage, with melanocytes being a major exception. The secretory lysosomes of T and NK cells are enclosed by an external membrane and contain internal vesicles and electron-dense cores of tightly packed proteins. In addition to proteins representative of conventional lysosomes such as lysosomal hydrolases and lysosomal membrane proteins (LAMPs), secretory lysosomes of cytotoxic cells contain proteins responsible for target cell elimination, including granzymes, perforin and FasL. Other proteins, for example the proteoglycan serglycin, facilitate the storage and packaging of these compounds, and also serve accessory functions during the directed exocytosis of the cytolytic effectors.

Upon recognition of a target cell, secretory lysosomes are transported to the site of intercellular contact, where they fuse with the plasma membrane and release the soluble cytotoxic effectors into the forming cytotoxic immunological synapse. This degranulation event locally exposes transmembrane components of the granules on the cell surface. Thus, these organelles facilitate the safe storage, regulated secretion and target restricted action of the different executive, accessory and regulatory components of cell-mediated cytotoxicity. As mentioned, FasL is a membrane component of secretory lysosomes in T and NK cells. Unlike other lysosomal proteins, however, FasL sorting to the lysosomal compartment is mediated by its proline-rich domain. Therefore, proteins containing SH3 or WW domains are presumably engaged in FasL storage and transport.

FasL can also be released as a membrane component of microvesicles that are about 100–200 nm in diameter. As secretory lysosomes contain internal vesicles and thus have the characteristics of multivesicular bodies, degranulation would also release such microvesicles (Lettau et al. 2007).

In accordance with its intracellular storage, stimulation of T and NK cells with phorbol ester and calcium ionophore results in a biphasic surface expression of FasL, with a first maximum after only 10 min and a late phase of surface expression after about 2 h. We showed that the rapid surface appearance of FasL relies on the actin-dependent degranulation of the intracellular storage granules, whereas the second peak requires de novo synthesis (Lettau et al. 2004).

Although the storage of FasL in secretory lysosomes is well-accepted in the scientific community, a very recent report convincingly showed that at least in murine $CD8^+$ T cells, intracellularly stored FasL might be located in structures different from cytolytic granules/secretory lysosomes. In their system, He and

Fig. 2 (continued) ADAM10-mediated shedding, leading to the release of soluble FasL (sFasL). The remaining membrane-bound N-terminal fragment of FasL is released into the cytosol by the protease SPPL2a. Subsequently, this small fragment may enter the nucleus and regulate the activity of target genes (Figure modified from Lettau et al. 2008).

coworkers observed a rapid release of FasL independent of the degranulation of cytolytic granules/secretory lysosomes. However, this compartment has not been further characterised (He and Ostergaard 2007). Future studies will clarify whether T cells or different T cell subpopulations (and NK cells) employ different compartments for the intracellular storage of FasL or whether secretory lysosomes themselves are far more heterogenous and versatile with respect to their cargo than initially anticipated. In this regard, the recent analysis of the proteome of enriched lysosomes from human NK cells revealed a high degree of similarity in the overall protein composition, with clear cell line-specific differences in functionally relevant proteins such as perforin or granzymes (Schmidt et al. 2008).

FasL surface expression is also modulated by metalloproteases. In different cellular systems, matrix metalloproteases facilitate the release of a soluble form of FasL by cleavage within an extracellular cleavage site and thus also decrease its surface expression. The cleavage site has been mapped to Ser126/127 in human FasL and to Lys129/Gln130 in the murine system. It is thus located N-terminal of the region that is essential for trimerisation, indicating that soluble FasL (sFasL) may form functional trimers. However, the cleaved soluble form of FasL has been described to mediate pro- as well as anti-apoptotic and neutrophil chemotactic effects presumably depending on the local micromilieu. Serum levels of sFasL have been associated with the progress of numerous diseases such as the asymptomatic phase of HIV infection, rheumatoid arthritis and different malignancies.

FasL is cleaved by matrix metalloproteinase-7 (MMP-7, matrilysin) in some cells, for example glandular epithelial cells or a specialised population of murine epithelial cells. Furthermore, MMP-7 may facilitate tumour survival and resistance to cytotoxic drugs (i.e. doxorubicin) by cleaving FasL and thereby reducing its death-promoting activity. Besides FasL, MMP-7 also cleaves the receptor Fas. The released soluble form of Fas then antagonises the effects of cytotoxic FasL and thus contributes to the resistance of tumour cells to FasL-induced apoptosis (Janssen et al. 2003; Linkermann et al. 2005).

Since the 'a-disintegrin-and-metalloproteinase' ADAM17 (TACE) was identified to specifically cleave TNF-α, and since the TNF-homologous portion of FasL is similarly processed and shed, a closely related protease was suggested. In fact, we recently demonstrated that also in T cells, FasL is constitutively cleaved by the closely related protease ADAM10. We showed that shedding decreased the FasL death-promoting activity in the context of cell-mediated cytotoxicity and of AICD (Schulte et al. 2007). Moreover, ADAM10 may cooperate with the protease SPPL2a (signal peptide peptidase-like 2a) to release a small unstable fragment that mainly consists of the intracellular domain of FasL. This intracellular remnant may then enter the nucleus and specifically modulate the activity of target genes. This mechanism with marked similarities to the Notch-signalling pathway might provide an alternate route for FasL reverse signalling (Kirkin et al. 2007).

Apart from shedding, the pro-apoptotic activity of FasL is further modulated by its association with lipid rafts. Lipid rafts are membrane compartments rich in cholesterol, glycosphingolipids and saturated phospholipids. An important feature of these plasma membrane microdomains is the selective recruitment or exclusion of signalling molecules, thereby regulating or modulating their activity. The death receptor-mediated induction of apoptosis has been shown to be modulated by lipid rafts. As an example, Fas associates with microdomains of the plasma membrane and this localisation influences Fas-mediated cell death in T cells. Moreover, FasL has been described to localise to lipid rafts both in transfectants and primary T cells and this association positively modulates its death-promoting activity. Also in this scenario, the proline-rich domain is indispensable for the recruitment to lipid rafts (Cahuzac et al. 2006). However, the molecular basis for the selective recruitment to membrane microdomains remains unclear to date.

5 FasL-Interacting Proteins

Assuming that the intracellular proline-rich domain (PRD) of FasL enables interactions with proteins containing proline binding modules such as SH3 or WW domains, over the past years, we identified and functionally characterised several FasL-interacting proteins (Janssen et al. 2003; Lettau et al. 2008). We initially showed that, for example SH3 domains of several kinases including the Src-type kinases Src, Fyn, Lyn, Lck, Hck, Fgr and Abl, and the p85 adapter subunit of PI3 kinase precipitate FasL. Since most of these kinases play a major role in T cell activation, these interactions might at least in part account for the postulated reverse signal transduction capacity of FasL by facilitating the cross-talk of the Fas/FasL system with the T cell receptor (Janssen et al. 2003) (see Sect. 6). Apart from reverse signalling, the phosphorylation of FasL (e.g. by Src kinases) has recently been implicated in the sorting of the death factor to the ‘inner vesicles’ that form part of secretory lysosomes (Zuccato et al. 2007).

Moreover, FasL has been shown to interact with numerous adapter proteins. Upon cotransfection, proteins of the so-called pombe cdc15 homology family (Chitu and Stanley 2007) including the CD2-interacting protein 1 (CD2BP1), the formin-binding protein 17 (FBP17), the cdc42-interacting protein 4 (CIP4) and the protein kinase c and casein kinase substrate in neurons 1–3 (PACSIN 1–3) affect the subcellular localisation of FasL. While FasL is expressed on the plasma membrane upon overexpression in non-haematopoietic cells, cotransfection of FBP17, the PACSINs 1–3, CD2BP1 or CIP4 results in an intracellular retention of FasL (Qian et al. 2006). FasL also interacts with the adapter protein Nck. A main function of Nck is to link receptor tyrosine kinases or associated proteins to the machinery of actin reorganisation. In this context, Nck is a central adapter protein also in the TCR-mediated rearrangement of the actin cytoskeleton. With regard to the regulation of FasL expression, Nck is required for the recruitment of FasL and FasL-associated vesicles to the T cell/target cell interface, thus facilitating the local surface expression of FasL within the cytotoxic immunological synapse (Lettau et al. 2006).

6 Reverse Signalling by FasL

In recent years, several members of the TNF family including CD27 Ligand (CD27L), 4-1BB Ligand (4-1BBL), OX40 Ligand (OX40L), CD30 Ligand (CD30L), CD40 Ligand (CD40L), LIGHT, TRANCE, TRAIL, TNF and FasL were found to be implicated in the modulation of cellular activation in different systems, leading to the assumption that most TNF family members are capable of a reverse or retrograde signal transduction. Thus, it has been suggested that CD27L, 4-1BBL, OX40L, CD30L, CD40L and FasL affect the activation of peripheral T lymphocytes by transducing costimulatory signals. Several ligands of the TNF family seem to have growth promoting functions, for example CD27L, which augments the proliferation of PHA-stimulated T cells. Similarly, CD30L ligation results in an increased metabolic activity, proliferation and cytokine production of TCR-stimulated cells, whereas in freshly isolated neutrophils, CD30L ligation induces an increased IL-8 production. CD40L-stimulation augments phosphorylation of Lck and PLCγ, and subsequent activation of the mitogen-activated protein kinase (MAPK) p38 and JNK. Furthermore, it was shown that $CD40^+$ transfectants costimulate anti-CD3 induced T-cell proliferation in an IL-2 dependent manner. In this context, CD40L seems to provide a signal for the selective expansion of $CD4^+$ T cells after interaction with CD40-expressing APC (Cayabyab et al. 1994). Nevertheless, in the case of CD40L the molecular basis for its reverse signal transduction capability is still somehow enigmatic, since this molecule has a very short cytoplasmic tail without any known enzymatic activity.

A selective enhancement of TCR-triggered $CD4^+$ T cell proliferation has also been described for TRAIL-ligation. Furthermore, besides its role in the induction of apoptotic cell death in cells expressing HVEM or LTα/β receptor, LIGHT has been implicated in reverse signal transduction in T cells by enhancing proliferation, cytokine production and cytotoxic activity (Sun and Fink 2007). Stimulation of membrane-bound TNF-α with anti-TNFα antibodies costimulates T cells to secrete cytokines and upregulate adhesion molecules. In this context, $CD4^+$ and $CD8^+$ T lymphocytes again react differently. While the mTNF-α-stimulation of $CD4^+$ T cells leads to the downregulation of TH2 cells, $CD8^+$ T lymphocytes increase their cytotoxic potential. Interestingly, mTNF-α-triggering by soluble TNF-Rs at the same time desensitises macrophages and monocytes against LPS by inhibiting the production of multiple cytokines responsible for activation (Sun and Fink 2007). Because of their individual expression profiles, other TNF family members have been mainly investigated in the context of B cell activation. For example, OX40L was found to enhance T cell-dependent proliferation and immunoglobulin secretion in murine splenic B cells (Stuber et al. 1995). A similar costimulation of B cells can be induced by 4-1BBL-ligation. Interestingly, in monocytes 4-1BBL-ligation induces secretion of IL-6, IL-8 and TNF-α (Sun and Fink 2007).

As mentioned, within its unique intracellular part, the FasL contains several motifs that argue for a multi-facetted signalling capacity. These include a potential nuclear localisation site, the exceptional proline-rich domain (PRD), the casein kinase I (CKI) phosphorylation motif and several tyrosine residues. A similar CKI substrate motif is in fact present in the cytoplasmic tails of six members of the TNF family. Although the molecular mechanisms and functional consequences have not been completely unravelled, at least in the cases of FasL and TNF-α, serine phosphorylation of that motif was implicated in reverse signal transduction (Sun and Fink 2007).

The PRD (aa 37–70) within the intracellular part of FasL is absolutely unique within the TNF family and is of special interest as a protein–protein interaction site. As mentioned (see Sect. 5), most of the PRD-interacting adapter proteins analysed in more detail so far are regulators of FasL storage and transport. Interestingly, before the FasL-interaction was described for the PCH family member CD2BP1, this protein had been implicated in synapse formation and T cell activation by facilitating actin reorganisation upon interaction with an APC (Badour et al. 2003). Moreover, CD2BP1 can interact with the phosphatase PTP-PEST, which is involved in negative regulation of lymphocyte activation (Yang and Reinherz 2006). Thus, besides its role for the subcellular localisation of FasL, CD2BP1 might also contribute to reverse signalling by recruiting PTP-PEST (Baum et al. 2005).

TCR ligation results in the activation of T cell-specific Src kinases Fyn and Lck, which phosphorylate immunoreceptor tyrosine-based activation motifs (ITAMs) within TCR-associated ζ chains to serve as docking sites for the Syk-type kinase ZAP-79 (Fig. 3). Subsequently, active ZAP-70 phosphorylates an array of regulators of the membrane-proximal activation complex, including the 'linker for the activation of T cells' (LAT). Phosphorylated LAT can then recruit different SH2 domain containing adapter proteins such as Gads/SLP-76 and thus initiate the Grb2/SOS/Ras/MAPK signalling cascade and the activation of phospholipase C γ1 (PLCγ1). Resting lymphocytes require a costimulatory signal for full activation, otherwise T cells fail to produce cytokines like IL-2, stop proliferation and undergo apoptosis. Costimulation in turn can be negatively modulated by the expression of inhibitory receptors. The prototypic costimulatory signal is transmitted through the costimulatory receptor CD28 and results in an activation of the PI3 kinase/Akt pathway. Together with the 'inducible costimulator' (ICOS) they form the class of disulfide-linked homodimers that bind to different members of the B7 family of surface molecules. It is important to note that several members of the TNF receptor family such as CD40, OX40, CD30, CD95 meanwhile form a novel class of costimulatory molecules (Frauwirth and Thompson 2002). Finally, the different signalling cascades initiated by the TCR and costimulatory receptors merge at the level of transcription to alter the protein profile required for proliferation, differentiation or induction of anergy or apoptosis.

The first evidence for a signalling capacity of FasL came from two independent studies on murine T cells from Fas(lpr)/FasL(gld) mutant mice. The first report described the costimulatory capacity of FasL in CTL lines generated from B6 wildtype, lpr- and gld-mice and suggested that FasL ligation might be required as

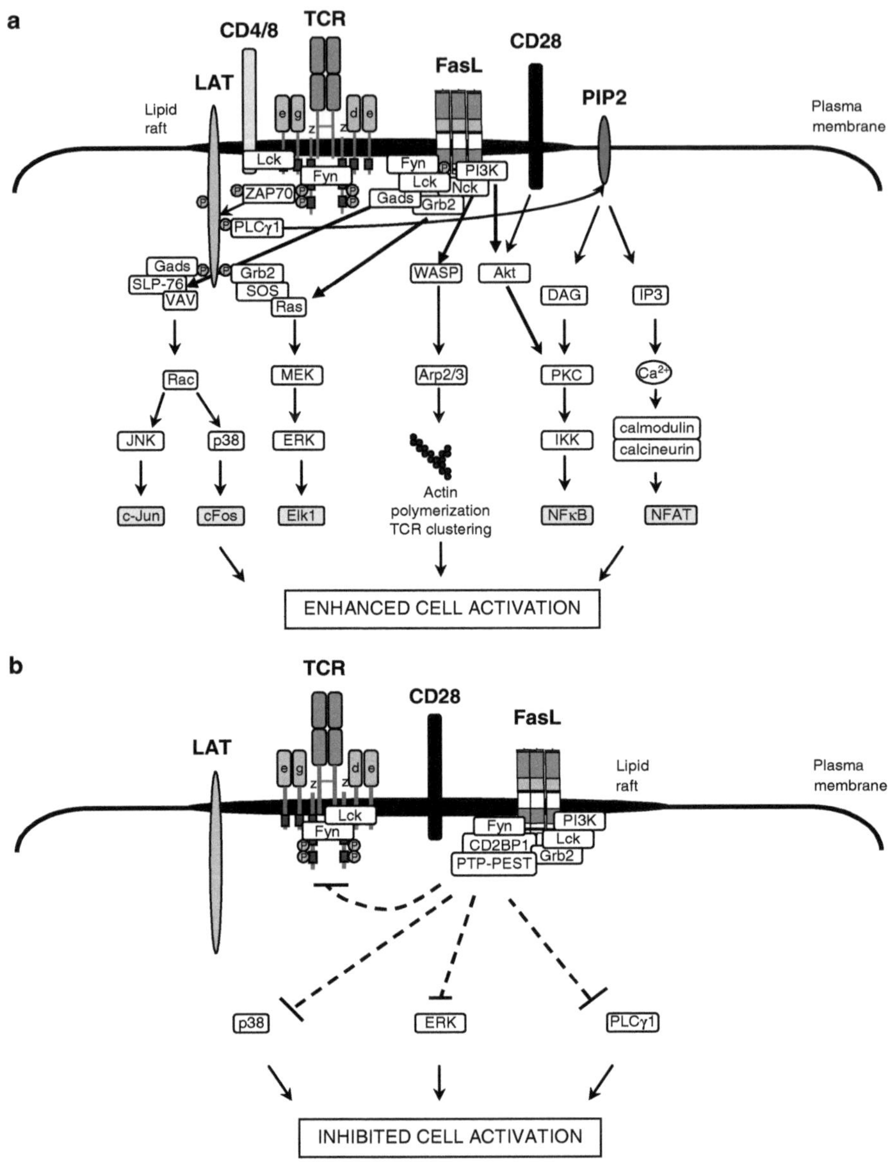

Fig. 3 FasL, an enhancer or inhibitor of T cell activation. The TNF family member FasL has been implicated in so-called reverse signal transduction. In this context, FasL affects TCR activation by positive (**a**) or negative (**b**) costimulation. (**a**) FasL enhances the TCR signal by activating several important signalling cascades, including the ERK-, p38 MAPK-, and JNK-pathways. In addition, Nck interacts with WASP, which in turn leads to cytoskeleton reorganisation, resulting in improved TCR clustering. (**b**) Under certain conditions, FasL blocks T cell activation. One explanation might be that recruitment of CD2BP1 and PTP-PEST inhibits p38, PLC, and ERK, but the exact mechanism of this proposal has not yet been investigated (Figure modified from Lettau et al. 2008).

a costimulatory signal to achieve optimal proliferation (Suzuki and Fink 1998). This 'FasL effect' was only seen in $CD8^+$ but not in $CD4^+$ T lymphocytes. It needed TCR coengagement and was independent of CD28. In contrast to these data, Desbarats et al. showed that FasL engagement on freshly isolated murine $CD4^+$ T cells inhibited TCR-stimulated cell proliferation and blocked IL-2 production (Desbarats et al. 1998).

Follow-up studies provided more evidence for a positive costimulation by FasL on the proliferation of mature murine CTL and during thymocyte maturation (Boursalian and Fink 2003; Suzuki et al. 2000). In terms of signalling, differences in the phosphorylation patterns induced by TCR ligation alone or in the presence of FasL-ligating FasFc fusion proteins were reported in murine CTLs. In this experimental model, phosphorylation of Akt, ERK and FasL itself was increased by FasL costimulation. Furthermore, FasL translocated to lipid raft platforms, and this association was supposed to be essential for the recruitment of adaptor molecules to enable a FasL-to-TCR cross-talk. Moreover, at the level of transcription factors, the nuclear translocation of NFAT and activation of the AP-1 complex were enhanced upon FasL-crosslinking, leading to an elevated production of IFN-γ (Sun et al. 2006).

Employing Jurkat T cells that overexpress a fusion, construct of FasL with an N-terminal myristylation motif (that tethers the molecule to the membrane) and the intracellular part of FasL provided first direct evidence that the cytoplasmic domain is sufficient to mediate TCR-costimulation. Interestingly, the casein kinase substrate motif and the PRD proved to be important for reverse signalling, since mutations within the CKI motif also interfered with NFAT translocation, indicating a more complex scenario for FasL signalling besides PRD-mediated alterations. Nevertheless, the deletion of amino acids 45–54 within the PRD abrogated ERK activation via the PI3 kinase pathway and blocked costimulation. At the same time, however, FasL-induced apoptosis on Fas-expressing cells was not influenced (Sun et al. 2007).

Apparently, these data somehow do not really match with the data by Julie Desbarats and colleagues described earlier (Desbarats et al. 1998). However, our own experiments on freshly isolated peripheral human T cells clearly show that TCR/CD3/CD28-induced proliferation and activation are completely blocked by immobilised (but not soluble) FasFc fusion protein or anti-FasL pAb at early stages of TCR signal initiation (Paulsen et al., Int. Immunol., in revision). In our hands, the degree of cross-linking does not account for the observed differences in the FasL signal, as FasFc fusion protein presented as a dimer or further cross-linked by anti-human IgGFc show similar effects. The inhibition of TCR signalling by FasL occurs at early stages, since activation-induced TCR internalisation and MAPK activation are blocked, whereas stimulation with phorbol ester and calcium ionophore is unaffected by FasL engagement. Whether the observed differences reflect the use of different cell types (murine/human, activated/resting, $CD4^+$/$CD8^+$) is still open. Nevertheless, although the exact FasL-dependent signalling events or cascades that interfere with TCR signal initiation are still not clear, there is increasing evidence that T cell activation is definitely modulated by the Fas/FasL system (Fig. 3).

References

Badour K, Zhang J, Shi F, McGavin MK, Rampersad V, Hardy LA, Field D, Siminovitch KA (2003) The Wiskott-Aldrich syndrome protein acts downstream of CD2 and the CD2AP and PSTPIP1 adaptors to promote formation of the immunological synapse. Immunity 18:141–154

Baum W, Kirkin V, Fernandez SB, Pick R, Lettau M, Janssen O, Zornig M (2005) Binding of the intracellular Fas ligand (FasL) domain to the adaptor protein PSTPIP results in a cytoplasmic localization of FasL. J Biol Chem 280:40012–40024

Blott EJ, Bossi G, Clark R, Zvelebil M, Griffiths GM (2001) Fas ligand is targeted to secretory lysosomes via a proline-rich domain in its cytoplasmic tail. J Cell Sci 114:2405–2416

Boursalian TE, Fink PJ (2003) Mutation in fas ligand impairs maturation of thymocytes bearing moderate affinity T cell receptors. J Exp Med 198:349–360

Brunner T, Wasem C, Torgler R, Cima I, Jakob S, Corazza N (2003) Fas (CD95/Apo-1) ligand regulation in T cell homeostasis, cell-mediated cytotoxicity and immune pathology. Semin Immunol 15:167–176

Cahuzac N, Baum W, Kirkin V, Conchonaud F, Wawrezinieck L, Marguet D, Janssen O, Zornig M, Hueber AO (2006) Fas ligand is localized to membrane rafts, where it displays increased cell death-inducing activity. Blood 107:2384–2391

Cayabyab M, Phillips JH, Lanier LL (1994) CD40 preferentially costimulates activation of CD4+ T lymphocytes. J Immunol 152:1523–1531

Chitu V, Stanley ER (2007) Pombe Cdc15 homology (PCH) proteins: coordinators of membrane-cytoskeletal interactions. Trends Cell Biol 17:145–156

Desbarats J, Duke RC, Newell MK (1998) Newly discovered role for Fas ligand in the cell-cycle arrest of CD4+ T cells. Nat Med 4:1377–1382

Frauwirth KA, Thompson CB (2002) Activation and inhibition of lymphocytes by costimulation. J Clin Invest 109:295–299

He JS, Ostergaard HL (2007) CTLs contain and use intracellular stores of FasL distinct from cytolytic granules. J Immunol 179:2339–2348

Igney FH, Krammer PH (2005) Tumor counterattack: fact or fiction? Cancer Immunol Immunother 54:1127–1136

Janssen O, Sanzenbacher R, Kabelitz D (2000) Regulation of activation-induced cell death of mature T-lymphocyte populations. Cell Tissue Res 301:85–99

Janssen O, Qian J, Linkermann A, Kabelitz D (2003) CD95 ligand – death factor and costimulatory molecule? Cell Death Differ 10:1215–1225

Kirkin V, Cahuzac N, Guardiola-Serrano F, Huault S, Luckerath K, Friedmann E, Novac N, Wels WS, Martoglio B, Hueber AO, Zornig M (2007) The Fas ligand intracellular domain is released by ADAM10 and SPPL2a cleavage in T-cells. Cell Death Differ 14:1678–1687

Krammer PH, Arnold R, Lavrik IN (2007) Life and death in peripheral T cells. Nat Rev Immunol 7:532–542

Lettau M, Paulsen M, Kabelitz D, Janssen O (2008) Storage, expression and function of Fas Ligand, the key death factor of immune cells. Curr Med Chem 15(17):1684–1696

Lettau M, Qian J, Linkermann A, Latreille M, Larose L, Kabelitz D, Janssen O (2006) The adaptor protein Nck interacts with Fas ligand: guiding the death factor to the cytotoxic immunological synapse. Proc Natl Acad Sci USA 103:5911–5916

Lettau M, Schmidt H, Kabelitz D, Janssen O (2007) Secretory lysosomes and their cargo in T and NK cells. Immunol Lett 108:10–19

Lettau M, Qian J, Kabelitz D, Janssen O (2004) Activation-dependent FasL expression in T lymphocytes and Natural Killer cells. Signal Transduction 4:206–211

Li-Weber M, Krammer PH (2003) Function and regulation of the CD95 (APO-1/Fas) ligand in the immune system. Semin Immunol 15:145–157

Linkermann A, Qian J, Lettau M, Kabelitz D, Janssen O (2005) Considering Fas ligand as a target for therapy. Expert Opin Ther Targets 9:119–134

Newell MK, Desbarats J (1999) Fas ligand: receptor or ligand? Apoptosis 4:311–315
Niederkorn JY (2006) See no evil, hear no evil, do no evil: the lessons of immune privilege. Nat Immunol 7:354–359
Qian J, Chen W, Lettau M, Podda G, Zornig M, Kabelitz D, Janssen O (2006) Regulation of FasL expression: a SH3 domain containing protein family involved in the lysosomal association of FasL. Cell Signal 18:1327–1337
Russell JH, Ley TJ (2002) Lymphocyte-mediated cytotoxicity. Annu Rev Immunol 20:323–370
Schmidt H, Gelhaus C, Nebendahl M, Lettau M, Watzl C, Kabelitz D, Leippe M, Janssen O (2008) 2-D DIGE analyses of enriched secretory lysosomes reveal heterogeneous profiles of functionally relevant proteins in leukemic and activated human NK cells. Proteomics 8:2911–2925
Schulte M, Reiss K, Lettau M, Maretzky T, Ludwig A, Hartmann D, de SB, Janssen O, Saftig P (2007) ADAM10 regulates FasL cell surface expression and modulates FasL-induced cytotoxicity and activation-induced cell death. Cell Death Differ 14:1040–1049
Stuber E, Neurath M, Calderhead D, Fell HP, Strober W (1995) Cross-linking of OX40 ligand, a member of the TNF/NGF cytokine family, induces proliferation and differentiation in murine splenic B cells. Immunity 2:507–521
Suda T, Okazaki T, Naito Y, Yokota T, Arai N, Ozaki S, Nakao K, Nagata S (1995) Expression of the Fas ligand in cells of T cell lineage. J Immunol 154:3806–3813
Sun M, Ames KT, Suzuki I, Fink PJ (2006) The cytoplasmic domain of Fas ligand costimulates TCR signals. J Immunol 177:1481–1491
Sun M, Fink PJ (2007) A new class of reverse signaling costimulators belongs to the TNF family. J Immunol 179:4307–4312
Sun M, Lee S, Karray S, Levi-Strauss M, Ames KT, Fink PJ (2007) Cutting edge: two distinct motifs within the Fas ligand tail regulate Fas ligand-mediated costimulation. J Immunol 179:5639–5643
Suzuki I, Fink PJ (1998) Maximal proliferation of cytotoxic T lymphocytes requires reverse signaling through Fas ligand. J Exp Med 187:123–128
Suzuki I, Martin S, Boursalian TE, Beers C, Fink PJ (2000) Fas ligand costimulates the in vivo proliferation of CD8+ T cells. J Immunol 165:5537–5543
Yang H, Reinherz EL (2006) CD2BP1 modulates CD2-dependent T cell activation via linkage to protein tyrosine phosphatase (PTP)-PEST. J Immunol 176:5898–5907
Zuccato E, Blott EJ, Holt O, Sigismund S, Shaw M, Bossi G, Griffiths GM (2007) Sorting of Fas ligand to secretory lysosomes is regulated by mono-ubiquitylation and phosphorylation. J Cell Sci 120:191–199

Impact of TNF-R1 and CD95 Internalization on Apoptotic and Antiapoptotic Signaling

Stefan Schütze and Wulf Schneider-Brachert

Abstract Internalization of cell surface receptors has long been regarded as a pure means to terminate signaling via receptor degradation. A growing body of information points to the fact that many internalized receptors are still in their active state and that signaling continues along the endocytic pathway. Thus endocytosis orchestrates cell signaling by coupling and integrating different cascades on the surface of endocytic vesicles to control the quality, duration, intensity, and distribution of signaling events. The death receptors tumor necrosis factor-receptor 1 (TNF-R1) and CD95 (Fas, APO-1) are known not only to signal for cell death via apoptosis but are also capable of inducing antiapoptotic signals via transcription factor NF-κB induction or activation of the proliferative mitogen-activated protein kinase (MAPK)/ERK (extracellular signal-regulated kinase) protein kinase cascades, resulting in cell protection and tissue regeneration. A clue to the understanding of these contradictory biological phenomena may arise from recent findings which reveal a regulatory role of receptor internalization and intracellular receptor trafficking in selectively transmitting signals, which lead either to apoptosis or to the survival of the cell.

In this chapter, we discuss the dichotomy of pro- and antiapoptotic signaling of the death receptors TNF-R1 and CD95. First, we will address the role of lipid rafts and post-translational modifications of death receptors in regulating the formation of receptor complexes. Then, we will discuss the role of internalization in determining the fate of the receptors and subsequently the specificity of signaling events. We propose that fusion of internalized TNF-receptosomes with trans-Golgi vesicles should be recognized as a novel mechanism to transduce death signals along the endocytic route. Finally, the lessons learnt from the strategy of adenovirus to escape

S. Schütze (✉)
Institute of Immunology, University Hospital of Schleswig-Holstein, Campus Kiel, Michaelisstr. 5, D-24105, Kiel, Germany
e-mail: schuetze@immunologie.uni-kiel.de

W. Schneider-Brachert
Institute for Medical Microbiology and Hygiene, University of Regensburg, Franz-Josef-Strauss-Allee 11, D-93053 Regensburg, Germany

Results Probl Cell Differ, doi:10.1007/400_2008_23

apoptosis by targeting death receptor internalization demonstrate the biological significance of TNF receptor compartmentalization for immunosurveillance.

1 Pathways of Receptor Internalization

It is now widely accepted that many internalized receptors are still in their active state and that signaling by these receptors continues along the endocytic pathway (McPherson et al. 2001; Sorkin and Von Zastrow 2002; Teis and Huber 2003; Miaczynska et al. 2004). The compartmentalization of signaling pathways would ensure the precise spatial and temporal regulation to evoke a unique, receptor-specific response.

The classical model of signal transduction involves cell surface receptors that are activated after binding to their ligands and transmit intracellular signals by recruitment of adaptor proteins to the intracytoplasmic tail to generate secondary messengers.

Receptors and ligands can internalize from the cell surface by different routes. The best known mechanism of endocytosis is clathrin-mediated endocytosis (CME). The first step in CME involves the selective recruitment of transmembrane receptors and their bound ligands into specialized membrane microdomains (termed clathrin-coated pits (CCPs)). Several adaptor protein complexes participate in this process, initiated by the binding of the adaptor protein complex-2 to the plasma membrane through its lipid-binding domains and to specific transport sequences within the intracellular sequence of the activated receptor. The best-characterized endocytosis motifs of cargo proteins are the tyrosine-based YXXΦ motif (where Y represents tyrosine, X any amino acid, and Φ a bulky hydrophobic amino acid) and dileucine-based motifs. Interaction of AP2 with the GTPase dynamin then forms rings around the neck of budding vesicles, which results in membrane fission and generation of free clathrin-coated vesicles (CCVs). After uncoating during intracellular trafficking, CCVs fuse with early endosomes from which ligand–receptor complexes are sorted to various intracellular compartments such as trans-Golgi vesicles or late endosomes (also known as multivesicular bodies (MVBs)). Receptor complexes that are destined for degradation then fuse with lysosomes and are proteolysed.

In addition, clathrin-independent pathways also play important roles in endocytosis (reviewed in Le Roy and Wrana (2005), Glebov et al. (2006), and Mayor and Pagano (2007)). One form of clathrin-independent endocytosis is based on cholesterol and sphingolipid-enriched membrane domains (lipid rafts) and special membrane invaginations (caveolae), termed RCE (raft/caveolar endocytosis). Raft association tends to concentrate specific proteins within the plasma membrane microdomains, thereby affecting receptor signaling pathways (recently reviewed in Lajoie and Nabi (2007)). In addition, several different dynamin-independent endocytosis mechanisms distinct from both CCPs and caveolae have been identified (recently reviewed in Parton and Simons (2007), Simons and Toomre (2000), and Helms and Zurzolo (2004)).

2 Apoptosis Signaling by Death Receptors

A subgroup of the TNF receptor family are the "death receptors" including TNF receptor-1 (TNF-R1), CD95 (Fas/APO-1), TNF-related apoptosis-inducing ligand (TRAIL) receptors (TRAIL-R1/DR4 and TRAIL R2/DR5), DR3, DR6, and p75 NTR. These death receptors share a "death domain" (DD), a conserved, 80-amino acid sequence in the cytoplasmic tail which is necessary for direct activation of the apoptotic program of the cells by some (TNF-R1, CD95, TRAIL-R1, TRAIL-R2) of these receptors. Ligand-activated TNF-R1 recruits the TNFR-associated death domain (TRADD) protein to the cytoplasmic DD of the receptor (Hsu et al. 1996a). In turn, TRADD functions as an assembly platform to diverge TNF-R1 signaling from the DD; interaction of TRADD with receptor interacting protein-1 (RIP-1) and TNFR-associated TRAF2 (TNF receptor associated protein-2) leads to the activation of the survival transcription factor nuclear factor-κB (NF-κB) and induction of the c-Jun *N*-terminal kinase (JNK) cascade (Hsu et al. 1996b). Alternatively, TRADD can recruit FADD (Fas-associated via death domain) to form the death-inducing signaling complex (DISC). In the case of CD95 and TRAIL receptors, the adaptor TRADD is not required for the recruitment of FADD; FADD directly binds by homotypic interaction through its own DD to the DD of CD95, TRAIL-R1, or TRAIL-R2. Subsequently, FADD recruits caspase-8 to the DISC, through its second functional domain, the death effector domain (DED).

Recruitment of caspase-8 to the DISC results in the activation of this initiator caspase by autoproteolytic self-activation. Then, cell death can follow two paths depending on the level of DISC formed. Activated caspase-8 directly cleaves caspase-3 to initiate the effector arm of the caspase cascade which leads to apoptosis (type I pathway (Scaffidi et al. 1998; Barnhart et al. 2003) characteristic of type I cells). However, if the activation of caspase-8 is not sufficient to mediate efficient caspase-3 activation, an amplification loop is required. Caspase-8 mediates cleavage of the proapoptotic Bcl-2 family member Bid. The truncated form, tBid, then translocates to the mitochondria to initiate apoptosome assembly (cytochrome c, Apaf-1, caspase-9, and ATP), which leads to the activation of caspase-3 and -7 (type II pathway, in type II cells). Both apoptotic pathways are tightly controlled by multiple agonistic and antagonistic regulators of caspase activation; at the level of the DISC, the inhibitory proteins (cFLIP, cIAP) can prevent recruitment of procaspase-8 by competitive binding to the DED of FADD (cIAP, $cFLIP_S$) or by inhibition of caspase-8 activation within the FLICE-inhibitory protein (FLIP)-containing DISC ($cFLIP_L$). Type II signaling is regulated by antiapoptotic Bcl-2 proteins such as Bcl-2 or Bcl-X_L, which inhibit the function of their proapoptotic counterparts (such as Bcl-2-associated X protein (BAX) and Bcl-2-antagonist of cell death (BAD)) to block tBid-mediated apoptosis, or by the inhibitory protein X inhibitor of apoptosis protein (XIAP) which prevents caspas-9 activation. These inhibitory proteins themselves are counteracted by mitochondrial proteins such as Smac Direct IAP Binding Protein with low isoeiectric point (DIABLO), which are released in response to apoptotic stimuli to guarantee caspase activation.

Micheau and Tschopp (2003) proposed a model in which TNF-R1 signaling involves assembly of two molecularly and spatially distinct signaling complexes that sequentially activate NF-kB and caspases. TNF-R1 recruits RIP-1, TRAF-2, and TRADD to form a signaling complex at the cell surface termed "complex I" within a few minutes of TNF binding. Complex I signals for NF-κB activation through recruitment of the I-κB kinase "signalosome" high molecular weight complex. At later time points and after TNF-R1 internalization, in this model, RIP-1, TRAF-2, and TRADD are modified by ubiquitinylation and dissociate from the receptor. Within the cytosol, this complex then recruits FADD and caspase-8 to a secondary signaling complex (termed "complex II").

Three recent reports of our group (Schneider-Brachert et al. 2004, 2006; Neumeyer et al. 2006) confirmed the existence of spatially distinct signaling complexes at the cell surface and intracellularly, as proposed (Harper et al. 2003; Micheau and Tschopp 2003). However, we showed that recruitment of RIP-1 and TRAF-2 to the cell surface TNF-R1 is sufficient to signal for NF-κB and could demonstrate that the DISC was found to be still associated together with the internalized TNF receptor in TNF-R1 receptosomes (Schneider-Brachert et al. 2004, 2006) (see below).

3 Role of Lipid Rafts in TNF-R1 Signaling

TNF was first described as a cytokine that exhibits antitumor effects in mouse models and is now recognized as a highly pleiotropic cytokine that elicits diverse cellular responses, which range from proliferation and differentiation to activation of apoptosis (Locksley et al. 2001; Wajant et al. 2003). The biological activities of TNF are mediated by two distinct cell surface receptors: TNF-R1 (also known as p55/60; CD120a) and TNF-R2 (also known as p75/80; CD120b). In contrast to TNF-R1, TNF-R2 does not contain a DD and cannot transmit apoptosis signals.

Binding of TNF initiates rapid clustering of the TNF-R1, followed by internalization of the ligand–receptor complex (Mosselmans et al. 1988; Bradley et al. 1993; Schütze et al. 1999; Schneider-Brachert et al. 2004, 2006). At the level of the plasma membrane, lipid rafts have been implicated to play a role in TNF-R1 signaling. TNF-R1 translocations to lipid rafts within 2 min of ligand binding have been observed in the human fibrosarcoma cell line HT1080, concomitant with the recruitment of the adaptor molecules RIP-1, TRADD, and TRAF2 (Legler et al. 2003). Disruption of lipid rafts by cholesterol depletion results in inhibition of Iκ-Ba phosphorylation, which is required for NF-κB activation (and subsequently transmission of survival signals) in response to TNF treatment, switching the TNF-response to induction of apoptosis.

However, the role of lipid rafts for TNF-R1 signaling still remains unclear and may depend on the cell type. In contrast to the HT1080 cell line (Legler et al. 2003), TNF-R1-induced apoptosis has been reported to depend on lipid rafts in the U937 myeloid cell line (Doan et al. 2004). In primary mouse macrophages, lipid rafts appear to be important for transducing TNF-R1 signaling to the MAPK/ERK pathway but not to NF-κB activation (Ko et al. 1999). A selective lipid raft dependency of TNF-R1 signaling to $p42^{mapk/erk2}$ was observed in primary mouse macrophages

(Doan et al. 2004), but in human airway smooth muscle cells NF-κB and MAPK activation by TNF was found to be independent of lipid rafts (Hunter and Nixon 2006). In the human endothelial cell line EA.hy926, TNF-R1-mediated activation of phosphatidyl-inositol 3-kinase (PI3K), but not of NF-κB, seems to originate from caveolae after interaction of TNF-R1 with caveolin-1 (D'Alessio et al. 2005). By contrast, disruption of lipid rafts in HT1080 fibrosarcoma blocked NF-κB activation and sensitized cells to apoptosis (Legler et al. 2003). Thus, redistribution of TNF-R1 into lipid rafts and nonraft regions of the plasma membrane seems to regulate diversity of signaling responses by TNF in various cell types, but the quality of signals transduced from lipid rafts varies significantly between different cell lines.

4 Role of Internalization in TNF-R1 Signaling

In various cell types, TNF-R1 endocytosis is mediated by CCP formation (Mosselmans et al. 1988; Bradley et al. 1993; Schütze et al. 1999; Schneider-Brachert et al. 2004, 2006) (Table 1). In human endothelial cells, TNF receptor endocytosis was linked to TNF-induced expression of NF-κB regulated genes, which encode the cell adhesion molecules endothelial leucocyte adhesion molecule-1 (ELAM-1), intracellular adhesion molecule-1(ICAM-1), and vascular cell adhesion molecule-1 (VCAM-1) (Bradley et al. 1993). Earlier reports suggested that TNF receptor internalization in other cells may play a role in mediating TNF cytotoxicity (Kull Jr and Cuatrecasas 1981, Pastorino et al. 1996). In U937 cells, it was demonstrated that selected TNF-R1 DD signaling pathways including the pathways that lead to apoptosis were dependent on TNF receptor internalization, while others were not (Schütze et al. 1999): blocking CCP formation by monodansyl cadaverine (MDC) inhibited activation of the endolysosomal acid sphingomyelinase (A-SMase) and JNK, as well as TNF-induced cell death. By contrast, the interaction of the adaptor molecules factor associated with neutral sphingomyelinase (FAN) and TRADD to TNF-R1 at the cell surface and the activation of plasma membrane-associated neutral sphingomyelinase (N-SMase) as well as the stimulation of proline-directed protein kinases (PDPK) were not influenced by inhibition of TNF-R1 internalization (Schütze et al. 1999). These findings point to a role of TNF-R1 internalization in transmitting proapoptotic signals from intracellular compartments, while nonapoptotic signaling occurs from TNF-R1 at the cell surface. As another example, in nonphagocytic cells TNF receptor internalization is required for TNF-induced production of reactive oxygen species (ROS) and activation of MAPKs and AKT/protein kinase B (PKB), but not for activation of NF-κB (Woo et al. 2006).

Thus, TNF-R1 compartmentalization seems to play an important role for selective internalization-dependent (cytotoxic, proapoptotic) and internalization-independent (mitogenic and proinflammatory) signaling pathways. On the basis of these observations, the endosomal compartment has to be recognized as a novel signaling organelle that is involved in selectively transmitting death signals from TNF-R1 (see Table 1).

Table 1 Role of TNF-R1 internalization for selective TNF signaling

Cell type	Internalization-dependent effects	Internalization-independent effects	Reference
L-M, murine tumorigenic fibroblasts	Tumor necrosis serum cytotoxicity	n.d.	Kull, Jr. and Cuatrecasas (1981)
Human myosarcoma cell line KYM	TNF-cytotoxicity	n.d.	Watanabe et al. (1988)
Human umbilical vein endothelial cells	TNF-induced ICAM-1 and VCAM-1- expression	n.d.	Bradley et al. (1993)
U937 cells	TNF-cytotoxicity, activation of A-SMase and JNK	Activation of N-SMase and PDP-kinase	Schütze et al. (1999)
U937 cells, murine fibroblasts, 70Z/3 murine pre-B cells	TNF-mediated apoptosis, activation of A-SMase, CTSD, TNF-R1 recruitment of TRADD and FADD, activation of caspase-8	Activation of N-SMase, TNF-R1 recruitment of RIP-1 and TRAF-2	Schneider-Brachert et al. (2004)
Murine fibroblasts	TNF-mediated apoptosis, activation of caspase-8 and caspase-9	Residual apoptosis, and activation of caspase-3 that is mediated by enhanced N-SMase activity	Neumeyer et al. (2006)
NIH 3T3 and C127 murine fibroblasts, human H1299 fibrosarcoma cells and HeLa cells	Adenovirus 14.7K selectively blocks TNF-R1 internalization, DISC-recruitment, activation of caspase-8 and apoptosis	14.7K does not block TNF-R1 recruitment of RIP-1 and TRAF-2, and activation of NF-kB	Schneider-Brachert et al. (2006)
HEK293 cells	ROS production, activation of MAP kinase, PI3K and PKB	Activation of NF-kB	Woo et al. (2006)

TNF tumor necrosis factor; *ICAM-1* Intracellular adhesion molecule-1; *VCAM-1* vascular cell adhesion molecule 1; *A-SMase* acid sphingomyelinase; *JNK* c-jun *N*-terminal protein kinase; *N-SMase* neutral sphingomyelinase; *PDP-kinase* proline-directed protein kinase; *CTSD* cathepsin D; *TRADD*, TNF receptor-associated death domain protein; *FADD* Fas-associated via death domain protein; *RIP-1* receptor interacting protein-1; *TRAF-2* TNF receptor associated protein-2; *DISC* death-inducing signaling complex; NF-κB nuclear factor κB; *ROS* reactive oxygen species; *MAP kinase* mitogen-activated protein kinase; *PKB* protein kinase B

An important question is how apoptotic signals are further transmitted from this intracellular compartment. As discussed above, TNF receptor triggering activates the endo-lysosomal enzyme A-SMase (Schütze et al. 1992; Wiegmann et al. 1994, 1999; Schwandner et al. 1998), generating the potent proapoptotic lipid second messenger ceramide (reviewed in Lin et al. (2000) and Morales et al. (2007)). A role for A-SMase in transmitting apoptotic signals of death receptors has been reported for TNF (Monney et al. 1998; Garcia-Ruiz et al. 2003; Heinrich et al. 2004), CD95 (Cifone et al. 1994; Herr et al. 1997; De Maria et al. 1998; Brenner et al. 1998), and TRAIL (Dumitru and Gulbins 2006; Thon et al. 2006).

The aspatate-protease cathepsin D (CTSD) is a direct downstream target for ceramide within the same endo-lysosomal compartment (Heinrich et al. 1999). The proapoptotic Bcl-2 protein family member Bid colocalizes with CTSD-positive vesicles and after TNF stimulation both CTSD and Bid are located in ROS5 positive early endosomes, implicating that Bid is located at the subcellular site of CTSD activation. After TNF-induced, ceramide-mediated translocation through the endosomal membrane, it was shown that CTSD cleaves Bid, leading to activation of caspase-9 and -3 (Heinrich et al. 2004).

Thus, endo-lysosomal proteases such as CTSD can be understood, like caspases, as a group of proteases that are activated in a cascade-like manner. Once released to the cytoplasm, these proteases may execute apoptosis either independent of caspases or might individually participate in different apoptotic or cell death signaling cascades by connecting the endosomal compartment to the classical apoptosis signaling pathways.

5 DISC Assembly Occurs at Internalized TNF Receptosomes

The key elements of the signaling pathways of apoptosis (the TNF-R1 adaptor proteins TRADD, FADD, and caspase-8) and of NF-κB (TRADD, RIP-1, and TRAFs) are well defined (reviewed in Chen and Goeddel (2002) and Wajant et al. (2003)). However, the molecular mechanisms that regulate the formation of the initial signaling complexes at the activated TNF-R1 to selectively transmit specific signal transduction events to various intracellular compartments are poorly understood.

We recently demonstrated the important role of internalized TNF-R1 as the essential platform for recruiting the DISC to the TNF-R1 receptosomes (Schneider-Brachert et al. 2004, 2006). Recruitment of the adaptor proteins TRADD, FADD, and caspase-8 to form the DISC occurred within 3 min after TNF stimulation, and DISC was still associated with TNF-R1 after 60 min. Inhibition of TNF-R1 internalization blocked DISC recruitment and apoptosis but still allowed the recruitment of RIP-1 and TRAF-2 to signal for NF-κB activation. We used a novel experimental approach: TNF receptors were labeled with biotin-TNF coupled to streptavidin-coated magnetic nanobeads, and intact TNF–TNFR complexes were isolated within their native membrane environment using a specialized magnetic device (Schütze and Tchikov 2008). Immunomagnetic isolation of morphologically intact vesicles revealed trafficking and maturation of TNF receptosomes along the endocytic pathway and fusion of TNF receptosomes with trans-Golgi membranes resulting in formation of multivesicular endosomes (Schneider-Brachert et al. 2004).

Deletion of the TNFR internalization domain (TRID) within the cytoplasmic tail of TNF-R1 or point mutations within the YQRW internalization motif resulted in complete elimination of TNF-R1 internalization. These mutations prevented the recruitment of TRADD, FADD, and caspase-8 and led to an almost entire inhibition of TNF-induced apoptosis (Schneider-Brachert et al. 2004). In cells that express TNF-R1 ΔTRID, a residual DISC-independent cell death was observed after prolonged incubation as a result of high local production of ceramide by overactivated

N-SMase. In adenovirus-infected cells or cells transduced with the adenoviral E3-14.7K, however, inhibition TNF-R1 endocytosis correlated with a complete blocking of TNF-induced apoptosis (Schneider-Brachert et al. 2006).

The DD of internalization-deficient TNF-R1 ΔTRID could still recruit RIP-1 and TRAF2 for activation of NF-κB even in the absence of TRADD as an assembly platform (Schneider-Brachert et al. 2004, 2006). Although it is generally accepted that TRADD is required for RIP-1 recruitment to TNF-R1, this observation is in line with previous reports on a direct interaction of RIP-1 with TNF-R1 and CD95 (Stanger et al. 1995; Hsu et al. 1996b). In addition, Zheng et al. (2006), and Jin and El Deiry (2006), showed that TRADD and RIP-1 can independently and competitively associate with TNF-R1. RIP-1 is essential for TNF-R1-mediated NF-κB activation, since cells derived from mice with a functional deletion of RIP-1 showed no NF-κB activation and were highly sensitive to the induction of apoptosis (Kelliher et al. 1998). RIP-1 then can interact with TRAF-2 (Hsu et al. 1996b), and TRAF-2 in turn mediates the physical interaction with the NF-κB-inducing I-kappa B kinase (IKK) complex (Devin et al. 2000).

How are the signaling events during endocytosis of TNF receptor complexes regulated? Since TRADD, RIP-1, and TRAF-2 are co-internalized with TNF-R1, NF-κB signaling has to be downregulated during TNF receptor endocytosis, allowing for full propagation of DISC-mediated proapoptotic signaling. The ubiquitin protein ligase CARP-2 was identified as a constitutive negative regulator of TNF-induced NF-κB activation (Liao et al. 2008). CARP-2 is localized to endocytic vesicles where it interacts with internalized TNF receptosomes and, together with A20, mediates ubiquitinylation and degradation of RIP-1 and subsequently downregulation of NF-κB signaling.

In summary, two temporary and spatially distinct TNF-R1 signaling complexes are formed with the capacity to either signal for NF-κB activation from the cell surface or for apoptosis from internalized receptosomes (see Fig. 1), indicating that TNF-R1 compartmentalization has an important role in the diversification of TNF-mediated biological responses. The reason for the discrepancy between our findings of TNF-R1-associated DISC and the results of Harper et al. (2003) and Micheau and Tschopp (2003) might be due to the different detergents used for solubilization of TNF-R1 from (clathrin-coated) endosomes and the fact that we used a different labeling and precipitation protocol. Furthermore, early DISC formation at TNF-receptosomes occurs rapidly and transiently and may be detectable only when the internalization is synchronized by prelabeling the receptors with TNF at 4°C followed by a rapid increase in the temperature to 37°C.

6 Regulation of CD95 Activation

CD95 is a key mediator of apoptosis (Li-Weber and Krammer 2003); however, ligation of CD95 can also mediate several nonapoptotic activities (recently discussed in Peter et al. (2007)). Thus, under certain conditions and in some cell types, CD95 ligation may also protect cells and regulate tissue regeneration and proliferation.

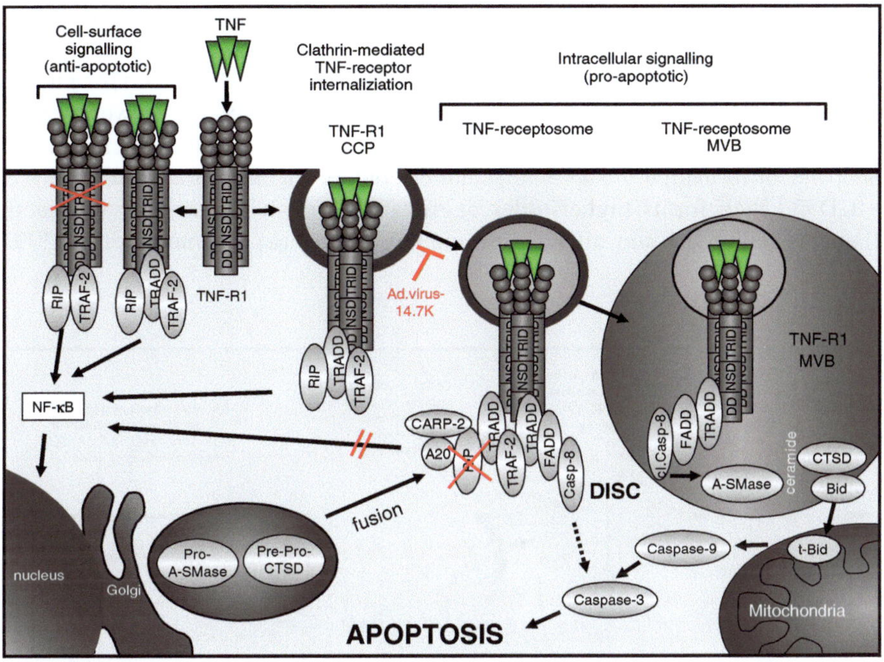

Fig. 1 Signaling from cell surface TNF-R1 or from internalized TNF-R1 decisive for NF-κB activation or induction of apoptosis. Ligand-activated TNF-R1 promotes activation of NF-κB via recruitment of TRADD, RIP-1, and TRAF-2 at the cell surface. When receptor internalization is blocked (by mutations within the TNF-receptor internalization domain (TRID) or by the adenovirus protein 14.7K), the recruitment of RIP-1 and TRAF2 to the cytoplasmic death domain (DD) of cell surface TNF-R1 is sufficient for NF-κB signaling. TNF-R1 is internalized via clathrin-dependent endocytosis within minutes, and NF-κB signaling is terminated by recruitment of the ubiquitin-E3 ligase CARP-2, together with A20, leading to the degradation of RIP-1. TRADD then recruits FADD and caspase-8 to TNF-R1 at the internalized receptosomes. Subsequently, caspase-8 is activated, which can induce caspase-3 activation. Along the endocytic pathway, TNF-receptosomes fuse with trans-Golgi vesicles that contain pro A-SMase and pre-pro CTSD to form multivesicular bodies (MVB). Within the MVB, activated caspase-8 stimulates the A-SMase/ceramide/CTSD cascade, which is capable of mediating apoptosis through cleavage of Bid and activation of caspase-9 and caspase-3

A clue to the understanding of these contradicting biological activities of CD95 may lie, as discussed earlier for the TNFR system, in the balance of the dynamics of CD95 membrane localization and internalization-dependent and internalization-independent pathways to signal for apoptosis and other functions.

A current view of the early events in CD95 signaling is depicted in Fig. 2.

Earlier work on the discovery of the CD95 DISC by Kischkel et al. (1995), Scaffidi et al. (1997), and Medema et al. (1997) showed that recruitment of FADD and caspase-8 occurs in a very rapid phase, within seconds. A further significant enhancement of p26/p28 and p18/p10 active caspase-8 was detected between 5 and 10 min after CD95 ligation, indicating two phases of caspase-8 activation. Also, within seconds after ligand binding, CD95 forms sodium dodecyl sulfate (SDS)- and mercaptoethanol-stable aggregates (Kischkel et al. 1995; Kamitani et al. 1997; Papoff et al. 1999; Algeciras-Schimnich et al. 2002; Feig et al. 2007). These highly aggregated CD95 forms of approximately 180 kDa in SDS–PAGE were termed $CD95^{hi}$ (Feig et al. 2007). SDS-stable $CD95^{hi}$ is likely the result of the caspase-8-independent formation of supramolecular oligomers (Henkler et al. 2005).

$CD95^{hi}$ then forms higher-order aggregates through interactions with actin filaments within 15 min after ligand binding (Algeciras-Schimnich et al. 2002;

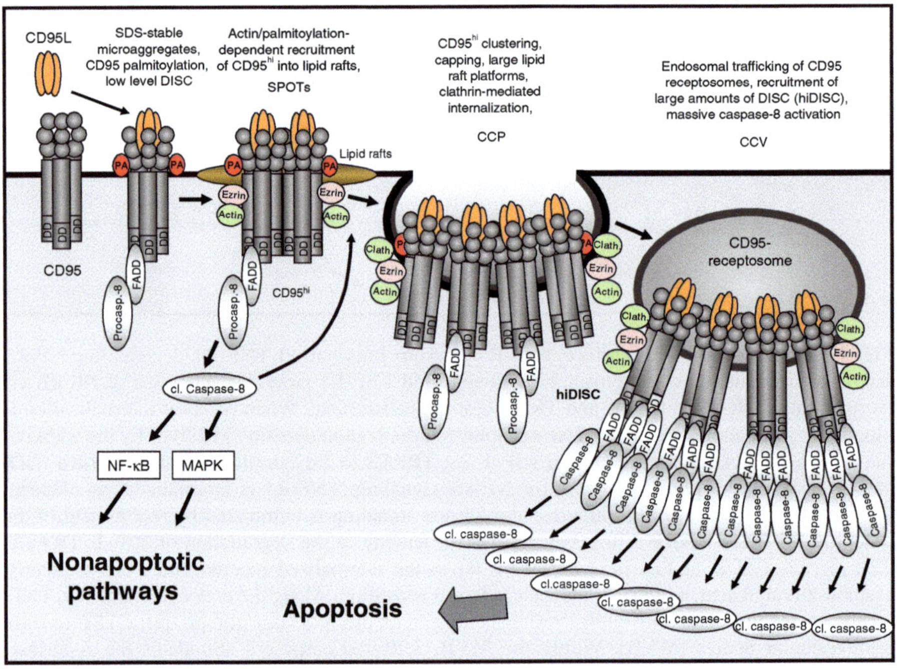

Fig. 2 CD95 compartmentalization linked to pro- and nonapoptotic signaling. Ligation of CD95 leads to rapid formation of SDS-stable microaggregates ($CD95^{hi}$), which translocate to lipid raft plasma membrane microdomains, a process that is regulated by palmitoylation (PA) and ezrin-mediated association of the receptor with the actin cytoskeleton. Low amounts of FADD and caspase-8 are recruited within this very early time frame and "signaling protein oligomerization transduction structures" (SPOTS) are then formed resulting in clustering of CD95 ("capping"), which depends on active caspase-8. This leads to the formation of large lipid raft platforms. At this stage, CD95 has the potential to activate nonapoptotic pathways by inducing activation of the mitogen-activated protein kinase (MAPK) and the transcription factor nuclear factor NF-κB, leading to cell proliferation and migration, but is unable to kill type I cells. Between 5 and 15 min after CD95 triggering, CD95 is internalized in a clathrin-dependent manner into endosomal compartments via ezrin-mediated actin filament association. During endosomal trafficking, DISC proteins (FADD and caspase-8) are massively recruited, resulting in strong caspase-8 activation within very high molecular weight structures of many megadaltons in size (hiDISC complexes) resulting in the propagation of apoptosis signaling

Henkler et al. 2005). These aggregates could be detected by fluorescence microscopy as "signaling protein oligomerization transduction structures" (SPOTS) in cells that die by the type I death pathway (type I cells) (Siegel et al. 2004).

Between 15 min and 1 h and after initial caspase-8 activation, a lateral segregation of CD95 complexes forming large clusters on one pole of the cells, termed "caps,", that correspond to internalized receptors, derived from hiDISC structures, was observed (Algeciras-Schimnich et al. 2002; Lee et al. 2006; Feig et al. 2007). However, small amounts of low molecular weight DISC also form at the cell surface prior to receptor internalization in type I cells or at the plasma membrane of type II cells but fail to internalize CD95. Notably, this low molecular weight DISC seems to be capable of mediating nonapoptotic signaling (Barnhart et al. 2004; Lee et al. 2006).

7 The Role of Lipid Rafts in CD95 Signaling

On the basis of experiments in cholesterol-depleted membranes, it has been postulated that CD95 apoptosis signals independently from rafts (Algeciras-Schimnich et al. 2002; Algeciras-Schimnich and Peter 2003; O'Reilly et al. 2004). However, several other studies have indicated a constitutive and inducible association of CD95 with lipid rafts (Henkler et al. 2005; Siegel et al. 2004; Grassme et al. 2001; Hueber et al. 2002; Scheel-Toellner et al. 2002; Muppidi et al. 2004; Eramo et al. 2004; Miyaji et al. 2005; Nakayama et al. 2006; Legembre et al. 2005; Chakrabandhu et al. 2007; Koncz et al. 2007). It was indeed thought that CD95 preassociation, lipid raft translocation, SPOT formation, DISC recruitment, and initiation of apoptosis signaling from lipid rafts are restricted to type I cells (Siegel et al. 2004; Eramo et al. 2004; Nakayama et al. 2006). In type II cells, CD95 was mostly found to be absent from lipid rafts. A number of reports showed that in type II cells lipid raft translocation of CD95 could be induced either by pretreatments with antitumor drugs (Gajate et al. 2004; Gajate and Mollinedo 2007), by T-cell receptor restimulation of activated $CD4^+$ T cells before binding of CD95L (Muppidi et al. 2004), or by coligation of the membrane receptor CD28 (Legembre et al. 2005, 2006). Since these treatments also resulted in amplification of CD95-mediated apoptosis, lipid raft localization of CD95 seems to be important for apoptosis signaling.

By comparing various type I and type II cells, the study of Eramo et al. (2004) demonstrated that in type II cells also CD95 translocated to lipid rafts, but with a much slower kinetic than in type I cells. At later times, CD95 was also able to internalize from these membrane structures and recruit the DISC, but to a much lesser extent. Thus, the differences between type I and type II cells may also be related to differences in the kinetics of CD95 ligand (CD95L)-induced CD95 translocation, internalization, and signaling. According to Feig et al. (2007), recruitment and activation of caspase-8 may occur in lipid rafts, but signaling for apoptosis is mediated from high molecular weight aggregates that are localized inside and outside lipid rafts.

The observation that the localization of CD95 to lipid raft microdomains and the formation of "caps" are enhanced by the action of ceramide generated by A-SMase

(Cremesti et al. 2001) points to another important regulatory mechanism in early CD95 activation. In WR19L lymphoid transfectants, CD95 translocation to lipid rafts, DISC formation, and apoptosis depended on the sphingomyelin content of the plasma membranes (Miyaji et al. 2005). These findings indeed suggest that CD95 is recruited to a special class of lipid raft microdomains that contain more sphingomyelin than others.

The facts that "cap" formation of CD95 represents internalized CD95 receptors (Lee et al. 2006) and that internalization as well as CD95-induced apoptosis is more pronounced in type I cells (Algeciras-Schimnich et al. 2002; Feig et al. 2007; Lee et al. 2006) point to an important role of CD95 internalization in the propagation of the apoptotic signal.

8 Impact of CD95 Internalization

Internalization of CD95 in response to agonistic antibody or CD95 ligand stimulation in various lymphoid and nonlymphoid type I cells was first described by Algeciras-Schimnich et al. (2002). These data confirmed that the role of receptor internalization for signaling in apoptosis is not restricted to TNF-R1 but is also valid for another member of the TNF death receptor family (Table 2).

A clathrin- and actin-filament-dependent CD95 internalization was observed between 5 and 15 min after stimulation. The connection between CD95 and actin filaments is mediated through the association of CD95 with ezrin, a membrane cytoskeletal crosslinker protein (Chakrabandhu et al. 2007; Parlato et al. 2000). Ezrin/actin cytoskeleton rearrangements were proposed to be regulated by the SH2-domain-containing tyrosine phosphatase SHP1 (Koncz et al. 2007). According to Algeciras-Schimnich et al. (2002), FADD- and caspase-8 recruitment preceded CD95 internalization in lymphoblasts and T cells, pointing toward the regulatory role of caspase-8 in CD95 internalization. CD95 internalization was not observed in type II cells (Jurkat T cells and CEM lymphoblasts), and disruption of the actin filaments by Latrunculin A had no effect on apoptosis sensitivity in these cells, suggesting that CD95 signaling in type II cells is different from that in type I cells and independent of actin (Algeciras-Schimnich et al. 2002).

Caspase-8 has a regulatory role in CD95 internalization in type I cells (Siegel et al. 2004). The SPOTS are formed in lipid rafts and preceded caspase-8 activation and CD95 internalization; thus, CD95 internalization most likely occurs from lipid raft microdomains. In the same lane, electron microscopy revealed that after CD95 stimulation lipid rafts aggregated in large clusters that were internalized in endosomal vesicles, and caspase-8 was processed and activated within these vesicles (Eramo et al. 2004).

A recent study by Lee et al. (2006) highlighted the role of receptor endocytosis for apoptosis and other nonapoptotic biological functions of CD95 in type I cells. According to Lee, internalization of CD95 was already detected after 3 min, as evident from the recruitment of endocytosis markers such as Rab4, EEA-1, and cathepsin D at isolated CD95 receptosomes (Lee et al. 2006; Feig et al. 2007). At this time, active caspase-8 (p43/41 and p18) was also detected at isolated CD95

Table 2 Impact of receptor internalization in CD95 signaling

Cell type	CD95 signaling from rafts	Reference
SKW 6.4B-lymphoblasts, H9 T cells, Burkitt's lymphoma, BL-60 transfectants	Formation of SDS-stable CD95 microaggregates, caspase-8-dependent and actin-dependent CD95 clustering, DISC-formation and internalization from non-rafts	Algeciras-Schimnich et al. (2002); Algeciras-Schimnich and Peter (2003)
SKW 6.4 and H9 cells, Jurkat cells, Cos-7 transfectants	Type I cells: FADD-mediated early formation of CD95 SPOTs in lipid rafts, caspase-8 dependent capping, and internalization of CD95 from lipid rafts.	Siegel et al. (2004)
JY B-cell line and HuT78 –cells, Jurkat cells, and CEM lympho-blasts	Type I cells: rapid caspase-8-independent CD95 translocation, rapid CD95 internalization, and DISC formation from lipid rafts. Type II cells: slow CD95 trans-location, delayed internalization, and less DISC formation compared to type I cells.	Eramo et al. (2004)
SKW 6.4, BJAB, ACHN, H9, MCF-7, primary T lymphocytes, Jurkat and HCT15 cells	Type I cells: clathrin-mediated CD95 internalization: DISC-proteins are recruited to endosomes (CD95 receptosomes). CD95 internalization-independent signaling: activation of ERK and NF-kB, cell motility and invasiveness after blocking CD95 internalization in type I cells, or stimulation with nonapoptotic anti APO-1 antibody	Lee et al. (2006)
HEK293 cells, L1210 mouse T-cell transfectants, peripheral T cells, Jurkat cells	CD95 palmitoylation targets CD95 to cytoskeleton-linked lipid rafts, mediates raft-dependent, ezrin-mediated cytoskeleton association to CD95, which is required for CD95 internalization, DISC-formation, and apoptosis.	Chakrabandhu et al. (2007)
SKW 6.4 cells, 293 T cells, NIH 3T3 cells	Formation of high molecular weight CD95 complexes (CD95 ^{hi}) preceeds CD95 internalization. Formation of SDS-stable megadalton-size DISC complexes (hiDISC) initiate apoptosis signaling. Most hiDISC is formed outside rafts. Palmitoylation of CD95 regulates CD95 aggregation, CD95 internalization, DISC formation, and apoptosis	Feig et al. (2007)
A20 murine B cells, murine spleen B cells	SHP1-regulates downmodulation of CD95-ezrin-actin linkage via Vav dephosphorylation: a fine-tuned switch-off mechanism to terminate CD95 internalization, DISC formation, and cell death	Koncz et al. (2007)

FADD Fas-associated via death domain protein; *DISC* death-inducing signaling complex; *SPOTs* signaling protein oligomerization transduction structures; *ERK* extracellular signal-related kinase; *SHP-1* SH2-homology containing protein-tyrosine phosphatase-1; *Vav* GDP/GTP nucleotide exchange factor regulated by tyrosine phosphorylation receptosomes. These activation profiles correspond to the second phase of enhanced caspase-8 activation between 5 and 10 min as described earlier by Kischkel et al. (1995), Scaffidi et al. (1997), and Medema et al. (1997). The isolation of internalizing receptosomes that contained magnetically labeled CD95 protein complexes (Schütze and Tchikov 2008; Lee et al. 2006) revealed that Rab4 and EEA-1 were readily detectable very early after stimulation and peaking at 10 min, consistent with the ability of CD95 to internalize in type I cells. Low levels of FADD were detected in CD95-containing membrane structures at basal levels, whereas maximal FADD recruitment in magnetically labeled CD95 receptosomes was observed at 30 min. Similar to FADD, caspase-8 and its intermediate cleavage products peaked at 10 min and could be detected in isolated receptosomes as late as 3 h following stimulation, suggesting that most of the caspase-8 activation occurred inside the cells and is located on endosomal and even lysosomal vesicles. Blocking CD95 internalization resulted in the inhibition of DISC recruitment and apoptosis. In contrast to type I cells, no significant increase in Rab4, EEA-1, or CTSD was observed in type II cells, confirming a lack of directional movement of CD95 into endosomal vesicles within 1 h. All these approaches demonstrated that DISC assembly occurs predominantly after CD95 was internalized and has entered an early endosomal compartment (see Fig. 2).

Since the physiological stimulus of CD95 is more likely to be membrane-bound (mCD95L) than soluble ligand (sCD95L), the intriguing question is whether CD95 internalization also takes place after stimulation of CD95 with mCD95L. In coculture experiments using cells that express noncleavable membrane CD95L, it could be shown that mCD95 induces internalization of CD95 and similar levels of caspase-8 activation as in stimulation with crosslinked sCD95L (Lee et al. 2006). In the same lane, binding of agonistic anti APO-1 antibodies to CD95 is required for CD95 aggregation but is no longer required for CD95 internalization, DISC formation, and caspase-8 activation (Feig et al. 2007).

Inhibition of CD95 internalization, which blocks signaling for apoptosis, enabled the induction of NF-κB activation and activation of ERK1/2 following CD95 engagement (Lee et al. 2006). These observations suggest that additional types of signaling occur independently of CD95 internalization. Indeed, treatment of CD95L-resistant MCF7 (FB) cells with anti APO-1 antibody or soluble CD95L that did not induce CD95 internalization increased tumor cell motility and invasiveness (Lee et al. 2006). These observations provided the molecular basis for the assumption that activation of nonapoptotic signaling pathways by CD95L, including MAPK and NF-κB signaling pathways, plays a role in the tumorigenesis of CD95-resistant tumors (Barnhart et al. 2004; Ahn et al. 2001). Alternative theories, however, postulate that proliferative signaling is mediated via the caspase-8 inhibitor FLIP to promote activation of NF-κB and ERK signaling pathways in various cell lines (Kataoka et al. 2000; Golks et al. 2006).

Thus, it appears that the dynamics of CD95 membrane localization and internalization plays a critical role to balance internalization-dependent apoptotic and internalization-independent nonapoptotic pathways to drive cellular cell death and other functions, respectively (Fig. 2).

9 Regulation of CD95 by Post-translational Modifications

The important and apparent decisive role of CD95 compartmentalization for selective CD95 signal transduction raises the question as to how the different steps in CD95 activation are regulated at the molecular level. It is well known that membrane proteins that are post-translationally modified by *N*-myristoylation and/or *S*-palmitoylation can be found in lipid rafts, whereas proteins that are modified by unsaturated fatty acids or prenyl groups are excluded from lipid rafts (reviewed in Smotrys and Linder (2004)). Indeed, human and murine CD95 was found to be palmitoylated at the membrane proximal Cys199 or Cys194, respectively (Feig et al. 2007; Chakrabandhu et al. 2007). Mutations in these sites or competition for CD95 palmitoylation prevented $CD95^{hi}$ complex formation and resulted in a marked reduction of CD95 translocation to lipid rafts, CD95L-induced CD95 internalization, DISC formation, and apoptosis. Thus, CD95 palmitoylation appears to play an essential role in the initiation of CD95 apoptotic signaling pathway.

10 Not on the Same TRAIL?

Two recent studies aimed to characterize the specific contribution of TRAIL receptor endocytosis for apoptosis signaling. Within the first 30 min after stimulation with labeled ligands, TRAIL-R1 and TRAIL-R2 were rapidly internalized, concomitant with the recruitment of FADD and caspase-8 (Austin et al. 2006; Kohlhaas et al. 2007). Ultrastructural analysis of the early phase of TRAIL endocytosis localized the labeled ligand at the cell surface within CCPs (Austin et al. 2006). However, a significant portion of TRAIL receptors was internalized by a non-clathrin-mediated pathway (Kohlhaas et al. 2007). Prolonged stimulation of TRAIL for 2 h induced caspase-mediated cleavage of the clathrin heavy chain (CHC), and the α subunit of adaptor protein-2 (AP2) terminated TRAIL receptor endocytosis (Austin et al. 2006). However, under the condition in which TRAIL receptor endocytosis was blocked, TRAIL-induced apoptosis signaling was not inhibited, but rather amplified. Taken together, these results suggest that TRAIL stimulated internalization of its cognate receptors proceeds along both clathrin-dependent and clathrin-independent pathways, but TRAIL-induced DISC formation, caspase activation, and apoptosis signaling also occur in the absence of clathrin-mediated TRAIL receptor endocytosis.

11 Viral Targeting of Death Receptors

In this section, we focus on molecular mechanisms by which adenovirus selectively manipulate death receptor internalization to inhibit apoptosis (Fig. 3). During their coevolution with the immune system, pathogens successfully adopt sophisticated strategies to counteract the innate and adaptive immune responses mounted by the infected host (Benedict et al. 2002). Many of the antiimmune mechanisms are directed against activation of the apoptotic pathway particularly by members of the TNF cytokine family (Benedict et al. 2003; Rahman and McFadden 2006).

The adenovirus early transcription unit 3 (E3) encodes several proteins to protect infected cells from death signals mediated by TNF, CD95L, and TRAIL (Lichtenstein et al. 2004b). Expression of the receptor internalization and degradation (RID) complex specifically downregulates presentation of CD95 and TRAIL-R1 on the cell surface within the first hours after adenovirus infection to prevent ligand-induced apoptosis (Shisler et al. 1997; Tollefson et al. 1998, 2001; Benedict

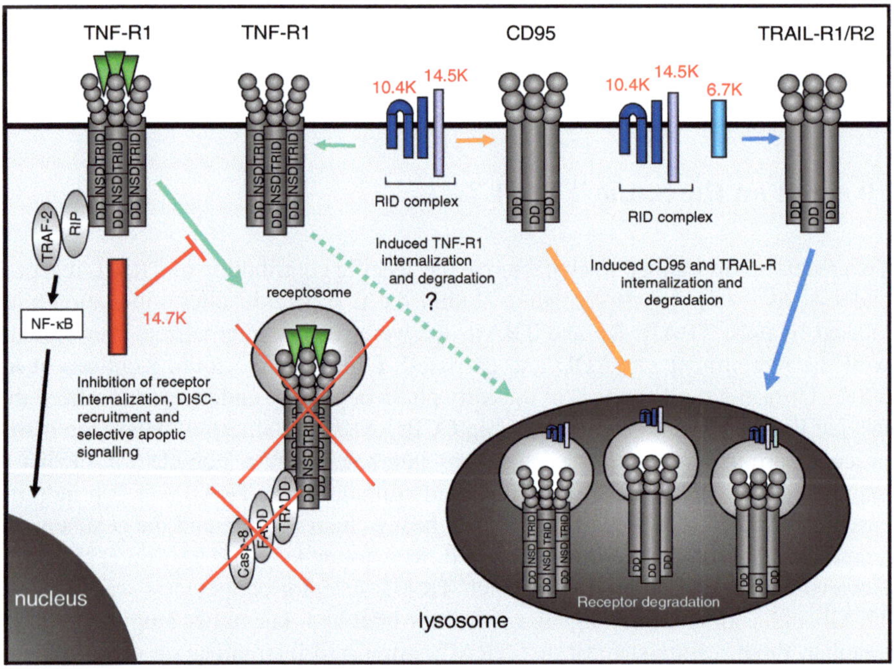

Fig. 3 Adenovirus targeting of death receptor endocytosis as a novel immune escape mechanism. The adenovirus early transcription region 3 (E3) encodes several proteins that interfere with apoptosis signaling by targeting TNF-R1, CD95, and TRAIL receptor internalization. The cytosolic E3-14.7 protein (14.7K) prevents internalization of the ligand-activated TNF-R1, thereby selectively inhibiting the assembly of the death-inducing complex (DISC) at intracellular TNF-R1 receptosomes. On the other hand, the recruitment of RIP-1 and TRAF-2 to the DD of TNF-R1 to mediate NF-κB activation is unaffected. Expression of E3-10.4K and E3-14.6K (termed RID complex for receptor internalization and degradation) can induce the ligand-independent receptor internalization of TNF-R1, CD95, and TRAIL receptor 1. In addition, RID together with an other E3-enconded protein, E3-6.7K, is necessary to remove TRAIL receptor 2 from the cell surface. The RID complex is expressed as a heterotrimer together with the E3-6.7K at the cell membrane and colocalizes with the internalized death receptors along the endocytic route until their lysosomal degradation

et al. 2001). In addition, internalization of TRAIL-R2 requires simultaneous expression of the RID complex together with another E3-encoded membrane protein, E3-6.7K (Lichtenstein et al. 2004a).

The molecular mechanism of RID-mediated receptor downregulation involves the ligand-independent recruitment of the AP2 complex to the intracellular receptor tail, which usually contains either an YxxΦ or a dileucin transport motif. Both RID proteins (RIDα and RIDβ) harbor distinct transport motifs within their intracellular tail to mediate ligand-independent death receptor endocytosis (Lichtenstein et al. 2002; Zanardi et al. 2003). Thus, RID pirates the clathrin-mediated endocytic route of receptor internalization to remove CD95 and TRAIL receptors from the cell surface, followed by their lysosomal degradation (Fig. 3). The RID complex also protects cells from TNF-mediated apoptosis; however, there is conflicting evidence as to whether this protection is mediated by forced TNF-R1 removal from the cell membrane (Shisler et al. 1997; Fessler et al. 2004; Chin and Horwitz 2005, 2006).

14.7K is a potent antiimmune protein encoded by the E3 region of adenoviruses, which was shown to protect from apoptosis induced by TNF, CD95L, and TRAIL (Lichtenstein et al. 2004b). The anti-TNF effect occurred independently of other adenovirus proteins and did not affect the level of TNF-R1 expression on the cell surface (Horton et al. 1991). Only recently was the molecular mechanism of 14.7K-mediated TNF resistance elucidated. Murine and human cells either stably expressing 14.7K or infected with adenovirus showed a significantly reduced rate of TNF-R1 internalization (Schneider-Brachert et al. 2006). It was demonstrated that 14.7K-mediated inhibition of TNF-R1 internalization resulted in a complete blockade in the recruitment of the DISC proteins. Notably, TNF-triggered recruitment of RIP-1 and TRAF-2 to the DD of TNF-R1 was not affected by 14.7K, providing evidence for its selective interference only with the apoptotic pathway. Thus, 14.7K inhibits TNF-induced apoptosis by a new molecular mechanism to escape immunosurveillance by targeting TNF-R1 endocytosis to selectively prevent DISC formation (Schneider-Brachert et al. 2006).

As illustrated above, interference with death receptor endocytosis is a sophisticated strategy to block the apoptotic response of the host. Understanding of the targeting of death-receptor internalization by viral proteins may provide a unique and powerful tool to develop novel strategies to combat death receptor–mediated diseases.

12 Conclusions

The current data suggest that internalization of TNF-R1 and CD95 is an important mechanism for the diversification of intracellular signaling, determining the biological outcomes after ligand binding. Clathrin-mediated internalization of both TNF-R1 and CD95 is a prerequisite for efficient recruitment of DISC proteins and apoptosis signaling from endosomal compartments. By contrast, antiapoptotic signaling through activation of NF-kB and MAPKs occurs independently of receptor internalization through RIP-1 and TRAF-2 binding to TNF-R1 or low-level recruitment of caspase-8 to CD95, respectively. The physiological relevance of TNF-R1 internalization becomes evident while considering infection by pathogens; for example, adenoviruses use the inhibition of TNF-R1 endocytosis to selectively prevent TNF-mediated apoptosis of infected cells, leaving other signaling events from the cell surface unaffected.

On the basis of these novel findings, it will be interesting to evaluate whether resistance of cells infected with other pathogens or the resistance of tumor cells against TNF- and CD95L-induced apoptosis is related to defects in internalization and intracellular trafficking pathways of TNF-R1 and CD95. Elucidation of the pathway(s) involved in TRAIL receptor internalization, which seems to be mediated independently of clathrin, remains another challenge. Once we know more about these events, pharmacological interference with the mechanisms involved in TNF-R1 and CD95 compartmentalization might be a promising strategy to break the resistance of infected cells or tumors against immunosurveillance and therapeutic interventions.

References

Ahn JH, Park SM, Cho HS, Lee MS, Yoon JB, Vilcek J, Lee TH (2001) Non-apoptotic signaling pathways activated by soluble Fas ligand in serum-starved human fibroblasts. Mitogen-activated protein kinases and NF-kappaB-dependent gene expression. J Biol Chem 276:47100–47106

Algeciras-Schimnich A, Peter ME (2003) Actin dependent CD95 internalization is specific for Type I cells. FEBS Lett 546:185–188

Algeciras-Schimnich A, Shen L, Barnhart BC, Murmann AE, Burkhardt JK, Peter ME (2002) Molecular ordering of the initial signaling events of CD95. Mol Cell Biol 22:207–220

Austin CD, Lawrence DA, Peden AA, Varfolomeev EE, Totpal K, De Maziere AM, Klumperman J, Arnott D, Pham V, Scheller RH, Ashkenazi A (2006) Death-receptor activation halts clathrin-dependent endocytosis. Proc Natl Acad Sci U S A 103:10283–10288

Barnhart BC, Alappat EC, Peter ME (2003) The CD95 type I/type II model. Semin Immunol 15:185–193

Barnhart BC, Legembre P, Pietras E, Bubici C, Franzoso G, Peter ME (2004) CD95 ligand induces motility and invasiveness of apoptosis-resistant tumor cells. EMBO J 23:3175–3185

Benedict CA, Norris PS, Prigozy TI, Bodmer JL, Mahr JA, Garnett CT, Martinon F, Tschopp J, Gooding LR, Ware CF (2001) Three adenovirus E3 proteins cooperate to evade apoptosis by tumor necrosis factor-related apoptosis-inducing ligand receptor-1 and -2. J Biol Chem 276:3270–3278

Benedict CA, Norris PS, Ware CF (2002) To kill or be killed: viral evasion of apoptosis. Nat Immunol 3:1013–1018

Benedict CA, Banks TA, Ware CF (2003) Death and survival: viral regulation of TNF signaling pathways. Curr Opin Immunol 15:59–65

Bradley JR, Johnson DR, Pober JS (1993) Four different classes of inhibitors of receptor-mediated endocytosis decrease tumor necrosis factor-induced gene expression in human endothelial cells. J Immunol 150:5544–5555

Brenner B, Ferlinz K, Grassme H, Weller M, Koppenhoefer U, Dichgans J, Sandhoff K, Lang F, Gulbins E (1998) Fas/CD95/Apo-I activates the acidic sphingomyelinase via caspases. Cell Death Differ 5:29–37

Chakrabandhu K, Herincs Z, Huault S, Dost B, Peng L, Conchonaud F, Marguet D, He HT, Hueber AO (2007) Palmitoylation is required for efficient Fas cell death signaling. EMBO J 26:209–220

Chen G, Goeddel DV (2002) TNF-R1 signaling: a beautiful pathway. Science 296:1634–1635

Chin YR, Horwitz MS (2005) Mechanism for removal of tumor necrosis factor receptor 1 from the cell surface by the adenovirus RIDalpha/beta complex. J Virol 79:13606–13617

Chin YR, Horwitz MS (2006) Adenovirus RID complex enhances degradation of internalized tumour necrosis factor receptor 1 without affecting its rate of endocytosis. J Gen Virol 87:3161–3167

Cifone MG, De Maria R, Roncaioli P, Rippo MR, Azuma M, Lanier LL, Santoni A, Testi R (1994) Apoptotic signaling through CD95 (Fas/Apo-1) activates an acidic sphingomyelinase. J Exp Med 180:1547–1552

Cremesti A, Paris F, Grassme H, Holler N, Tschopp J, Fuks Z, Gulbins E, Kolesnick R (2001) Ceramide enables fas to cap and kill. J Biol Chem 276:23954–23961

D'Alessio A, Al Lamki RS, Bradley JR, Pober JS (2005) Caveolae participate in tumor necrosis factor receptor 1 signaling and internalization in a human endothelial cell line. Am J Pathol 166:1273–1282

De Maria R, Rippo MR, Schuchman EH, Testi R (1998) Acidic sphingomyelinase (ASM) is necessary for fas-induced GD3 ganglioside accumulation and efficient apoptosis of lymphoid cells. J Exp Med 187:897–902

Devin A, Cook A, Lin Y, Rodriguez Y, Kelliher M, Liu Z (2000) The distinct roles of TRAF2 and RIP in IKK activation by TNF-R1: TRAF2 recruits IKK to TNF-R1 while RIP mediates IKK activation. Immunity 12:419–429

Doan JE, Windmiller DA, Riches DW (2004) Differential regulation of TNF-R1 signaling: lipid raft dependency of p42mapk/erk2 activation, but not NF-kappaB activation. J Immunol 172:7654–7660

Dumitru CA, Gulbins E (2006) TRAIL activates acid sphingomyelinase via a redox mechanism and releases ceramide to trigger apoptosis. Oncogene 25:5612–5625

Eramo A, Sargiacomo M, Ricci-Vitiani L, Todaro M, Stassi G, Messina CG, Parolini I, Lotti F, Sette G, Peschle C, De Maria R (2004) CD95 death-inducing signaling complex formation and internalization occur in lipid rafts of type I and type II cells. Eur J Immunol 34:1930–1940

Feig C, Tchikov V, Schütze S, Peter ME (2007) Palmitoylation of CD95 facilitates formation of SDS-stable receptor aggregates that initiate apoptosis signaling. EMBO J 26:221–231

Fessler SP, Chin YR, Horwitz MS (2004) Inhibition of tumor necrosis factor (TNF) signal transduction by the adenovirus group C RID complex involves downregulation of surface levels of TNF receptor 1. J Virol 78:13113–13121

Gajate C, Mollinedo F (2007) Edelfosine and perifosine induce selective apoptosis in multiple myeloma by recruitment of death receptors and downstream signaling molecules into lipid rafts. Blood 109:711–719

Gajate C, Canto-Janez E, Acuna AU, Amat-Guerri F, Geijo E, Santos-Beneit AM, Veldman RJ, Mollinedo F (2004) Intracellular triggering of Fas aggregation and recruitment of apoptotic molecules into Fas-enriched rafts in selective tumor cell apoptosis. J Exp Med 200:353–365

Garcia-Ruiz C, Colell A, Mari M, Morales A, Calvo M, Enrich C, Fernandez-Checa JC (2003) Defective TNF-alpha-mediated hepatocellular apoptosis and liver damage in acidic sphingomyelinase knockout mice. J Clin Invest 111:197–208

Glebov OO, Bright NA, Nichols BJ (2006) Flotillin-1 defines a clathrin-independent endocytic pathway in mammalian cells. Nat Cell Biol 8:46–54

Golks A, Brenner D, Krammer PH, Lavrik IN (2006) The c-FLIP-NH2 terminus (p22-FLIP) induces NF-kappaB activation. J Exp Med 203:1295–1305

Grassme H, Jekle A, Riehle A, Schwarz H, Berger J, Sandhoff K, Kolesnick R, Gulbins E (2001) CD95 signaling via ceramide-rich membrane rafts. J Biol Chem 276:20589–20596

Harper N, Hughes M, MacFarlane M, Cohen GM (2003) Fas-associated death domain protein and caspase-8 are not recruited to the tumor necrosis factor receptor 1 signaling complex during tumor necrosis factor-induced apoptosis. J Biol Chem 278:25534–25541

Heinrich M, Wickel M, Schneider-Brachert W, Sandberg C, Gahr J, Schwandner R, Weber T, Saftig P, Peters C, Brunner J, Krönke M, Schütze S (1999) Cathepsin D targeted by acid sphingomyelinase-derived ceramide. EMBO J 18:5252–5263

Heinrich M, Neumeyer J, Jakob M, Hallas C, Tchikov V, Winoto-Morbach S, Wickel M, Schneider-Brachert W, Trauzold A, Hethke A, Schütze S (2004) Cathepsin D links TNF-induced acid sphingomyelinase to Bid-mediated caspase-9 and -3 activation. Cell Death Differ 11:550–563

Helms JB, Zurzolo C (2004) Lipids as targeting signals: lipid rafts and intracellular trafficking. Traffic 5:247–254

Henkler F, Behrle E, Dennehy KM, Wicovsky A, Peters N, Warnke C, Pfizenmaier K, Wajant H (2005) The extracellular domains of FasL and Fas are sufficient for the formation of supramolecular FasL-Fas clusters of high stability. J Cell Biol 168:1087–1098

Herr I, Wilhelm D, Bohler T, Angel P, Debatin KM (1997) Activation of CD95 (APO-1/Fas) signaling by ceramide mediates cancer therapy-induced apoptosis. EMBO J 16:6200–6208

Horton TM, Ranheim TS, Aquino L, Kusher DI, Saha SK, Ware CF, Wold WS, Gooding LR (1991) Adenovirus E3 14.7K protein functions in the absence of other adenovirus proteins to protect transfected cells from tumor necrosis factor cytolysis. J Virol 65:2629–2639

Hsu H, Huang J, Shu HB, Baichwal V, Goeddel DV (1996a) TNF-dependent recruitment of the protein kinase RIP to the TNF receptor-1 signaling complex. Immunity 4:387–396

Hsu H, Shu HB, Pan MG, Goeddel DV (1996b) TRADD-TRAF2 and TRADD-FADD interactions define two distinct TNF receptor 1 signal transduction pathways. Cell 84:299–308

Hueber AO, Bernard AM, Herincs Z, Couzinet A, He HT (2002) An essential role for membrane rafts in the initiation of Fas/CD95-triggered cell death in mouse thymocytes. EMBO Rep 3:190–196

Hunter I, Nixon GF (2006) Spatial compartmentalization of tumor necrosis factor (TNF) receptor 1-dependent signaling pathways in human airway smooth muscle cells. Lipid rafts are essential for TNF-alpha-mediated activation of RhoA but dispensable for the activation of the NF-kappaB and MAPK pathways. J Biol Chem 281:34705–34715

Jin Z, El Deiry WS (2006) Distinct signaling pathways in TRAIL- versus tumor necrosis factor-induced apoptosis. Mol Cell Biol 26:8136–8148

Kamitani T, Nguyen HP, Yeh ET (1997) Activation-induced aggregation and processing of the human Fas antigen. Detection with cytoplasmic domain-specific antibodies. J Biol Chem 272:22307–22314

Kataoka T, Budd RC, Holler N, Thome M, Martinon F, Irmler M, Burns K, Hahne M, Kennedy N, Kovacsovics M, Tschopp J (2000) The caspase-8 inhibitor FLIP promotes activation of NF-kappaB and Erk signaling pathways. Curr Biol 10:640–648

Kelliher MA, Grimm S, Ishida Y, Kuo F, Stanger BZ, Leder P (1998) The death domain kinase RIP mediates the TNF-induced NF-kappaB signal. Immunity 8:297–303

Kischkel FC, Hellbardt S, Behrmann I, Germer M, Pawlita M, Krammer PH, Peter ME (1995) Cytotoxicity-dependent APO-1 (Fas/CD95)-associated proteins form a death-inducing signaling complex (DISC) with the receptor. EMBO J 14:5579–5588

Ko YG, Lee JS, Kang YS, Ahn JH, Seo JS (1999) TNF-alpha-mediated apoptosis is initiated in caveolae-like domains. J Immunol 162:7217–7223

Kohlhaas SL, Craxton A, Sun XM, Pinkoski MJ, Cohen GM (2007) Receptor-mediated endocytosis is not required for TRAIL-induced apoptosis. J Biol Chem 282:12831–12841

Koncz G, Kerekes K, Chakrabandhu K, Hueber AO (2007) Regulating Vav1 phosphorylation by the SHP-1 tyrosine phosphatase is a fine-tuning mechanism for the negative regulation of DISC formation and Fas-mediated cell death signaling. Cell Death Differ 15:494–503

Kull FC Jr, Cuatrecasas P (1981) Possible requirement of internalization in the mechanism of in vitro cytotoxicity in tumor necrosis serum. Cancer Res 41:4885–4890

Lajoie P, Nabi IR (2007) Regulation of raft-dependent endocytosis. J Cell Mol Med 11:644–653

Le Roy C, Wrana JL (2005) Clathrin- and non-clathrin-mediated endocytic regulation of cell signalling. Nat Rev Mol Cell Biol 6:112–126

Lee KH, Feig C, Tchikov V, Schickel R, Hallas C, Schütze S, Peter ME, Chan AC (2006) The role of receptor internalization in CD95 signaling. EMBO J 24:1009–1023
Legembre P, Daburon S, Moreau P, Ichas F, de Giorgi F, Moreau JF, Taupin JL (2005) Amplification of Fas-mediated apoptosis in type II cells via microdomain recruitment. Mol Cell Biol 25:6811–6820
Legembre P, Daburon S, Moreau P, Moreau JF, Taupin JL (2006) Modulation of Fas-mediated apoptosis by lipid rafts in T lymphocytes. J Immunol 176:716–720
Legler DF, Micheau O, Doucey MA, Tschopp J, Bron C (2003) Recruitment of TNF Receptor 1 to lipid rafts is essential for TNFalpha-mediated NF-kappaB activation. Immunity 18:655–664
Li-Weber M, Krammer PH (2003) Function and regulation of the CD95 (APO-1/Fas) ligand in the immune system. Semin Immunol 15:145–157
Liao W, Xiao Q, Tchikov V, Fujita K, Yang W, Wincovitch S, Garfield S, Conze D, El-Deiry WS, Schütze S, Srinivasula SM (2008) CARP-2 is an endosome-associated ubiquitin protein ligase for RIP and regulates TNF-induced NF-B activation. Curr Biol 18:641–649
Lichtenstein DL, Krajcsi P, Esteban DJ, Tollefson AE, Wold WS (2002) Adenovirus RIDbeta subunit contains a tyrosine residue that is critical for RID-mediated receptor internalization and inhibition of Fas- and TRAIL-induced apoptosis. J Virol 76:11329–11342
Lichtenstein DL, Doronin K, Toth K, Kuppuswamy M, Wold WS, Tollefson AE (2004a) Adenovirus E3-6.7K protein is required in conjunction with the E3-RID protein complex for the internalization and degradation of TRAIL receptor 2. J Virol 78:12297–12307
Lichtenstein DL, Toth K, Doronin K, Tollefson AE, Wold WS (2004b) Functions and mechanisms of action of the adenovirus E3 proteins. Int Rev Immunol 23:75–111
Lin T, Genestier L, Pinkoski MJ, Castro A, Nicholas S, Mogil R, Paris F, Fuks Z, Schuchman EH, Kolesnick RN, Green DR (2000) Role of acidic sphingomyelinase in Fas/CD95-mediated cell death. J Biol Chem 275:8657–8663
Locksley RM, Killeen N, Lenardo MJ (2001) The TNF and TNF receptor superfamilies: integrating mammalian biology. Cell 104:487–501
Mayor S, Pagano RE (2007) Pathways of clathrin-independent endocytosis. Nat Rev Mol Cell Biol 8:603–612
McPherson PS, Kay BK, Hussain NK (2001) Signaling on the endocytic pathway. Traffic 2:375–384
Medema JP, Scaffidi C, Kischkel FC, Shevchenko A, Mann M, Krammer PH, Peter ME (1997) FLICE is activated by association with the CD95 death-inducing signaling complex (DISC). EMBO J 16:2794–2804
Miaczynska M, Pelkmans L, Zerial M (2004) Not just a sink: endosomes in control of signal transduction. Curr Opin Cell Biol 16:400–406
Micheau O, Tschopp J (2003) Induction of TNF receptor I-mediated apoptosis via two sequential signaling complexes. Cell 114:181–190
Miyaji M, Jin ZX, Yamaoka S, Amakawa R, Fukuhara S, Sato SB, Kobayashi T, Domae N, Mimori T, Bloom ET, Okazaki T, Umehara H (2005) Role of membrane sphingomyelin and ceramide in platform formation for Fas-mediated apoptosis. J Exp Med 202:249–259
Monney L, Olivier R, Otter I, Jansen B, Poirier GG, Borner C (1998) Role of an acidic compartment in tumor-necrosis-factor-alpha-induced production of ceramide, activation of caspase-3 and apoptosis. Eur J Biochem 251:295–303
Morales A, Lee H, Goni FM, Kolesnick R, Fernandez-Checa JC (2007) Sphingolipids and cell death. Apoptosis 12:923–939
Mosselmans R, Hepburn A, Dumont JE, Fiers W, Galand P (1988) Endocytic pathway of recombinant murine tumor necrosis factor in L-929 cells. J Immunol 141:3096–3100
Muppidi JR, Tschopp J, Siegel RM (2004) Life and death decisions: secondary complexes and lipid rafts in TNF receptor family signal transduction. Immunity 21:461–465
Nakayama J, Ogawa Y, Yoshigae Y, Onozawa Y, Yonemura A, Saito M, Ichikawa K, Yamoto T, Komai T, Tatsuta T, Ohtsuki M (2006) A humanized anti-human Fas antibody, R-125224, induces apoptosis in type I activated lymphocytes but not in type II cells. Int Immunol 18:113–124

Neumeyer J, Hallas C, Merkel O, Winoto-Morbach S, Jakob M, Thon L, Adam D, Schneider-Brachert W, Schütze S (2006) TNF-receptor I defective in internalization allows for cell death through activation of neutral sphingomyelinase. Exp Cell Res 312:2142–2153

O'Reilly LA, Divisekera U, Newton K, Scalzo K, Kataoka T, Puthalakath H, Ito M, Huang DC, Strasser A (2004) Modifications and intracellular trafficking of FADD/MORT1 and caspase-8 after stimulation of T lymphocytes. Cell Death Differ 11:724–736

Papoff G, Hausler P, Eramo A, Pagano MG, Di Leve G, Signore A, Ruberti G (1999) Identification and characterization of a ligand-independent oligomerization domain in the extracellular region of the CD95 death receptor. J Biol Chem 274:38241–38250

Parlato S, Giammarioli AM, Logozzi M, Lozupone F, Matarrese P, Luciani F, Falchi M, Malorni W, Fais S (2000) CD95 (APO-1/Fas) linkage to the actin cytoskeleton through ezrin in human T lymphocytes: a novel regulatory mechanism of the CD95 apoptotic pathway. EMBO J 19:5123–5134

Parton RG, Simons K (2007) The multiple faces of caveolae. Nat Rev Mol Cell Biol 8:185–194

Pastorino JG, Simbula G, Yamamoto K, Glascott PA Jr, Rothman RJ, Farber JL (1996) The cytotoxicity of tumor necrosis factor depends on induction of the mitochondrial permeability transition. J Biol Chem 271:29792–29798

Peter ME, Budd RC, Desbarats J, Hedrick SM, Hueber AO, Newell MK, Owen LB, Pope RM, Tschopp J, Wajant H, Wallach D, Wiltrout RH, Zornig M, Lynch DH (2007) The CD95 receptor: apoptosis revisited. Cell 129:447–450

Rahman MM, McFadden G (2006) Modulation of tumor necrosis factor by microbial pathogens. PLoS Path 2:e4

Scaffidi C, Medema JP, Krammer PH, Peter ME (1997) FLICE is predominantly expressed as two functionally active isoforms, caspase-8/a and caspase-8/b. J Biol Chem 272:26953–26958

Scaffidi C, Fulda S, Srinivasan A, Friesen C, Li F, Tomaselli KJ, Debatin KM, Krammer PH, Peter ME (1998) Two CD95 (APO-1/Fas) signaling pathways. EMBO J 17:1675–1687

Scheel-Toellner D, Wang K, Singh R, Majeed S, Raza K, Curnow SJ, Salmon M, Lord JM (2002) The death-inducing signalling complex is recruited to lipid rafts in Fas-induced apoptosis. Biochem Biophys Res Commun 297:876–879

Schneider-Brachert W, Tchikov V, Neumeyer J, Jakob M, Winoto-Morbach S, Held-Feindt J, Heinrich M, Merkel O, Ehrenschwender M, Adam D, Mentlein R, Kabelitz D, Schütze S (2004) Compartmentalization of TNF receptor 1 signaling: internalized TNF receptosomes as death signaling vesicles. Immunity 21:415–428

Schneider-Brachert W, Tchikov V, Merkel O, Jakob M, Hallas C, Kruse ML, Groitl P, Lehn A, Hildt E, Held-Feindt J, Dobner T, Kabelitz D, Krönke M, Schütze S (2006) Inhibition of TNF receptor 1 internalization by adenovirus 14.7K as a novel immune escape mechanism. J Clin Invest 116:2901–2913

Schütze, S. and Tchikov, V. (2008) Immunomagnetic isolation of TNF-receptosomes. Methods in Enzymology 442: 101–123

Schütze S, Potthoff K, Machleidt T, Berkovic D, Wiegmann K, Krönke M (1992) TNF activates NF-kappa B by phosphatidylcholine-specific phospholipase C-induced "acidic" sphingomyelin breakdown. Cell 71:765–776

Schütze S, Machleidt T, Adam D, Schwandner R, Wiegmann K, Kruse ML, Heinrich M, Wickel M, Krönke M (1999) Inhibition of receptor internalization by monodansylcadaverine selectively blocks p55 tumor necrosis factor receptor death domain signaling. J Biol Chem 274:10203–10212

Schwandner R, Wiegmann K, Bernardo K, Kreder D, Krönke M (1998) TNF receptor death domain-associated proteins TRADD and FADD signal activation of acid sphingomyelinase. J Biol Chem 273:5916–5922

Shisler J, Yang C, Walter B, Ware CF, Gooding LR (1997) The adenovirus E3-10.4K/14.5K complex mediates loss of cell surface Fas (CD95) and resistance to Fas-induced apoptosis. J Virol 71:8299–8306

Siegel RM, Muppidi JR, Sarker M, Lobito A, Jen M, Martin D, Straus SE, Lenardo MJ (2004) SPOTS: signaling protein oligomeric transduction structures are early mediators of death receptor-induced apoptosis at the plasma membrane. J Cell Biol 167:735–744

Simons K, Toomre D (2000) Lipid rafts and signal transduction. Nat Rev Mol Cell Biol 1:31–39
Smotrys JE, Linder ME (2004) Palmitoylation of intracellular signaling proteins: regulation and function. Annu Rev Biochem 73:559–587
Sorkin A, Von Zastrow M (2002) Signal transduction and endocytosis: close encounters of many kinds. Nat Rev Mol Cell Biol 3:600–614
Stanger BZ, Leder P, Lee TH, Kim E, Seed B (1995) RIP: a novel protein containing a death domain that interacts with Fas/APO-1 (CD95) in yeast and causes cell death. Cell 81:513–523
Teis D, Huber LA (2003) The odd couple: signal transduction and endocytosis. Cell Mol Life Sci 60:2020–2033
Thon L, Mathieu S, Kabelitz D, Adam D (2006) The murine TRAIL receptor signals caspase-independent cell death through ceramide. Exp Cell Res 312:3808–3821
Tollefson AE, Hermiston TW, Lichtenstein DL, Colle CF, Tripp RA, Dimitrov T, Toth K, Wells CE, Doherty PC, Wold WS (1998) Forced degradation of Fas inhibits apoptosis in adenovirus-infected cells. Nature 392:726–730
Tollefson AE, Toth K, Doronin K, Kuppuswamy M, Doronina OA, Lichtenstein DL, Hermiston TW, Smith CA, Wold WS (2001) Inhibition of TRAIL-induced apoptosis and forced internalization of TRAIL receptor 1 by adenovirus proteins. J Virol 75:8875–8887
Wajant H, Pfizenmaier K, Scheurich P (2003) Tumor necrosis factor signaling. Cell Death Differ 10:45–65
Watanabe N, Kuriyama H, Sone H, Neda H, Yamauchi N, Maeda M, Niitsu Y (1988) Continuous internalization of tumor necrosis factor receptors in a human myosarcoma cell line. J Biol Chem 263:10262–10266
Wiegmann K, Schütze S, Machleidt T, Witte D, Krönke M (1994) Functional dichotomy of neutral and acidic sphingomyelinases in tumor necrosis factor signaling. Cell 78:1005–1015
Wiegmann K, Schwandner R, Krut O, Yeh WC, Mak TW, Krönke M (1999) Requirement of FADD for tumor necrosis factor-induced activation of acid sphingomyelinase. J Biol Chem 274:5267–5270
Woo CH, Kim TH, Choi JA, Ryu HC, Lee JE, You HJ, Bae YS, Kim JH (2006) Inhibition of receptor internalization attenuates the TNFalpha-induced ROS generation in non-phagocytic cells. Biochem Biophys Res Commun 351:972–978
Zanardi TA, Yei S, Lichtenstein DL, Tollefson AE, Wold WS (2003) Distinct domains in the adenovirus E3 RIDalpha protein are required for degradation of Fas and the epidermal growth factor receptor. J Virol 77:11685–11696
Zheng L, Bidere N, Staudt D, Cubre A, Orenstein J, Chan FK, Lenardo M (2006) Competitive control of independent programs of tumor necrosis factor receptor-induced cell death by TRADD and RIP1. Mol Cell Biol 26:3505–3513

Ubiquitination and TNFR1 Signaling

Ken-ichi Fujita and Srinivasa M. Srinivasula

Abstract Death receptors are a subset of the tumor necrosis factor receptor (TNFR) family of proteins and share a characteristic cytoplasmic motif called the "death domain." In addition to mediating cell death, these receptors regulate cell proliferation, inflammatory responses, and tumor progression. Receptor occupancy triggers the assembly of several cytoplasmic molecules into distinct complexes, each initiating separate signaling events leading to different biological responses. Post-translational modifications involving ubiquitin, a peptide of 76 amino acids, regulate events at nearly all stages of signaling. All ubiquitin chains function as docking platforms for molecules with specific recognition motifs that either propagate the signal or target the protein for proteasomal degradation. Moreover, enzymes with ubiquitin thioesterase activity (deubiquitinating enzymes, or DUBs) reverse modifications by removing the ubiquitin chains, allowing ubiquitin editing at the molecular level. Ubiquitin protein ligases (E3s), DUBs, and signaling molecules with ubiquitin recognition motifs control TNFR1 mediated cell death and activation of NF-κB and JNK. Here, we discuss the current understanding of how these proteins regulate TNFR1 signaling.

1 Introduction

1.1 General

Death receptor (DR) proteins, TNF-R1 (Tumor necrosis factor alpha-receptor-1/DR1/p55/p60/CD120a), Fas (APO-1/DR2/CD95), DR3 (APO-3/TRAMP/WSL1), TRAIL-R1 (TNF related apoptosis-inducing ligand-receptor 1/APO-2/DR4), TRAIL-R2 (KILLER/DR5/TRICK2), DR6, ectodysplasin A receptor (EDAR) and nerve growth factor receptor (NGFR) form a subset of Type-I plasma membrane

Ken-ichi Fujita and S.M. Srinivasula (✉)
Laboratory of Immune Cell Biology, Center for Cancer Research,
National Cancer Institute, National Institutes of Health, Building 37, Room 3016A,
9000 Rockville Pike, Bethesda, MD, 20892-4256, USA
e-mail: fujitak@mail.nih.gov, srinivsr@mail.nih.gov

Results Probl Cell Differ, DOI 10.1007/400_2009_18

TNF receptor super family (Dempsey et al. 2003). These receptors are characterized by the presence of a death domain (DD) of 80 amino acids in the intracellular portion that acts upon activation, as a platform to assemble signaling complexes (Lavrik et al. 2005). Death ligands play a critical role in apoptosis, a process of programmed cell death (PCD) that is essential for normal organismal development. In addition to apoptosis, DRs mediate a variety of biological processes in different tissues that control development, differentiation, and immune responses (MacEwan 2002). Dysregulation of DR mediated signaling was reported in a host of pathogenic diseases including arthritis, Crohn's disease, sepsis, cerebral malaria, diabetes, osteoporosis, allograft rejection, and autoimmune diseases such as multiple sclerosis, rheumatoid arthritis, and inflammatory bowel diseases. Unchecked death-ligand responses lead in some cases to tumorogenesis. Receptor occupancy leads to trimerization of the receptor and recruitment of many cytosolic proteins that form a signaling complex at the intracellular tail end involving the DD. Most of the proteins of this higher-order complex contain distinct functional domains such as DD, death effector domain (DED), kinase domain, caspase recruiting domain (CARD), really interesting gene (RING) some of which being protein–protein interaction motifs help these molecules to get recruited to the complex. Close proximity of catalytically active molecules is believed to result in extensive modifications, such as phosphorylation and ubiquitination, of many components of the complex and these events are the key for propagation of the signal.

Ubiquitination has emerged as an important protein modification in various biological processes, including degradation of proteins, receptor endocytosis, DNA repair, gene transcription, virus budding, the cell cycle, inflammation, and immune responses (Haglund and Dikic 2005). Many of these functions are dependent on the action of ubiquitin protein ligases (E3), which catalyze the transfer of ubiquitin to a protein substrate, leading to distinct types of ubiquitin modifications of the target proteins. Ubiquitination is a reversible process as ubiquitin or polyubiquitin chains can be cleaved from its protein substrates by deubiquitinating enzymes (DUBs). Covalently attached ubiquitin chains confer further specificity to the macromolecular interactions, and help to control the duration and the intensity of signaling. While proteins that are linked with polyubiquitin molecules via K48-chains are targeted by proteasome complexes for degradation, K63 or head to tail linked ubiquitin moieties were shown to facilitate signal transmission to the downstream complexes (Fig. 1c).

Many DR pathways share critical cytosolic molecules and function in a conserved fashion. For example, the death inducing signaling complexes (DISC) initiated by Fas, tumor necrosis factor receptor 1 (TNFR1), and TRAIL receptors share Fas associated death domain (FADD) and caspase-8 and activate downstream caspase pathways in a similar manner. Some of the DRs like TNFR1, mediate the activation of a critical transcription factor nuclear factor-kappa B (NF-κB) and c-JUN kinase (JNK). Here, we discuss the role ubiquitination plays in TNFR1 mediated cell death, NF-κB, JNK signaling. We will begin by describing the hierarchical cascade of ubiquitin conjugation and TNFR1 signaling, and will then include recent advances in the mechanisms of action of individual molecules involved in signaling via ubiquitination (Table 1).

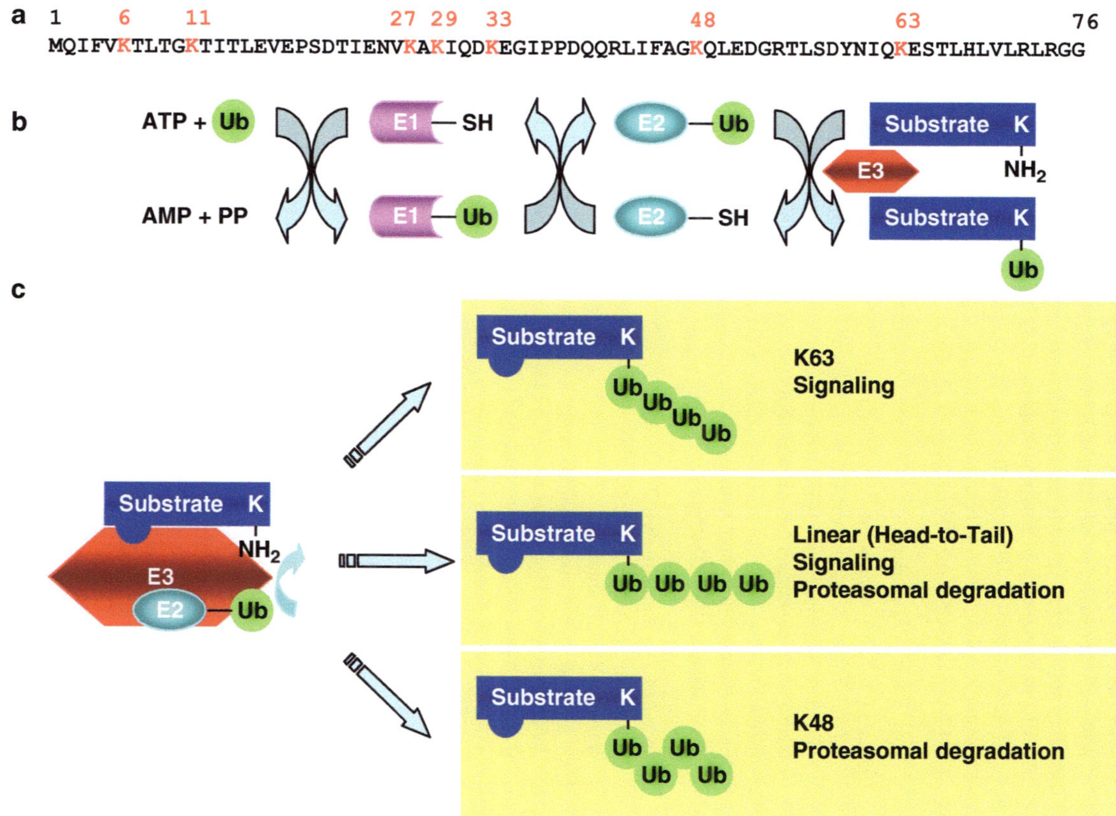

Fig. 1 Over view of the ubiquitin conjugation system. (**a**) Amino acid sequence of ubiquitin protein is shown. The positions of lysine residues (in *red*) are numbered. (**b**) Ubiquitin is activated by E1 in an energy-dependent manner, and then the activated ubiquitin is conjugated to an E2, and is then transferred to a specific lysine residue on the protein substrate with the help of an E3. (**c**) A specific set of E2 and E3 complex coordinate the formation of linkages (K63, or K48, or linear) between multiple ubiquitin molecules on the substrate. While substrates marked with K48 chains are targeted for proteasomal degradation, those marked with K63 propagate signaling. Presently, it is believed that the linear ubiquitin chains can promote both protein degradation and signaling

Table 1 Summary of gene ablation phenotypes of proteins involved in TNF signaling

Protein	KO in mice	Reference
A20	Premature lethality with severe inflammation and cachexia. Mice are highly sensitive to TNFα A20-deficient MEFs in response to TNFα show prolonged activation of NF-κB and JNK	Lee et al. (2000)
RIP1	Lethality at early embryonic stage. RIP1-deficient MEFs in response to TNFα show impaired NF-κB, but normal JNK signaling	Kelliher et al. (1998)
TAX1BP1	Premature lethality with severe inflammation and cardiac valvulitis. TAX1BP1-deficient MEFs in response to TNFα show prolonged activation of NF-κB and JNK	Shembade et al. (2007), Iha et al. (2008)
Itch	Premature lethality with pulmonary interstitial inflammation, develop diverse immunological disorders. Itch-deficient MEFs in response to TNFα show prolonged NF-κB and JNK activation	Hustad et al. (1995), Fang et al. (2002), Shembade et al. (2008)
NEMO	Lethality at embryonic stage with liver degeneration. Heterozygous female mice develop skin lesions similar to the human incontinentia pigmenti. NEMO-deficient MEFs in response to TNFα treatment show impaired NF-κB signaling	Rudolph et al. (2000), Schmidt-Supprian et al. (2000)
CYLD	Male mice are sterile owing to defects in seminiferous tubular organization. CYLD-deficient macrophages in response to TNFα treatment show increased JNK activation	Wright et al. (2007), Bignell et al. (2000), Simonson et al. (2007)
TRAF2	Lethality at embryonic stage. TRAF2-deficient MEFs in response to TNFα treatment show increased JNK activation	Yeh et al. (1997)
XIAP	Viable. MEFs without the RING domain of XIAP in response to TNFα treatment show increased caspase-3 activity	Schile et al. (2008)
cIAP-1	Viable. Acute loss in tumor cells leads to TNFα-mediated apoptosis	Conze et al. (2005), Wu et al. (2007)
cIAP-2	Viable. Acute loss in tumor cells leads to TNFα-mediated apoptosis	Conte et al. (2006), Wu et al. (2007)
ABIN-1	Lethality at embryonic stage with fetal liver apoptosis. ABIN-1-deficient MEFs in response to TNFα show hypersensitivity to apoptosis	Oshima et al. (2009)

1.2 Ubiquitination

Ubiquitination is a reversible posttranslational modification by which ubiquitin, a highly conserved peptide of 76 amino acids, is covalently attached to substrates by an isopeptide bond between the C-terminal glycine (Gly 76) of ubiquitin and a specific lysine residue on the substrate (Weissman 2001) (Fig. 1a). Attachment of a single ubiquitin molecule is known as monoubiquitination and addition of several single ubiquitin molecules to different lysine residues results in multiubiquitination. Both these modifications are generally implicated in cellular functions such as DNA repair and endosomal sorting (Bonifacino and Traub 2003; Gao and Karin 2005). Alternatively, ubiquitin itself contains seven lysine residues (K6, K11, K27, K29, K33, K48, and K63), and all these lysines in an interactive process known as polyubiquitination can possibly form bonds with other ubiquitin molecules (Fig. 1a). Ubiquitin chains linked via K48 and K63 are best characterized thus far and, as discussed earlier, these modifications target conjugated proteins for either proteasomal degradation (molecules with K48 chains), or to establish stable protein–protein interactions (molecules with K63 chains) (Gao and Karin 2005; Haglund and Dikic 2005). One reason for this selectivity appears to be the differences in the conformations with the ubiquitin chains conjugated via K63 linkages, which are much more extended than the ones linked via K48 chains. A notion that polyubiquitin chains can be formed with both kinds of linkages increasing the complexity even further. It is generally believed that proteins with specialized Ub-binding domains (UBDs) that form diverse structural folds interact in vivo with Ub chains or ubiquitinylated proteins with different affinities and this allows formation of specific Ub-mediated networks. This concept is supported by the evidence that a specific ubiquitin-binding protein, NEMO,is associated with polyUb much more strongly than with monoUb (Lo et al. 2009). Moreover, modifications of proteins with multiple ubiquitin moieties attached in head to tail fashion have also been reported, although many details of this kind of modifications remain unknown (Kirisako et al. 2006; Tokunaga et al. 2009).

Ubiquitinations occur in an energy intensive multistep process involving ubiquitin-activating enzyme (E1), ubiquitin-conjugating enzyme (E2 or UBC) and ubiquitin protein ligating enzyme (E3). In most organisms, in an initial step the C-terminal glycine (G76) of ubiquitin is covalently linked, in an ATP-dependent manner, to ubiquitin-activating enzyme (Gao and Karin 2005; Weissman 2001) (Fig. 1b). The activated ubiquitin is subsequently transferred to a cysteine thus forming E2-ubiquitin thioester linkage on an E2 or an UBC. A member of the ubiquitin protein ligases then recruits the ubiquitin-conjugated E2 and facilitates the transfer of the ubiquitin to specific lysine residue on the substrate (Fang and Weissman 2004; Gao and Karin 2005; Haglund and Dikic 2005). The hierarchal nature of the ubiquitin conjugation is appreciated by the fact that most eukaryotes possess only one E1, scores of E2s, and hundreds of E3s. Each E3 in conjunction with one or a few E2s selectively targets substrates for modification. E3 ligases can be divided broadly in to three categories: homologous to E6-AP carboxyl terminus (HECT), really interesting new gene (RING) finger, and U-box domain-containing proteins (Gao and Karin 2005). While structurally related RING finger and U-box domains mediate the transfer of ubiquitin to the substrate by facilitating the interaction between an E2 and the target protein, in case of HECT E3s the interaction with an E2-uniquitin results in the conjugation of uniquitin to a specific cysteine residue of the HECT domain which subsequently is transferred to interacting substrates. Multiple E3s, as components of a ubiquitin ligase complex, are known to facilitate the assembly of linear ubiquitin chains (Kirisako et al. 2006). In addition, proteins with zinc finger domains (example, A20) were shown to facilitate ubiquitination (Wertz et al. 2004). Other factors that regulate ubiquitin-mediated signaling are molecules with ubiquitin thioesterase, also known deubiquitinating (DUB) activity. These proteins with ubiquitin-specific processing protease activity that removes ubiquitin moieties from proteins serve as ubiquitin-editing enzymes. Although many details remain to be elucidated at least two proteins, A20 and CYLD, are reported to negatively regulate TNFR1 signaling via DUB activity (Brummelkamp et al. 2003; Kovalenko et al. 2003; Trompouki et al. 2003; Wertz et al. 2004).

1.3 TNFR1 Signaling

1.3.1 NF-κB and JNK Activation

Tumor Necrosis Factor alpha (TNFα) was first identified as an endotoxin-induced serum factor that promotes necrosis in tumor cells (Carswell et al. 1975). Numerous subsequent studies have demonstrated that TNFα, in addition to stimulating cell death, elicits many biological responses that are critical for cellular homeostasis (Dempsey et al. 2003; MacEwan 2002). Among the two receptors (TNFR1 and TNFR2) that TNFα uses to exert its function, TNFR2 lacks the DD and is less well characterized (MacEwan 2002). Hence, this article focuses exclusively on TNFR1 and discusses TNFR1 signaling pathways that control cell survival via NF-κB activation, and cell death via JNK or caspase activation (Fig. 2a, b). TNFR1 signaling involves sequential formation of two complexes, known as complex I, and DISC (or complex II) (Micheau and Tschopp 2003; Schneider-Brachert et al. 2004). Complex I contains TNFR1, TNF-receptor-associated death domain protein (TRADD), RIP1, and the RING domain-containing molecules TNF-receptor-associated factor 2 (TRAF2), cellular IAP1 (cIAP-1), and cIAP-2. The assembly of complex I that takes place immediately after stimulation at the plasma membrane triggers the activation of IκB kinase (IKK) complex that consists of two catalytically active kinase subunits (IKKα and IKKβ) and a regulatory subunit IKKγ (Baud and Karin 2001; Dempsey et al. 2003; Hayden and Ghosh 2008). NF-κB represents a group of structurally related and evolutionarily conserved family of transcriptional factors that control diverse biological responses (Ghosh and Karin 2002; Karin et al. 2002). In resting cells, most of the NF-κB is sequestered in the cytoplasm by a family of inhibitory proteins known as IκB. Stimulation with TNFα triggers signal transduction pathways that ultimately leads to the phosphorylation of specific serine residues on IκB proteins resulting in their ubiquitination and degradation, and nuclear translocation of NF-κB where it affects the expression of target genes (Fig. 2a). Modification of IκB by ubiquitin and its degradation were the first ubiquitination events reported in TNFR1 signaling and have since then been widely reviewed in the literature (Chen and Greene 2004; Ghosh and Karin 2002; Perkins 2006), and thus are not discussed here. Both RIP1 and NEMO are essential for animal development,

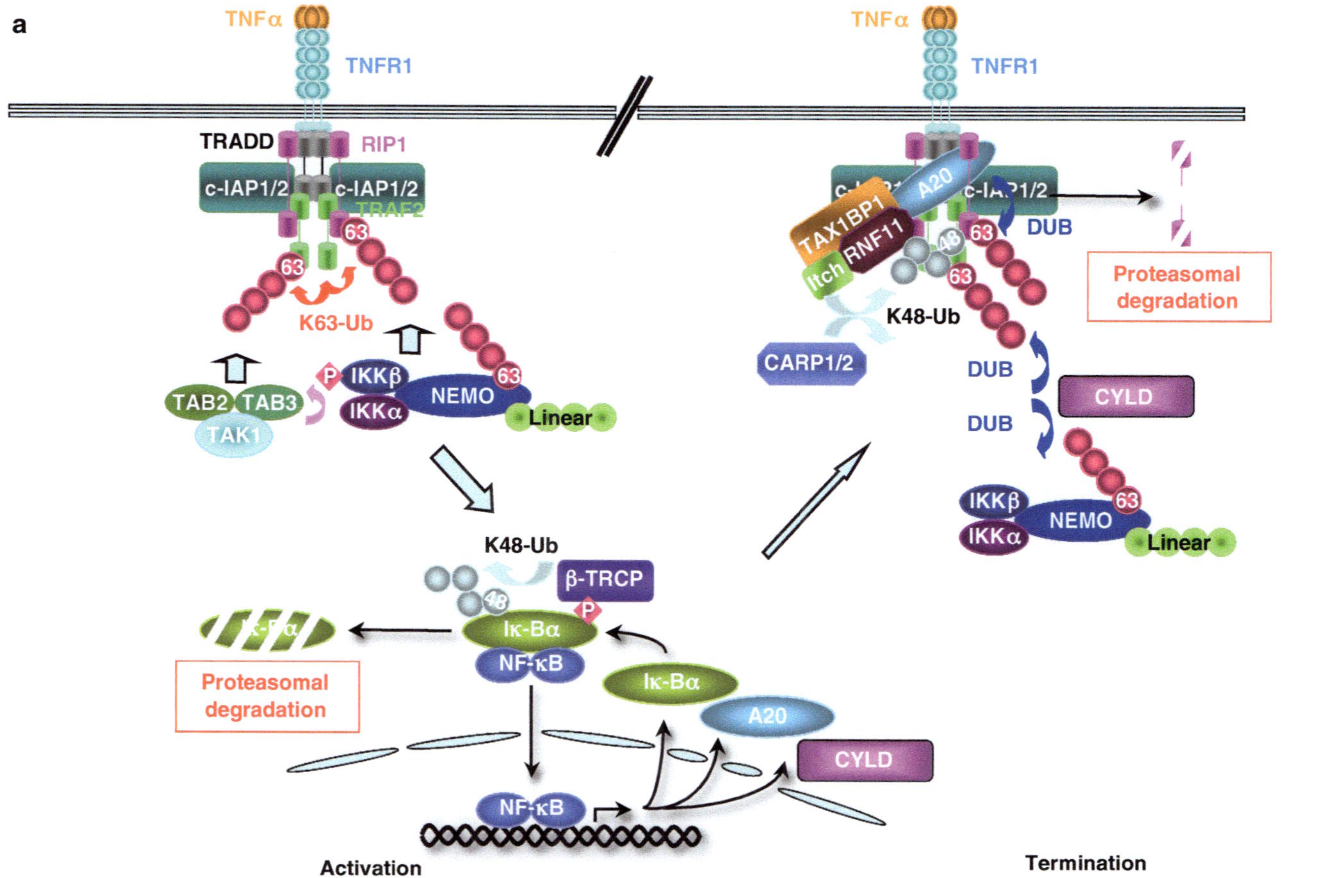

Fig. 2 TNFR1 signaling. Engagement of cell surface receptor TNFR1 with TNFα triggers the recruitment of several cytosolic molecules into distinct complexes, each leading to diverse biological responses. The complex formed at the plasma membrane includes TNFR1, TRADD, TRAF2, RIP1, cIAP-1/2 and it either activates NF-κB (**a**), or reassembles in the cytosol to include FADD and caspase-8 that promotes the activation of caspase cascade resulting in apoptosis (**b**).

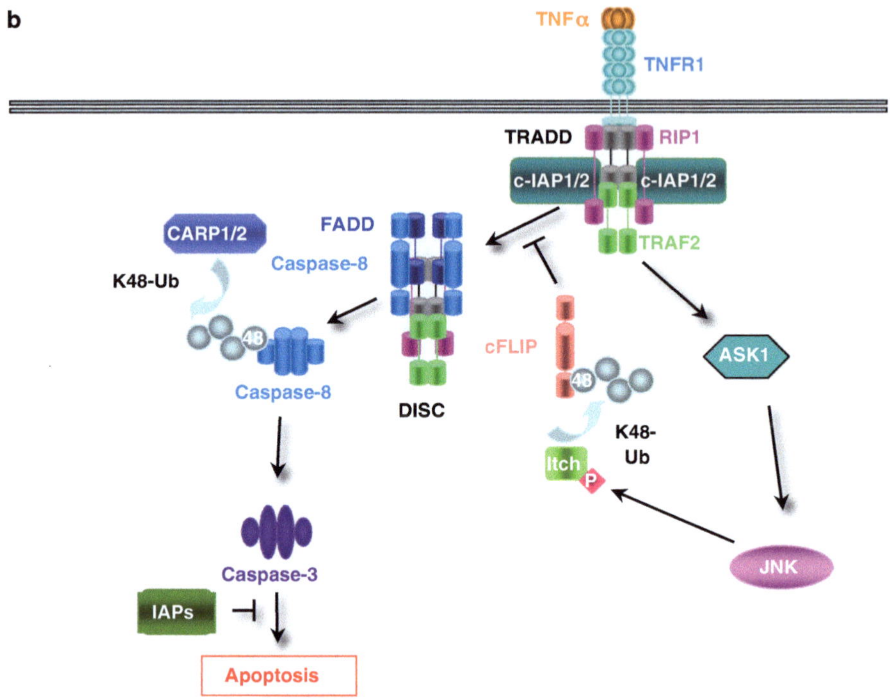

Fig. 2 (continued) (**a**) Components of the receptor complex including RIP1 and TRAF2 are ubiquitinated, and the K-63 ubiquitin chains recruit TAK1- and NEMO-containing complexes, leading to IKK activation and IκBα phosphorylation, resulting in the degradation of IκBα and nuclear translocation of NF-κB. In a negative feed back loop NF-κB upregulates the expression of IκBα, A20, and CYLD. While A20 in complex with TAX1BP1, Itch and RNF11 terminates signaling by targeting RIP1 for degradation, CYLD exerts its function via DUB activity. Constitutively expressed CARP1, and CARP2 proteins also limit NF-κB activation by targeting RIP1. In addition, linear- and K63-Ub chains on NEMO are known to contribute to the regulation of NF-κB signaling. (**b**) The RING domain-containing proteins XIAP, cIAP-1, cIAP-2, Itch, CARP1, and CARP2 regulate apoptosis by acting at different stages of signaling. While CARP1 and CARP2 can act as negative regulators by promoting caspase-8 degradation, JNK substrate Itch functions as a positive regulator by ubiquitinating and removing cFLIP, thus facilitating caspase-8 recruitment to the DISC. IAPs function by acting as inhibitors of caspase-3, caspase-7, and caspase-9, or at the level of the receptor complex

and TNFα-induced NF-κB activation. TNFα stimulation of cells that are deficient in either RIP1 or NEMO failed to activate NF-κB (Kelliher et al. 1998; Rudolph et al. 2000; Schmidt-Supprian et al. 2000; Ting et al. 1996).

Although TRAF2 possesses the E3 activity, TRAF2-deficiency had minimal effect on NF-κB activation and this was due to the functional redundancy by another member of the TRAF family, TRAF5 (Au and Yeh 2007; Tada et al. 2001; Yeh et al. 1997). Though no formal proof is available, TRAF2 in collaboration with an E2 dimer UEV1/UBC13 is believed to promote K63-linked ubiquitin chains on RIP1, as absence of TRAF2 had caused a decrease in the ubiquitination of RIP1

(Au and Yeh 2007; Lee et al. 2004; Wertz et al. 2004). In addition to its function as an E3 for RIP1 in NF-κB activation, TRAF2 also regulates TNFα-induced activation of JNK and overexpression TRAF2 increases JNK activity (Au and Yeh 2007; Meylan and Tschopp 2005). Moreover, apoptosis signal-regulated kinase 1 (ASK1) is activated in response to TNFα treatment, or TRAF2 overexpression, and a dominant negative form of ASK1 inhibited both TNFα- and TRAF2-induced JNK activation (Hoeflich et al. 1999; Ichijo et al. 1997; Nishitoh et al. 1998) suggesting that TRAF2 acts via ASK1 activation. Importantly, deletion of ASK1 resulted in defective JNK activation and concomitantly cell death upon TNFα treatment (Tobiume et al. 2001). While ASK1 functions downstream of TRAF2 in JNK pathway, TGFβ-activating kinase 1 (TAK1) connects TRAF2 and IKK signaling pathway. TAK1 complex consists of its regulator TAB2, which directly phosphorylates IKKβ of the IKK complex, thereby activating NF-κB pathway (Tobiume et al. 2001). Interestingly, the novel zinc finger (NZF) domain of TAB2 binds preferentially K63-linked ubiquitin chains on ubiquitinated RIP1, and ubiquitin modification of RIP1 is required for TAK1 and IKK activations (Ea et al. 2006; Kanayama et al. 2004) (Fig. 2a, b). The functional significance of the ubiquitin sensor motifs on NEMO and other molecules will be discussed in detail later.

1.3.2 Apoptosis

Apoptosis or PCD is a physiologically controlled cell suicide process that plays an essential role in homeostasis of multicellular organisms by the elimination of unwanted, damaged, or infected cells (Jacobson et al. 1997; Steller 1995). Dysregulation of apoptosis is implicated in the pathogenesis of a wide spectrum of diseases including neurological disorders such as Alzheimer's and Parkinson's diseases, and autoimmunity (Esposito et al. 2007; Gorman 2008; Martin 2001; Reich et al. 2008). Suppression of apoptotic programs often plays a key role in the survival of tumor cells. Stimulation of members of the DR family by death ligands results in the recruitment of several cytosolic molecules to the DD and activation of downstream caspase cascade (Baud and Karin 2001; Lavrik et al. 2005; MacEwan 2002; Schutze et al. 2008) (Fig. 2b). In spite of the differences in composition and kinetics of the assembly of the DISC, all DRs initiate the activation of apoptotic cascade by recruiting the DD-containing adaptor molecule FADD and caspase-8 as components of the DISC (Fig. 2b). Caspases are central components of apoptotic machinery and belong to a family of cysteine-dependent aspartate-specific proteases that are synthesized as inactive zymogens (procaspases) (Chang and Yang 2000; Yuan 2006). Cleavage of procaspase into two unequal subunits triggers the formation of a heterotetramer containing two copies of each that are functionally active. Recruitment of procaspase 8 in to the DISC is believed to result in increased local concentration, thus leading to its processing (Boatright et al. 2003). Apoptosis can be induced by stimulation with the members of the TNF family of ligands, including TNFα, Fas ligand (FasL, also known as CD95L or APO1L) and TNF-related apoptosis-inducing ligand (TRAIL, also known as APO2L). These "death ligands" bind to their cognate "death receptors" thus resulting in the formation of the DISC and the activation of the executionar caspases like caspase 3. One of the cellular inhibitor of apoptosis, cellular FLICE inhibitory protein (c-FLIP) possess, like caspase-8 and caspase-10, a DED and a functionally inactive caspase-like domain. The apoptotic inhibitor c-FLIP exerts its function by interfering with the recruitment of caspase-8 to the DISC. While c-FLIP itself had no E3 function, its cellular levels are regulated by TNFα in an ubiquitin-dependent manner, thus affecting caspase-8 activation and apoptosis. The RING-domain containing inhibitors of apoptosis (IAPs) also control the death ligand-mediated apoptosis by inhibiting the protease activity of caspase-9, caspase-7 and caspase-3. Moreover, TNFα activated JNK facilitates caspase activation and apoptosis in a ubiquitination-dependent manner. TNFR1-mediated activation of NF-κB, JNK and apoptotic processes is controlled at different stages of propagation by several molecules involving complex ubiquitin modifications. For simplicity we will describe the current knowledge of these molecules individually and discuss the mechanism of their function.

2 Regulators of Ubiquitination

2.1 A20

NF-κB family of transcriptional factors control the expression of numerous genes involved in inflammatory responses. Activation of NF-κB by TNFα is transient and genetic and biochemical evidence shows that the ubiquitin editing protein, A20, plays a key role in limiting the duration of NF-κB signaling (Liu et al. 2005). A20, a cytoplasmic protein with an N-terminal ovarian tumor (OTU) domain and seven zinc fingers, negatively regulates TNFα-induced activation of NF-κB (Rothe et al. 1995). Overwhelming evidence supports an indispensable function for A20 that acts by targeting RIP1 in a negative feedback mechanism.

A physiological role for A20 in restricting NF-κB and inflammation in vivo was provided by A20 knockout mice that died prematurely from multi-organ inflammation (Lee et al. 2000). Importantly, a similar role for A20 in humans is supported by the observed correlation between multiple polymorphisms in the A20 region and autoimmune diseases like Crohn's disease, systemic lupus erythomatosus, and rheumatoid arthritis (Wellcome Trust Case Control Consortium 2007; Graham et al. 2008; Musone et al. 2008; Plenge et al. 2007; Thomson et al. 2007). Recently, inactive genetic deletions and somatic mutations in A20 have been reported in marginal zone B-cell lymphomas (Chanudet et al. 2009; Novak et al. 2009). TNFα treatment dramatically induced A20 transcription in all tissues, and this rapid increase in A20 expression is essential for limiting NF-κB responses. In addition, A20 may also control apoptosis and JNK responses (Coornaert et al. 2009). Notwithstanding the deregulation of Toll-like receptor (TLR) signaling, cells from A20-deficient mice failed to downregulate TNFα, but not interleukin-1β (IL-1β) signaling (Lee et al. 2000; Wertz et al. 2004). Treatment of wild type MEFs with TNFα resulted in rapid IκBα degradation, and NF-κB nuclear translocation, followed by re-accumulation of IκBα protein and decreased nuclear NF-κB by 60 min. In contrast, loss in IκBα protein and the presence of NF-κB in the nucleus persisted in A20-deficient MEFs until 180 min. Moreover, direct measurements showed prolonged IKK activation and synthesis of IκBα mRNA in A20 –/– MEFs confirming the continued TNF signaling in A20 null cells (Lee et al. 2000; Werner et al. 2005). Similarities in the activation kinetics of IKK activation until 60 min between A20 positive and null cells indicate the requirement of de novo synthesis of A20 for its inhibitory function. Importantly, A20 null mice exposed to low doses of TNFα (0.1 mg kg^{-1}) died within 2 h, where as wild type mice survived high doses (0.4 mg kg^{-1}) (Lee et al. 2000).

Initial reports proposed a two sequential step mechanism involving distinct ubiquitin modulating activities to explain the inhibitory function of A20. The N-terminal OTU domain of A20 exhibits DUB activity and removes K63-linked ubiquitin chains from RIP1, an essential mediator of the proximal TNFR1 signaling complex (Wertz et al. 2004). The K63 polyubiquitin chains on RIP1 act as anchoring moieties for downstream signaling complexes such as IKK complex (Ea et al. 2006; Wu et al. 2006). This step is essential for the activation of IKK that triggers IκBα degradation, nuclear translocation of NF-κB transcriptional factor, and induction of target gene expression. However, to sufficiently turn off NF-κB signaling removal of K63-linked ubiquitin chains by A20 OTU is followed by the proteasomal degradation of the receptor-associated RIP1. Interestingly, the C-terminal Zn finger motifs of A20 also possess ubiquitin ligase activity and this function facilitated the ligation of K48-linked ubiquitin chains on RIP1 leading to its degradation (Wertz et al. 2004). While reconstitution of A20 –/– MEFs with wild type A20 promoted degradation of TNFR1-associated RIP1 over time, the loss of ubiquitinated RIP1 was blocked in cells complimented with either OTU, or ZnF4 inactive mutants. These studies confirmed separate DUB and ubiquitin protein ligase functions of A20 domains, and participation of both activities in regulating TNFR1 signaling. There is a growing body of evidence that several additional molecules, at least in certain tissues, play important roles in A20-mediated negative regulation as components of a quaternary complex known as A20 ubiquitin editing complex.

2.2 *TAX1BP1*

TAX1BP1 (also known as TXBP151 or T6BP) is an 86 kDa negative regulator of NF-κB, and its affinity to A20 was first unraveled in a yeast two-hybrid screen (Gachon et al. 1998; Jin et al. 1997). TAX1BP1 also associates with RIP1, and in TAX1BP1-null fibroblasts, as in case of A20 –/– MEFs, RIP1 remained hyperubiquitinated upon TNFα treatment indicating that TAX1BP1 might affect TNFα-induced NF-κB signaling by influencing A20-RIP1 complex formation (Iha et al. 2008; Shembade et al. 2007). This is indeed supported by the finding that suppression of TAX1BP1 (with siRNA) expression prevented stimulus-dependent RIP1-A20 complex assembly. Furthermore, overexpression of TAX1BP1 was unable to inhibit RIP1 signaling in A20-deficient MEFs, suggesting the requirement of A20 for TAX1BP1 inhibitory function. The lack of any known E3 activity of TAX1BP1, and the requirement of A20 implicates TAX1BP1 as an adaptor molecule that recruits A20 to RIP1-containing complex. Interestingly, a novel ubiquitin-binding Zn finger (UBZ) in TAX1BP1 that binds to purified K48 and K63 polyubiquitin molecules has been reported (Iha et al. 2008). So it is conceivable that TAX1BP1 recognizes ubiquitinated-signaling molecules such as RIP1 via UBZ, and recruits A20 for deubiquitination. In spite of the similarities in the kinetics of TNF signaling in cultured A20- and TAX1BP1-null fibroblasts, large differences exist between the phenotypes of A20 and TAX1BP1 knockout mice. While A20-deficient mice died young due to massive inflammation in multiple organs, TAX1BP1 KO mice showed age dependent inflammation-induced, valve disease (cardiac valvulitis) with infiltration of macrophages and T-cells in the myocardium and cardiac valves (Iha et al. 2008). The phenotype correlated with increased in vivo NF-κB activation - with elevated IκBα mRNA, and proinflammatory cytokines IL-1β, IL-2, and TNFα - in heart valve tissues of KO mice. The tissue-restricted phenotype of TAX1BP1-deficient mice could be due to the redundancy with other A20 adaptor molecules. Another study that generated TAX1BP1-deficient mutant mice by gene-trap strategy had reported that TAX1BP1 homozygous mutant (termed m/m) mice were not viable (Shembade et al. 2007). The difference between these studies could be the result of various technical and scientific approaches used to generate the mice and/or genetic backgrounds. Moreover, SV40 large T antigen transformed MEFs from TAX1BP1 m/m mice, showed persistent JNK activity, which was not observed in primary MEFs from KO mice. Despite these differences, the kinetics of NF-κB activation in fibroblasts from both TAX1BP1 m/m and KO mice was consistent, and these results collectively demonstrate a pivotal role for TAX1BP1 in limiting the TNFα-induced NF-κB signaling.

2.3 RNF11

RNF11, a RING-H2 finger domain-containing protein, was identified as a molecule overexpressed in the breast, colon, and pancreas tumors, and was proposed to play a regulatory role in protein ubiquitinylation by interacting with ubiquitin conjugating enzymes and ubiquitin protein ligases (Burger et al. 1998; Seki et al. 1999; Subramaniam et al. 2003). Notwithstanding its function as a promoter of transforming growth factor-beta (TGF-β) signaling, a potential role for RNF11 in NF-κB activation was suggested as it pulled numerous regulators of TNF signaling as binding partners in a yeast two-hybrid screen (Azmi and Seth 2005; Colland et al. 2004; Li and Seth 2004; Subramaniam et al. 2003). These proteins include A20 binding and inhibitor of NF-κB (ABIN)-1, NEMO, ITCH and TAX1BP1, molecules with either known E3 activity, or ubiquitin-binding function that are critical for cytokine-induced NF-κB signaling. Recent reports proposed that RNF11 negatively regulates TNFR1-signaling by acting together with RIP1, A20 and TAX1BP1 (Shembade et al. 2009). Both endogenous and exogenous RNF11 interacted with RIP1 and A20 in TNFα-stimulated, but not untreated cells. Interaction between endogenous molecules, in bone marrow-derived macrophages (BMDMs) and MEFs was observed as early as in 30 min of stimulation. Interestingly, this association appears to have occurred not at the receptor level, but in the cytosol. Unlike the association of RNF11 with TAX1BP1 or A20 that lasted for an hour, the interaction with RIP1 was prolonged, and could be observed even after 2 h of stimulation. Suppression of RNF11 expression (with siRNA) in monocyte cell line THP-1 led to TNFα-mediated specific enhancement of NF-κB reporter activity, prolonged IκBα degradation, and persistent JNK phosphorylation compared with a scrambled control siRNA (Shembade et al. 2009). Moreover, RNF11 knockdown also resulted in increased expression of NF-κB genes IL-6, IκB α, and A20. The enhanced NF-κB signaling observed was strikingly similar to what has been reported in cells deficient in individual components of A20-editing complex that function by targeting RIP1 indicating that RNF11 might also function as an another subunit of this complex. In agreement with such an idea, the suppression of RNF11 expression impaired the recruitment of A20 to RIP1, and elevated the levels of ubiquitinated RIP1 in TNFα-stimulated cells. Correspondingly, in the absence of RNF11, A20 failed to trigger the degradation of endogenous RIP1 in MEFs, or to inhibit TNFα-induced activation of NF-κB. The RING domain and the PPXY motif are required for binding of RNF11 to A20 and limiting NF-κB signaling as single point mutations in RNF11, a mutation in either the RING domain (C99A) or the PPXY motif (Y40A), abrogated the interaction between A20 and RNF11 and the inhibitory effect of RNF11 on TNF signaling (Shembade et al. 2009). Although these results suggested an essential function for RNF11 in limiting cytokine-induced NF-κB activation, generation and characterization of RNF11-deficient mice will be required to understand the biological function of RNF11 in TNF signaling.

2.4 Itch

Itch is a HECT E3 ligase with WW domains, and the loss of its function in mice leads to constant itching of the skin with inflammatory disorders including hyperplasia in the lymph nodes and spleen. Itch-deficient mice died by 35 weeks because of pulmonary interstitial inflammation (Fang et al. 2002; Hustad et al. 1995). The deleterious autoimmune-like disease phenotype is mediated by lymphocytes as the deficiency of Itch in recombinase-activating gene 1 (Rag1) null background that lacks mature B and T lymphocytes, did not cause severe inflammation (Shembade et al. 2008). $CD4^+$ T-cells from Itch −/− mice in response to TNFα showed, unlike cells from wild type mice, persistent phosphorylation and degradation of IκBα suggesting a negative regulatory role for Itch in TNFα-mediated NF-κB signaling. In agreement with such an idea, lack of Itch in MEFs also impaired the termination of NF-κB DNA binding as measured by electrophoretic mobility shift assay (EMSA). Consistent with these results Itch −/− MEFs, unlike wild type MEFs that showed only transient activation, exhibited persistent activation of IKK and JNK in response to TNFα treatment. Moreover, the defect in Itch null cells appears to be specific for pathways controlled by cytokines as stimulation with agonistic antibodies to CD3 and CD28 elicited normal NF-κB responses in T-cells from both wild type and Itch-null mice (Fang et al. 2002; Shembade et al. 2008). While reconstitution of Itch −/− MEFs with wild type Itch restored transient IKK activation, overexpression of an Itch mutant with impaired E3 ligase activity (Itch C830A) failed to inhibit TNFα-induced NF-κB activation suggesting that Itch helps to maintain the optimal duration of responses in TNFα-stimulated cells in an E3 ligase-dependent manner. Consistent with the enhanced NF-κB activation, the polyubiquitin chains on RIP1 in Itch-deficient MEFs were found to be conjugated via K63-linkages. This finding is in contrast to the K48-linked polyubiquitin chains observed on RIP1 in TNFα treated Itch +/+ MEFs indicating that Itch might promote K48-linked ubiquitination of RIP1. In fact, in overexpression experiments with hemagglutinin-tagged ubiquitin variants exogenous Itch triggered the K-48 linked ubiquitination, and degradation of RIP1 (Shembade et al. 2008). These functional effects of Itch on TNF signaling appear to overlap with that of A20 and TAX1BP1 molecules. Surprisingly, suppression of A20 expression with siRNA impaired Itch-induced RIP1 degradation suggesting that A20 and Itch might cooperate in promoting RIP1 degradation. Results from several experiments supported such a notion. For example, stimulus-dependent association between RIP1 and A20 requires Itch and in conjunction with such a role for Itch, A20 failed to inhibit TNFα-induced expression of NF-κB luciferase reporter gene in Itch-deficient MEFs. These results collectively suggest that Itch E3 ligase activity play an indispensable role in terminating TNFα-induced NF-κB responses as a component of A20-containing complex (Fig. 2a).

Interestingly, Itch also plays a decisive role in determining the cell fate in response to TNFα. As mentioned above, binding of TNFα to TNFR1 triggers formation of signaling complexes that promote cell survival pathways by activating NF-κB, and cell death pathways by activating JNK (Bradley and Pober 2001; Deng et al. 2003). The complex formed at the plasma membrane internalizes and during the endocytic journey it reassembles to form DISC by recruiting FADD and caspase-8, leading to caspase activation and apoptosis (Micheau and Tschopp 2003; Schneider-Brachert et al. 2004; Schutze et al. 2008). It was established that NF-κB upregulates the expression of cellular FLICE Inhibitory Protein long isoform (c-FLIP$_L$), a negative regulator of proximal caspases (Micheau et al. 2001). The amount of c-FLIP$_L$ determines the fate of cell as it prevents the recruitment of caspase-8 to FADD and its activation (Muppidi et al. 2004). Lack of c-FLIP$_L$ renders cells sensitive to TNFα-induced apoptosis even in the absence of NF-κB activity indicating that induction of c-FLIP$_L$ is the primary mechanism by which NF-κB prevents activation of caspase-8 and apoptosis during early stages of TNFR1 signaling (Yeh et al. 2000). However, TNFα activated JNK enhances cell death by promoting the ubiquitination and degradation of c-FLIP$_L$ via Itch phosphorylation (Chang et al. 2006). Treatment with TNFα and CHX resulted in the loss of c-FLIP$_L$ protein in wild type, but not JNK1-deficient hepatocytes, and the loss was prevented by treatment with proteasomal inhibitor MG-132 demonstrating that TNFα induces proteasomal degradation of c-FLIP$_L$ in a JNK1-dependent manner (Chang et al. 2006). JNK1 did not phosphorylate c-FLIP$_L$ directly, but facilitated its degradation via phosphorylation-dependent activation of Itch. Incubation of Itch-deficient fibroblasts with TNFα plus CHX resulted in, compared to wild type fibroblasts, minimal loss of c-FLIP$_L$ protein, and very little cleavage of caspase-8. Reconstitution of Itch –/– cells with wild type, but not a mutant form of Itch that is refractory to JNK1-induced phosphorylation (AA-Itch), restored the loss of c-FLIP$_{L.}$ Itch associated directly with c-FLIP$_{L,}$ but not with other isoform c-FLIP$_s$, and as a result coexpression of Itch and activated JNK1 specifically promoted the ubiquitination and degradation of c-FLIP$_L$. Importantly, the absence of Itch protected mice from TNFα plus GalN (D-galactosamine)-induced liver damage and morbidity. Thus, the ubiquitin modifications mediated by the E3 activity of Itch play a key role in determining the biological outcome of TNFα treatment.

2.5 CYLD

CYLD is a cytoplasmic protein with three cytoskeletal-associated protein–glycine-conserved (CAP-GLY) domains that functions as a tumor suppressor. Mutations in the *CYLD* gene have been associated with familial cylindromatosis, also known as turban tumor syndrome, an autosomal-dominant condition that predisposes affected individuals to the development of multiple skin tumors (Almeida et al. 2008; Bignell et al. 2000). The tumors, which are believed to arise from the secretory glands and the hair follicles, are called cylindromas because of their characteristic microscopic architecture. Several studies have revealed that CYLD possess DUB function and with this thioesterase activity CYLD can remove from the target proteins, polyubiquitin chains that are conjugated via either K63- or K48-linkages (Brummelkamp et al. 2003; Kovalenko et al. 2003; Trompouki et al. 2003). Although many details remain to be elucidated, CYLD appears to regulate multiple signaling events of immune system including TNFR1-mediated NF-κB and JNK pathways (Glittenberg and Ligoxygakis 2007; Reiley et al. 2004; Simonson et al. 2007). Initial studies have suggested that CYLD negatively regulates activation of NF-κB by removing K63-linked polyubiquitin chains from TRAF2 and NEMO, both positive regulators of NF-κB. Evidence from CYLD-deficient mice and over expression experiments supported a physiological role for CYLD in the negative regulation of TNF signaling. Overexpression of CYLD inhibited activation of NF-κB induced either by TNFα, or by several members of the TNFR1 complex including TNFR1, TRAF2, TRADD and RIP1, and conversely, suppression of CYLD expression with siRNA enhanced TNFα-induced NF-κB signaling (Kovalenko et al. 2003). Absence of CYLD in primary keratinocytes led to rapid decrease in IκBα level and higher activation of NF-κB activation. A physiological role for CYLD in animal developmental processes was demonstrated by recent studies using CYLD knockout mouse. CYLD-deficient male mice are sterile owing to defects in seminiferous tubular organization, and decreased apoptosis during the early stages of spermatogenesis (Wright et al. 2007). CYLD binds and negatively regulates RIP1 ubiquitination and function indicating that in germ cells CYLD controls RIP1/NF-κB signaling. Accordingly, the spermatogonia of CYLD-null mice showed increased polyubiquitinated RIP1 and this concomitantly associated with enhanced basal NF-κB activity. These results establish that in testicular cells CYLD regulates NF-κB activation by removing signaling polyubiquitin chains on RIP1.

Furthermore, TNFα treatment of peritoneal macrophages from CYLD null mice resulted in, an increase in the levels of ubiquitinated TRAF2 and phospho-JNK suggesting that in mammalian cells in addition to NF-κB, CYLD negatively regulates JNK signaling (Glittenberg and Ligoxygakis 2007; Simonson et al. 2007). In contrast, evidence from Drosophila studies indicate that CYLD functions as a positive regulator of TNFα-induced JNK signaling and apoptosis (Tsichritzis et al. 2007). In fly, ectopic expression of Egr (Eiger, the Drosophila ortholog of TNFα) in the developing eye triggers apoptosis via JNK signaling resulting in a reduction in the adult eye size. The small eye phenotype was totally suppressed by the deletion of Drosophila CYLD (dCYLD) indicating that the activity of dCYLD is required for Egr-induced activation of JNK and apoptosis. The effect was due to the loss of dCYLD function as the suppression of the small eye phenotype under dCYLD null background could be rescued by the expression of wild type dCYLD. Importantly, mutant dCYLD without the DUB functional domains had no effect on the suppression of the Egr eye phenotype implying that the DUB activity is required for dCYLD function in Egr-induced cell death. In Drosophila dTRAF2 is the adaptor protein that mediates Egr-induced JNK signaling. dCYLD appears to regulate JNK activation positively by binding and deubiquitinating dTRAF2, and promoting the accumulation of dTRAF2 protein in vivo. Biochemical analysis of the protein extracts from dCYLD-null flies showed marked increase in ubiquitinated dTRAF2 and substantial reduction in dTRAF2 protein levels. Interestingly, positive modulation of JNK signaling by CYLD was reported in some cell types of mammals as well with thymocytes from CYLD-deficient mice exhibiting decreased JNK activity (Reiley et al. 2004). Thus, these results collectively indicate that CYLD regulates TNF signaling in a DUB-dependent, cell-type specific manner and this process is evolutionarily conserved.

2.6 CARPs

TNFR1 signaling complex formed at the plasma membrane that is essential for the activation of NF-κB is known to internalize immediately after formation, and fuse with intracellular membrane vesicles to form multivesicular endosomes (MVE), and this event is believed to be critical to attenuate NF-κB activation and diversification of TNF signaling (Schutze et al. 2008). Caspase-8 and -10 associated RING proteins (CARP)-1 and CARP-2 are E3 ligases with a phospholipid-binding motif known as FYVE (conserved in Fab1p/YOTB/Vac1p/EEA1) domain that is known to target the protein endocytic membrane vesicles (McDonald and El-Deiry 2004). Both CARP1 and CARP2 were shown to associate constitutively with intracellular compartments that include vesicles that are positive to endosomal proteins Rab 5 and Rab 11 (Araki et al. 2003; Coumailleau et al. 2004; Liao et al. 2008). Analysis of purified TNF-receptosomes (internalized TNFR1 signaling complexes) revealed that endocytosed receptor-containing MVE associated with CARP-containing vesicles as early as 5 min after internalization. At this stage, CARP proteins appeared to target the receptor-associated RIP1 for ubiquitination and proteasome-mediated degradation (Liao et al. 2008, 2009). Endogenous CARP2 interacted with endogenous RIP1 in TNFα-stimulated cells and owing to the ability of CARP2 to enhance RIP1 degradation, endogenous interaction between these proteins could be observed only when proteasomal degradation was inhibited. Moreover, CARP proteins promoted K48-linked polyubiquitination and degradation of RIP1 in vivo. Suppression of CARP protein expression, using small hairpin RNA (shRNA), stabilized the receptor-associated ubiquitinated RIP1 and enhanced NF-κB reporter activity. Conversely, overexpression of CARP wildtype, but not a RING mutant that unable to function as an active E3 protein ligase, decreased IKK activation, IκBα degradation and cytokine IL-6 secretion in TNFα-stimulated cells (Liao et al. 2008, 2009). Interestingly, CARP proteins initially identified in a yeast two-hybrid screen as molecules that interact with procaspase-8 and -10 (McDonald and El-Deiry 2004). In coexpression experiments both proteins were shown to promote the degradation of caspases. Suppression with shRNA of either CARP1 or CARP2 enhanced TNFα-induced caspase activity and cell death. These results demonstrate that both CARP1 and CARP2 are E3 ligases that target obligatory signaling intermediates of TNF signaling and limit both cell survival and cell death pathways. Given the similarities in substrate affinity and tissue distribution, functional redundancy may exist between CARP1 and CARP2 proteins. CARP2-deficient mice indeed exhibited no overt phenotype, and MEFs derived from these mice showed no changes in TNFα-induced apoptosis or NF-κB activation (Ahmed et al. 2009). Analysis of mice that lack expression of both CARP1 and CARP2 may be required to understand the physiological function of these endosomal-associated E3 molecules.

3 Ubiquitin Sensors

3.1 NEMO

Ubiquitin chains act as a nexus for the assembly, and thereby promoting the activation of downstream cytosolic molecules. Studies have shown that in TNFα-stimulated cells RIP1 acquires polyubiquitin chains and this event is crucial for the activation of IKK and NF-κB. Results from transfection experiments using ubiquitin mutants that possess either K63 or K48 for ubiquitin conjugation indicated that both kinds of linkages are involved, albeit at different stages, in RIP1 ubiquitination (Ea et al. 2006; Li et al. 2006; Wertz et al. 2004; Wu et al. 2006; Zhang et al. 2000). This was further confirmed by evidence from a study using linkage-specific antibodies that recognized polyubiquitin chains on RIP1 that are linked through K63 in the early stages and chains that are conjugated with K48-linkages at later stages of signaling (Newton et al. 2008). Moreover, reconstitution of RIP1-deficient cells with mutated RIP1 (K377R mutation that disrupts stimulus dependent K63-polyubiquitination of RIP1) substantially reduced the TNFα-induced NF-κB activation suggesting that K63-linked ubiquitin chains on RIP1 are critical for mediating NF-κB signaling (Ea et al. 2006; Li et al. 2006). This also led to the idea that K63-linked ubiquitin chains serve as a platform for oligomerization of NEMO-containing IKK signaling complex. A clue that NEMO might sense ubiquitin chains came from a yeast two-hybrid screen in which NEMO pulled out endogenous ubiquitin precursors, tandem diubiquitin (di-Ub) or triubiquitin (tri-Ub) molecules (Wu et al. 2006). Deletion mapping analysis of NEMO revealed that the domain encompassing coiled–coil (CC2) and Leucine-Zipper (LZ) motifs was necessary and sufficient for its interaction with ubiquitin. In vitro binding assays demonstrated that this interaction is specific for di-Ub or K63-linked ubiquitin chains as GST-NEMO had shown little affinity towards mono-Ub or K48-linked polyUb molecules (Lo et al. 2009; Rahighi et al. 2009). Importantly, mutations that occur naturally in the ubiquitin-binding region have been shown to cause immunodeficiency and ectodermal dysplasia in humans as a result of impaired NF-κB activation (Wu et al. 2006). Reduced NF-κB activity in NEMO-deficient cells that express mutant NEMO also correlated with decreased affinity of mutant protein towards polyubiquitinated RIP1. Consistent with this observation GST-NEMO mutant protein, unlike wildtype, did not associate with K63-linked polyubiquitin molecules in vitro. Interestingly, regions with high homology to NEMO CC2-LZ are identified in other proteins such as ABIN 1–3 and Optineurin, and this conserved motif was referred as the ubiquitin-binding in ABIN and NEMO (UBAN) domain (Wagner et al. 2008; Wu et al. 2006; Zhu et al. 2007). It is now generally believed that docking of the UBAN with polyubiquitin chains is the key for TNFα-induced NF-κB and apoptosis events. It was also proposed that Optineurin or ABIN proteins affect NF-κB signaling by interfering with the association of the UBAN of NEMO to the K63-linked Ub chains.

While initial studies focused exclusively on the importance of K63-linkages to TNF signaling, subsequent structural, biochemical and functional experiments provided evidence that the UBAN binding to linear ubiquitin chains also play essential roles in NF-κB pathway (Rahighi et al. 2009). Quantitative isothermal titration measurements using the UBAN of NEMO detected nearly 100-fold more affinity towards tandem di-Ub than K63-linked di-Ub (with dissociation constants 1.4 μM vs. 131 μM) indicating that linear ubiquitin chains may physiologically be more relevant than once thought (Lo et al. 2009). On the other hand, low affinity of NEMO to K63-linked di-Ub in vitro is not surprising given that NEMO is reported to interact preferentially with longer chain of K63-linked ubiquitin molecules

(Wu et al. 2006). Details of intrinsic conformational differences between linear di-Ub and K63-linked di-Ub that contribute to differential NEMO binding are available. Structural and biochemical evidence suggested that the UBAN of NEMO forms a parallel coiled-coil dimer, and it employs distinct surface patches for its interaction with individual ubiquitin moieties of tandem di-Ub and K63-linked di-Ub (Lo et al. 2009; Rahighi et al. 2009). Amino acids of the UBAN that are the key for its association with linear di-Ub, but not with K63-linked di-Ub have been identified (Rahighi et al. 2009). As expected changes in these amino acids that disrupted the association of the UBAN with linear tetraubiquitin had only minimal effect on interaction between K63-linked tetraubiquitin and the UBAN. Interestingly, reconstitution of MEFs from NEMO knockout mice with mutants that can still bind to K63-linked tetraubiquitin in vitro, had failed to restore the activation of NF-κB, or to protect cells from apoptosis upon TNFα treatment. These results revealed that both K63 and linear ubiquitin chains play an important role in TNFα-induced NF-κB signaling.

Although many details remain unknown, genetic and biochemical evidence in support of a biological role for linear ubiquitin chains is available. Linear ubiquitin chain assembly complex (LUBAC) is an E3 complex consisting of two RING finger proteins HOIL-1L and HOIP that facilitates conjugation of linear polyubiquitin chains on substrates (Kirisako et al. 2006). Hepatocytes from mice-deficient for HOIL-1 exhibited impaired TNFα-induced activation of NF-κB and increased apoptosis. However, according to initial reports the target for LUBAC-mediated linear polyubiquitination appears to be NEMO, but not RIP1 (Tokunaga et al. 2009). LUBAC associated with NEMO in vivo and the Zn finger domains of HOIP and HOIL-1L are indispensable for NEMO binding. Furthermore, the UBAN domain of NEMO is essential for LUBAC interaction and coexpression in NEMO-deficient N-1 cells of LUBAC with NEMO lacking the UBAN domain failed to activate NF-κB. LUBAC promoted polyubiquitination of NEMO in vitro, but not IKKα/β, and induced polyubiquitination of endogenous NEMO in vivo. LUBAC-mediated modification of NEMO in the presence of ubiquitin mutant with no free lysine residues, i.e., ubiquitin in which all lysine residues were mutated to arginine indicated that LUBAC might promote linkage between the C-terminus of one ubiquitin and the N-terminal amino group of another ubiquitin (linear linkages). In agreement with such an idea when blocked ubiquitin in which all amino groups including the N-terminal α-amino group of ubiquitin were methylated was used in these assays no polyubiquitination of NEMO was detected (Tokunaga et al. 2009). Moreover, an antibody that specifically detected linear ubiquitin chains reacted with immunoprecipitates of NEMO from denatured extracts of cells expressing LUBAC and NEMO. Mass spectroscopy and mutant analyzes showed that Lys 285 and Lys 309 of NEMO are the acceptors of polyubiquitylation mediated by LUBAC. These results collectively demonstrate that NEMO recognition of both K63-linked and linear ubiquitin chains are required for NF-κB signaling (Haas 2009).

3.2 ABIN-1

Increasing evidence also suggest a role for ubiquitin sensing proteins in the regulation of TNFα-induced apoptotic pathways. ABIN-1 (A20 binding and inhibitor of NF-κB) also known as NEF-associated factor 1 (NAF1) was originally found as an interactor of the HIV-1 NEF protein in a yeast two-hybrid screen (Fukushi et al. 1999). Not-withstanding earlier reports that suggested that ABIN-1 binds A20 and inhibits of NF-κB, a recent study using ABIN-1 null cells have revealed only a subtle role for ABIN-1 in limiting NF-κB signaling (Heyninck et al. 1999; Oshima et al. 2009). This study, however, found ABIN-1 as an ubiquitin sensor with an essential role in embryonic development. Mice deficient for ABIN-1 died during embryogenesis with lower hematocrits (anemia), and hypocellularity in liver due to apoptosis. The failure to protect embryonic fetal livers from apoptosis is the result of dysregulation of TNF signaling as TNF-deficiency rescued ABIN-1 null embryos, thus confirming physiological role for ABIN-1 in TNF signaling. In agreement with this ABIN-1 deficient cells exhibited increased sensitivity to TNF-induced apoptosis. ABIN-1 appears to exert its function independent of A20, which is also known to protect cells against apoptosis, as ABIN-1 blocked cell death in A20 null cells (Oshima et al. 2009). Interestingly, ABIN-1 protects cells from death by interfering with the assembly of the DISC components. ABIN-1 recruited to FADD in a stimulus-dependent manner, and in cells where ABIN-1 expression was suppressed, greater amounts of endogenous FADD associated with caspase-8 indicating that ABIN-1 affects the interaction between FADD and caspase-8. ABIN-1 also contains a region that shared homology with the UBAN of NEMO, In GST-pull down assays ABIN-1, like NEMO, bound to K63-linked polyubiquitin chains with a preference for chains of three ubiquitin moieties or more (Oshima et al. 2009). ABIN-1 proteins harboring mutations of the conserved amino acids residues (QQ477/478 EE or F482S) failed to associate with K63 ubiquitin chains. Importantly, expression of either of these ubiquitin-binding deficient ABIN-1 mutants, but not wild type, failed to protect ABIN-1 null fibroblasts from TNFα-induced apoptosis. In agreement with this wild type, but not the mutant ABIN-1 (QQ477/478 EE) blocked the association between endogenous FADD and caspase-8 proteins. These results provide evidence that polyubiquitin modifications, probably of the DISC components regulate TNFα-induced caspase-8 activation and apoptosis by recruiting ABIN-1. It would be interesting to know whether removal of the polyubiquitin chains is necessary for the assembly of stable DISC. The finding that the action of CYLD, a K63- deubiquitinating enzyme, is required to form a caspase-8-activating complex consisting of FADD and caspase-8 supports such a notion (Wang et al. 2008).

3.3 IAPs

IAPs with ubiquitin protein ligase activity, namely X-linked IAP (XIAP), cIAP-1 and cIAP-2 are known to affect TNFα-signaling (Srinivasula and Ashwell 2008).

These proteins, when present at higher concentrations, protect cells from death against several onslaughts including stimulation by death ligands. Although initial structural and functional studies indicated that IAPs could inhibit cell death by binding and inhibiting caspases independent of the RING, genetic evidence is now available in favor of a physiological role for ubiquitin protein ligase activity (Vaux and Silke 2005). For example, in Drosophila the E3 function of Drosophila IAP1 appears to promote ubiquitination of the Drosophila caspase DRONC and this E3 activity is required to block apoptosis. Mice in which the RING motif of XIAP was inactivated by gene targeting showed, in some cells types, increased caspase activity and apoptosis (Schile et al. 2008). Previous studies that were based mostly on overexpression and in vitro experiments, have reported that the RING domain promotes autoubiquitination and degradation of IAPs (Yang et al. 2000). Consistent with these findings removal of the RING domain of XIAP in mice that impaired the E3 activity, stabilized XIAP protein in apoptotic thymocytes (Schile et al. 2008). Moreover, fibroblasts and embryonic stem cells that express XIAP without the RING domain had impaired ubiquitination of active caspase-3 and increased caspase-3 activity. Importantly, these cells were strongly sensitive to TNFα-induced apoptosis thus, demonstrating the contribution of the E3 activity of XIAP to TNF signaling (Schile et al. 2008).

The proteins cIAP-1 and cIAP-2 were originally identified as components of TNFR1 complex, yet no definitive evidence for a physiological role for these molecules TNFR1 signaling is forthcoming (Rothe et al. 1995; Shu et al. 1996; Conte et al. 2006; Conze et al. 2005). Primary human umbilical vein endothelial cells (HUVEC) deficient in both cIAP-1 and cIAP-2 showed limited TNFα-induced NF-κB activation, leading to the notion that cIAPs might regulate TNF signaling by acting at the proximal level (Santoro et al. 2007). Recent findings that acute deficiency of cIAPs leads to TNFR1-mediated cell death further supported this notion (Gaither et al. 2007; Petersen et al. 2007; Varfolomeev et al. 2007; Vince et al. 2007). IAP-antagonists (known as SMAC mimetics) bind and trigger the E3-mediated autoubiquitination and degradation of cIAP-1 and cIAP-2. Loss of cIAP proteins resulted in increased NF-κB activation leading to enhanced autocrine TNFα production and cell death in some tumor cell lines (Wu et al. 2007). Absence of cIAP also affected the assembly of TNFR1 signaling complex directly. Treatment of cell lines with IAP antagonists resulted in the formation of a death complex consisting of RIP1, FADD, and caspase-8, and cIAP degradation required for the assembly of this death complex. More studies are needed to understand the mechanistic details by which IAPs contribute to the signaling complex formation. In this context, the recent identification of ubiquitin-binding motifs in IAP proteins is intriguing as this motif enables IAPs, like NEMO and ABIN-1, to participate in Ub-dependent TNF signaling processes (Blankenship et al. 2009; Gyrd-Hansen et al. 2008). Sequence analysis and structural algorithms identified an evolutionarily conserved ubiquitin-associated (UBA) domain that consists of three tightly packed α-helices in XIAP, cIAP-1 and cIAP-2 in a region N-terminus to the RING domain. In vitro binding experiments revealed interactions by XIAP and cIAP-2 with K63-linked Ub molecules and by cIAP-1 with both K63- and K48-polyubiquitins. Point mutations in the conserved amino acid residues in the UBA domain abrogated the interaction, thus underlying the importance of the UBA for polyUb binding. Furthermore, formation of higher-order oligomeric forms of IAPs is required for this interaction as deletion mutants of XIAP or cIAP-1 that do not dimerize, failed to bind with ubiquitin molecules. Immortalized MEFs from cIAP-1 knockout mice are sensitive to TNFα-induced apoptosis, and expression of wild type cIAP-1, but not the polyUb-binding defective mutant, restored resistance to TNFα. Importantly, in zebrafish zcIAP-1 function is required for endothelial cell survival, and zcIAP-1 null mutants showed vascular defects, described in the literature as tomato mutant phenotype, as a result of increased caspase-8 dependent apoptosis. This phenotype could be rescued by the expression wild type zcIAP-1. Interestingly, mutant zcIAP-1 that does not bind ubiquitin molecules failed to rescue the tomato mutant phenotype. These results suggest that the Ub-binding is a critical cIAP-1 function in endothelial survival in vertebrates (Blankenship et al. 2009). Although many details remain unknown, evidence thus far indicate that IAP E3 activity and ubiquitin-binding function contribute to TNFα-mediated NF-κB signaling and apoptosis.

4 Conclusions

Modification of proteins by ubiquitin has emerged as a critical signal for the regulation of a variety of biological processes including TNFR1 signaling. It is now revealed that ubiquitin modifications of key signaling components affect, at multiple steps of the pathways, protein stability, localization, and a network of protein–protein interactions. It is important to know whether the ubiquitin-editing events such as the replacement of K63-linked ubiquitin chains involved in signal propagation with degradative K48 chains are more common, and whether such an exchange can occur between two activating K63 and linear ubiquitin chains. Given the possibility that K63-linked and linear ubiquitin chains promote distinct functions, knowledge of molecules that facilitate ubiquitin-editing might be the key for our understanding of the diversification of TNFR1 signaling. One of the many other questions that need attention is how the activation of E3 ligases is regulated, and the role played by phosphorylation in the E3 activation and its substrate recognition. The existence of large number of E3 ligases and the likelihood that ubiquitin moieties can be conjugated via more than one kind of linkages (mixed chains) make answers to these questions both complex and exciting.

Acknowledgments We apologize to those researchers whose contributions have not been cited due to space limitations. The research was supported by the Intramural Research Program of the NIH, Center for Cancer research, National Cancer Institute. We thank Jonathan Ashwell for helpful discussions.

References

Ahmed AU, Moulin M, Coumailleau F, Wong WW, Miasari M, Carter H, Silke J, Cohen-Tannoudji M, Vince JE, Vaux DL (2009) CARP2 deficiency does not alter induction of NF-kappaB by TNFalpha. Curr Biol 19:R15–17 author reply R17–19

Almeida S, Maillard C, Itin P, Hohl D, Huber M (2008) Five new CYLD mutations in skin appendage tumors and evidence that aspartic acid 681 in CYLD is essential for deubiquitinase activity. J Invest Dermatol 128:587–593

Araki K, Kawamura M, Suzuki T, Matsuda N, Kanbe D, Ishii K, Ichikawa T, Kumanishi T, Chiba T, Tanaka K, Nawa H (2003) A palmitoylated RING finger ubiquitin ligase and its homologue in the brain membranes. J Neurochem 86:749–762

Au PY, Yeh WC (2007) Physiological roles and mechanisms of signaling by TRAF2 and TRAF5. Adv Exp Med Biol 597:32–47

Azmi P, Seth A (2005) RNF11 is a multifunctional modulator of growth factor receptor signalling and transcriptional regulation. Eur J Cancer 41:2549–2560

Baud V, Karin M (2001) Signal transduction by tumor necrosis factor and its relatives. Trends Cell Biol 11:372–377

Bignell GR, Warren W, Seal S, Takahashi M, Rapley E, Barfoot R, Green H, Brown C, Biggs PJ, Lakhani SR et al (2000) Identification of the familial cylindromatosis tumour-suppressor gene. Nat Genet 25:160–165

Blankenship JW, Varfolomeev E, Goncharov T, Fedorova AV, Kirkpatrick DS, Izrael-Tomasevic A, Phu L, Arnott D, Aghajan M, Zobel K et al (2009) Ubiquitin binding modulates IAP antagonist-stimulated proteasomal degradation of c-IAP1 and c-IAP2(1). Biochem J 417:149–160

Boatright KM, Renatus M, Scott FL, Sperandio S, Shin H, Pedersen IM, Ricci JE, Edris WA, Sutherlin DP, Green DR, Salvesen GS (2003) A unified model for apical caspase activation. Mol Cell 11:529–541

Bonifacino JS, Traub LM (2003) Signals for sorting of transmembrane proteins to endosomes and lysosomes. Annu Rev Biochem 72:395–447

Bradley JR, Pober JS (2001) Tumor necrosis factor receptor-associated factors (TRAFs). Oncogene 20:6482–6491

Brummelkamp TR, Nijman SM, Dirac AM, Bernards R (2003) Loss of the cylindromatosis tumour suppressor inhibits apoptosis by activating NF-kappaB. Nature 424:797–801

Burger A, Li H, Zhang XK, Pienkowska M, Venanzoni M, Vournakis J, Papas T, Seth A (1998) Breast cancer genome anatomy: correlation of morphological changes in breast carcinomas with expression of the novel gene product Di12. Oncogene 16:327–333

Carswell EA, Old LJ, Kassel RL, Green S, Fiore N, Williamson B (1975) An endotoxin-induced serum factor that causes necrosis of tumors. Proc Natl Acad Sci USA 72:3666–3670

Chang HY, Yang X (2000) Proteases for cell suicide: functions and regulation of caspases. Microbiol Mol Biol Rev 64:821–846

Chang L, Kamata H, Solinas G, Luo JL, Maeda S, Venuprasad K, Liu YC, Karin M (2006) The E3 ubiquitin ligase itch couples JNK activation to TNFalpha-induced cell death by inducing c-FLIP(L) turnover. Cell 124:601–613

Chanudet E, Ye H, Ferry J, Bacon CM, Adam P, Muller-Hermelink HK, Radford J, Pileri SA, Ichimura K, Collins VP et al (2009) A20 deletion is associated with copy number gain at the TNFA/B/C locus and occurs preferentially in translocation-negative MALT lymphoma of the ocular adnexa and salivary glands. J Pathol 217:420–430

Chen LF, Greene WC (2004) Shaping the nuclear action of NF-kappaB. Nat Rev Mol Cell Biol 5:392–401

Colland F, Jacq X, Trouplin V, Mougin C, Groizeleau C, Hamburger A, Meil A, Wojcik J, Legrain P, Gauthier JM (2004) Functional proteomics mapping of a human signaling pathway. Genome Res 14:1324–1332

Conte D, Holcik M, Lefebvre CA, Lacasse E, Picketts DJ, Wright KE, Korneluk RG (2006) Inhibitor of apoptosis protein cIAP2 is essential for lipopolysaccharide-induced macrophage survival. Mol Cell Biol 26:699–708

Conze DB, Albert L, Ferrick DA, Goeddel DV, Yeh WC, Mak T, Ashwell JD (2005) Posttranscriptional downregulation of c-IAP2 by the ubiquitin protein ligase c-IAP1 in vivo. Mol Cell Biol 25:3348–3356

Coornaert B, Carpentier I, Beyaert R (2009) A20: central gatekeeper in inflammation and immunity. J Biol Chem 284:8217–8221

Coumailleau F, Das V, Alcover A, Raposo G, Vandormael-Pournin S, Le Bras S, Baldacci P, Dautry-Varsat A, Babinet C, Cohen-Tannoudji M (2004) Over-expression of Rififylin, a new RING finger and FYVE-like domain-containing protein, inhibits recycling from the endocytic recycling compartment. Mol Biol Cell 15:4444–4456

Dempsey PW, Doyle SE, He JQ, Cheng G (2003) The signaling adaptors and pathways activated by TNF superfamily. Cytokine Growth Factor Rev 14:193–209

Deng Y, Ren X, Yang L, Lin Y, Wu X (2003) A JNK-dependent pathway is required for TNFalpha-induced apoptosis. Cell 115:61–70

Ea CK, Deng L, Xia ZP, Pineda G, Chen ZJ (2006) Activation of IKK by TNFalpha requires site-specific ubiquitination of RIP1 and polyubiquitin binding by NEMO. Mol Cell 22:245–257

Esposito E, Di Matteo V, Di Giovanni G (2007) Death in the substantia nigra: a motor tragedy. Expert Rev Neurother 7:677–697

Fang S, Weissman AM (2004) A field guide to ubiquitylation. Cell Mol Life Sci 61:1546–1561

Fang D, Elly C, Gao B, Fang N, Altman Y, Joazeiro C, Hunter T, Copeland N, Jenkins N, Liu YC (2002) Dysregulation of T lymphocyte function in itchy mice: a role for itch in TH2 differentiation. Nat Immunol 3:281–287

Fukushi M, Dixon J, Kimura T, Tsurutani N, Dixon MJ, Yamamoto N (1999) Identification and cloning of a novel cellular protein Naf1, Nef-associated factor 1, that increases cell surface CD4 expression. FEBS Lett 442:83–88

Gachon F, Peleraux A, Thebault S, Dick J, Lemasson I, Devaux C, Mesnard JM (1998) CREB-2, a cellular CRE-dependent transcription repressor, functions in association with Tax as an activator of the human T-cell leukemia virus type 1 promoter. J Virol 72:8332–8337

Gaither A, Porter D, Yao Y, Borawski J, Yang G, Donovan J, Sage D, Slisz J, Tran M, Straub C et al (2007) A Smac mimetic rescue screen reveals roles for inhibitor of apoptosis proteins in tumor necrosis factor-alpha signaling. Cancer Res 67:11493–11498

Gao M, Karin M (2005) Regulating the regulators: control of protein ubiquitination and ubiquitin-like modifications by extracellular stimuli. Mol Cell 19:581–593

Ghosh S, Karin M (2002) Missing pieces in the NF-kappaB puzzle. Cell 109 Suppl:S81–96

Glittenberg M, Ligoxygakis P (2007) CYLD: a multifunctional deubiquitinase. Fly (Austin) 1:330–332

Gorman AM (2008) Neuronal cell death in neurodegenerative diseases: recurring themes around protein handling. J Cell Mol Med 12:2263–2280

Graham RR, Cotsapas C, Davies L, Hackett R, Lessard CJ, Leon JM, Burtt NP, Guiducci C, Parkin M, Gates C et al (2008) Genetic variants near TNFAIP3 on 6q23 are associated with systemic lupus erythematosus. Nat Genet 40:1059–1061

Gyrd-Hansen M, Darding M, Miasari M, Santoro MM, Zender L, Xue W, Tenev T, da Fonseca PC, Zvelebil M, Bujnicki JM et al (2008) IAPs contain an evolutionarily conserved ubiquitin-binding domain that regulates NF-kappaB as well as cell survival and oncogenesis. Nat Cell Biol 10:1309–1317

Haas AL (2009) Linear polyubiquitylation: the missing link in NF-kappaB signalling. Nat Cell Biol 11:116–118

Haglund K, Dikic I (2005) Ubiquitylation and cell signaling. EMBO J 24:3353–3359

Hayden MS, Ghosh S (2008) Shared principles in NF-kappaB signaling. Cell 132:344–362

Heyninck K, De Valck D, Vanden Berghe W, Van Criekinge W, Contreras R, Fiers W, Haegeman G, Beyaert R (1999) The zinc finger protein A20 inhibits TNF-induced NF-kappaB-dependent gene expression by interfering with an RIP- or TRAF2-mediated transactivation signal and directly binds to a novel NF-kappaB-inhibiting protein ABIN. J Cell Biol 145:1471–1482

Hoeflich KP, Yeh WC, Yao Z, Mak TW, Woodgett JR (1999) Mediation of TNF receptor-associated factor effector functions by apoptosis signal-regulating kinase-1 (ASK1). Oncogene 18:5814–5820

Hustad CM, Perry WL, Siracusa LD, Rasberry C, Cobb L, Cattanach BM, Kovatch R, Copeland NG, Jenkins NA (1995) Molecular genetic characterization of six recessive viable alleles of the mouse agouti locus. Genetics 140:255–265

Ichijo H, Nishida E, Irie K, ten Dijke P, Saitoh M, Moriguchi T, Takagi M, Matsumoto K, Miyazono K, Gotoh Y (1997) Induction of apoptosis by ASK1, a mammalian MAPKKK that activates SAPK/JNK and p38 signaling pathways. Science 275:90–94

Iha H, Peloponese JM, Verstrepen L, Zapart G, Ikeda F, Smith CD, Starost MF, Yedavalli V, Heyninck K, Dikic I et al (2008) Inflammatory cardiac valvulitis in TAX1BP1-deficient mice through selective NF-kappaB activation. EMBO J 27:629–641

Jacobson MD, Weil M, Raff MC (1997) Programmed cell death in animal development. Cell 88:347–354

Jin DY, Teramoto H, Giam CZ, Chun RF, Gutkind JS, Jeang KT (1997) A human suppressor of c-Jun N-terminal kinase 1 activation by tumor necrosis factor alpha. J Biol Chem 272:25816–25823

Kanayama A, Seth RB, Sun L, Ea CK, Hong M, Shaito A, Chiu YH, Deng L, Chen ZJ (2004) TAB2 and TAB3 activate the NF-kappaB pathway through binding to polyubiquitin chains. Mol Cell 15:535–548

Karin M, Cao Y, Greten FR, Li ZW (2002) NF-kappaB in cancer: from innocent bystander to major culprit. Nat Rev Cancer 2:301–310

Kelliher MA, Grimm S, Ishida Y, Kuo F, Stanger BZ, Leder P (1998) The death domain kinase RIP mediates the TNF-induced NF-kappaB signal. Immunity 8:297–303

Kirisako T, Kamei K, Murata S, Kato M, Fukumoto H, Kanie M, Sano S, Tokunaga F, Tanaka K, Iwai K (2006) A ubiquitin ligase complex assembles linear polyubiquitin chains. EMBO J 25:4877–4887

Kovalenko A, Chable-Bessia C, Cantarella G, Israel A, Wallach D, Courtois G (2003) The tumour suppressor CYLD negatively regulates NF-kappaB signalling by deubiquitination. Nature 424:801–805

Lavrik I, Golks A, Krammer PH (2005) Death receptor signaling. J Cell Sci 118:265–267

Lee EG, Boone DL, Chai S, Libby SL, Chien M, Lodolce JP, Ma A (2000) Failure to regulate TNF-induced NF-kappaB and cell death responses in A20-deficient mice. Science 289:2350–2354

Lee TH, Shank J, Cusson N, Kelliher MA (2004) The kinase activity of Rip1 is not required for tumor necrosis factor-alpha-induced IkappaB kinase or p38 MAP kinase activation or for the ubiquitination of Rip1 by Traf2. J Biol Chem 279:33185–33191

Li H, Seth A (2004) An RNF11: Smurf2 complex mediates ubiquitination of the AMSH protein. Oncogene 23:1801–1808

Li H, Kobayashi M, Blonska M, You Y, Lin X (2006) Ubiquitination of RIP is required for tumor necrosis factor alpha-induced NF-kappaB activation. J Biol Chem 281:13636–13643

Liao W, Xiao Q, Tchikov V, Fujita K, Yang W, Wincovitch S, Garfield S, Conze D, El-Deiry WS, Schutze S, Srinivasula SM (2008) CARP-2 is an endosome-associated ubiquitin ligase for RIP and regulates TNF-induced NF-kappaB activation. Curr Biol 18:641–649

Liao W, Fujita K, Xiao Q, Tchikov V, Yang W, Gunsor M, Garfield S, Goldsmith P, El-Deiry WS, Schutze S, Srinivasula SM (2009) Response: CARP1 regulates induction of NF-κB by TNF-α. Curr Biol 19:R17–R19

Liu YC, Penninger J, Karin M (2005) Immunity by ubiquitylation: a reversible process of modification. Nat Rev Immunol 5:941–952

Lo YC, Lin SC, Rospigliosi CC, Conze DB, Wu CJ, Ashwell JD, Eliezer D, Wu H (2009) Structural basis for recognition of diubiquitins by NEMO. Mol Cell 33:602–615

MacEwan DJ (2002) TNF receptor subtype signalling: differences and cellular consequences. Cell Signal 14:477–492

Martin LJ (2001) Neuronal cell death in nervous system development, disease, and injury (Review). Int J Mol Med 7:455–478

McDonald ER 3rd, El-Deiry WS (2004) Suppression of caspase-8- and -10-associated RING proteins results in sensitization to death ligands and inhibition of tumor cell growth. Proc Natl Acad Sci USA 101:6170–6175

Meylan E, Tschopp J (2005) The RIP kinases: crucial integrators of cellular stress. Trends Biochem Sci 30:151–159
Micheau O, Tschopp J (2003) Induction of TNF receptor I-mediated apoptosis via two sequential signaling complexes. Cell 114:181–190
Micheau O, Lens S, Gaide O, Alevizopoulos K, Tschopp J (2001) NF-kappaB signals induce the expression of c-FLIP. Mol Cell Biol 21:5299–5305
Muppidi JR, Tschopp J, Siegel RM (2004) Life and death decisions: secondary complexes and lipid rafts in TNF receptor family signal transduction. Immunity 21:461–465
Musone SL, Taylor KE, Lu TT, Nititham J, Ferreira RC, Ortmann W, Shifrin N, Petri MA, Kamboh MI, Manzi S et al (2008) Multiple polymorphisms in the TNFAIP3 region are independently associated with systemic lupus erythematosus. Nat Genet 40:1062–1064
Newton K, Matsumoto ML, Wertz IE, Kirkpatrick DS, Lill JR, Tan J, Dugger D, Gordon N, Sidhu SS, Fellouse FA et al (2008) Ubiquitin chain editing revealed by polyubiquitin linkage-specific antibodies. Cell 134:668–678
Nishitoh H, Saitoh M, Mochida Y, Takeda K, Nakano H, Rothe M, Miyazono K, Ichijo H (1998) ASK1 is essential for JNK/SAPK activation by TRAF2. Mol Cell 2:389–395
Novak U, Rinaldi A, Kwee I, Nandula SV, Rancoita PM, Compagno M, Cerri M, Rossi D, Murty VV, Zucca E, et al (2009) The NF-{kappa}B negative regulator TNFAIP3 (A20) is inactivated by somatic mutations and genomic deletions in marginal zone B-cell lymphomas. Blood 113:4918-4921
Oshima S, Turer EE, Callahan JA, Chai S, Advincula R, Barrera J, Shifrin N, Lee B, Yen B, Woo T et al (2009) ABIN-1 is a ubiquitin sensor that restricts cell death and sustains embryonic development. Nature 457:906–909
Perkins ND (2006) Post-translational modifications regulating the activity and function of the nuclear factor kappa B pathway. Oncogene 25:6717–6730
Petersen SL, Wang L, Yalcin-Chin A, Li L, Peyton M, Minna J, Harran P, Wang X (2007) Autocrine TNFalpha signaling renders human cancer cells susceptible to Smac-mimetic-induced apoptosis. Cancer Cell 12:445–456
Plenge RM, Cotsapas C, Davies L, Price AL, de Bakker PI, Maller J, Pe'er I, Burtt NP, Blumenstiel B, DeFelice M et al (2007) Two independent alleles at 6q23 associated with risk of rheumatoid arthritis. Nat Genet 39:1477–1482
Rahighi S, Ikeda F, Kawasaki M, Akutsu M, Suzuki N, Kato R, Kensche T, Uejima T, Bloor S, Komander D et al (2009) Specific recognition of linear ubiquitin chains by NEMO is important for NF-kappaB activation. Cell 136:1098–1109
Reich A, Spering C, Schulz JB (2008) Death receptor Fas (CD95) signaling in the central nervous system: tuning neuroplasticity? Trends Neurosci 31:478–486
Reiley W, Zhang M, Sun SC (2004) Negative regulation of JNK signaling by the tumor suppressor CYLD. J Biol Chem 279:55161–55167
Rothe M, Pan MG, Henzel WJ, Ayres TM, Goeddel DV (1995) The TNFR2-TRAF signaling complex contains two novel proteins related to baculoviral inhibitor of apoptosis proteins. Cell 83:1243–1252
Rudolph D, Yeh WC, Wakeham A, Rudolph B, Nallainathan D, Potter J, Elia AJ, Mak TW (2000) Severe liver degeneration and lack of NF-kappaB activation in NEMO/IKKgamma-deficient mice. Genes Dev 14:854–862
Santoro MM, Samuel T, Mitchell T, Reed JC, Stainier DY (2007) Birc2 (cIap1) regulates endothelial cell integrity and blood vessel homeostasis. Nat Genet 39:1397–1402
Schile AJ, Garcia-Fernandez M, Steller H (2008) Regulation of apoptosis by XIAP ubiquitin-ligase activity. Genes Dev 22:2256–2266
Schmidt-Supprian M, Bloch W, Courtois G, Addicks K, Israel A, Rajewsky K, Pasparakis M (2000) NEMO/IKK gamma-deficient mice model incontinentia pigmenti. Mol Cell 5:981–992
Schneider-Brachert W, Tchikov V, Neumeyer J, Jakob M, Winoto-Morbach S, Held-Feindt J, Heinrich M, Merkel O, Ehrenschwender M, Adam D et al (2004) Compartmentalization of TNF receptor 1 signaling: internalized TNF receptosomes as death signaling vesicles. Immunity 21:415–428

Schutze S, Tchikov V, Schneider-Brachert W (2008) Regulation of TNFR1 and CD95 signalling by receptor compartmentalization. Nat Rev Mol Cell Biol 9:655–662
Seki N, Hattori A, Hayashi A, Kozuma S, Sasaki M, Suzuki Y, Sugano S, Muramatsu MA, Saito T (1999) Cloning and expression profile of mouse and human genes, Rnf11/RNF11, encoding a novel RING-H2 finger protein. Biochim Biophys Acta 1489:421–427
Shembade N, Harhaj NS, Liebl DJ, Harhaj EW (2007) Essential role for TAX1BP1 in the termination of TNF-alpha-, IL-1- and LPS-mediated NF-kappaB and JNK signaling. EMBO J 26:3910–3922
Shembade N, Harhaj NS, Parvatiyar K, Copeland NG, Jenkins NA, Matesic LE, Harhaj EW (2008) The E3 ligase Itch negatively regulates inflammatory signaling pathways by controlling the function of the ubiquitin-editing enzyme A20. Nat Immunol 9:254–262
Shembade N, Parvatiyar K, Harhaj NS, Harhaj EW (2009) The ubiquitin-editing enzyme A20 requires RNF11 to downregulate NF-kappaB signalling. EMBO J 28:513–522
Shu HB, Takeuchi M, Goeddel DV (1996) The tumor necrosis factor receptor 2 signal transducers TRAF2 and c-IAP1 are components of the tumor necrosis factor receptor 1 signaling complex. Proc Natl Acad Sci USA 93:13973–13978
Simonson SJ, Wu ZH, Miyamoto S (2007) CYLD: a DUB with many talents. Dev Cell 13: 601–603
Srinivasula SM, Ashwell JD (2008) IAPs: what's in a name? Mol Cell 30:123–135
Steller H (1995) Mechanisms and genes of cellular suicide. Science 267:1445–1449
Subramaniam V, Li H, Wong M, Kitching R, Attisano L, Wrana J, Zubovits J, Burger AM, Seth A (2003) The RING-H2 protein RNF11 is overexpressed in breast cancer and is a target of Smurf2 E3 ligase. Br J Cancer 89:1538–1544
Tada K, Okazaki T, Sakon S, Kobarai T, Kurosawa K, Yamaoka S, Hashimoto H, Mak TW, Yagita H, Okumura K et al (2001) Critical roles of TRAF2 and TRAF5 in tumor necrosis factor-induced NF-kappa B activation and protection from cell death. J Biol Chem 276:36530–36534
Thomson W, Barton A, Ke X, Eyre S, Hinks A, Bowes J, Donn R, Symmons D, Hider S, Bruce IN et al (2007) Rheumatoid arthritis association at 6q23. Nat Genet 39:1431–1433
Ting AT, Pimentel-Muinos FX, Seed B (1996) RIP mediates tumor necrosis factor receptor 1 activation of NF-kappaB but not Fas/APO-1-initiated apoptosis. EMBO J 15:6189–6196
Tobiume K, Matsuzawa A, Takahashi T, Nishitoh H, Morita K, Takeda K, Minowa O, Miyazono K, Noda T, Ichijo H (2001) ASK1 is required for sustained activations of JNK/p38 MAP kinases and apoptosis. EMBO Rep 2:222–228
Tokunaga F, Sakata S, Saeki Y, Satomi Y, Kirisako T, Kamei K, Nakagawa T, Kato M, Murata S, Yamaoka S et al (2009) Involvement of linear polyubiquitylation of NEMO in NF-kappaB activation. Nat Cell Biol 11:123–132
Trompouki E, Hatzivassiliou E, Tsichritzis T, Farmer H, Ashworth A, Mosialos G (2003) CYLD is a deubiquitinating enzyme that negatively regulates NF-kappaB activation by TNFR family members. Nature 424:793–796
Tsichritzis T, Gaentzsch PC, Kosmidis S, Brown AE, Skoulakis EM, Ligoxygakis P, Mosialos G (2007) A Drosophila ortholog of the human cylindromatosis tumor suppressor gene regulates triglyceride content and antibacterial defense. Development 134:2605–2614
Varfolomeev E, Blankenship JW, Wayson SM, Fedorova AV, Kayagaki N, Garg P, Zobel K, Dynek JN, Elliott LO, Wallweber HJ et al (2007) IAP antagonists induce autoubiquitination of c-IAPs, NF-kappaB activation, and TNFalpha-dependent apoptosis. Cell 131:669–681
Vaux DL, Silke J (2005) IAPs, RINGs and ubiquitylation. Nat Rev Mol Cell Biol 6:287–297
Vince JE, Wong WW, Khan N, Feltham R, Chau D, Ahmed AU, Benetatos CA, Chunduru SK, Condon SM, McKinlay M et al (2007) IAP antagonists target cIAP1 to induce TNFalpha-dependent apoptosis. Cell 131:682–693
Wagner S, Carpentier I, Rogov V, Kreike M, Ikeda F, Lohr F, Wu CJ, Ashwell JD, Dotsch V, Dikic I, Beyaert R (2008) Ubiquitin binding mediates the NF-kappaB inhibitory potential of ABIN proteins. Oncogene 27:3739–3745
Wang L, Du F, Wang X (2008) TNF-alpha induces two distinct caspase-8 activation pathways. Cell 133:693–703

Weissman AM (2001) Themes and variations on ubiquitylation. Nat Rev Mol Cell Biol 2:169–178
Wellcome Trust Case Control Consortium (2007) Genome-wide association study of 14,000 cases of seven common diseases and 3000 shared controls. Nature 447:661–678
Werner SL, Barken D, Hoffmann A (2005) Stimulus specificity of gene expression programs determined by temporal control of IKK activity. Science 309:1857–1861
Wertz IE, O'Rourke KM, Zhou H, Eby M, Aravind L, Seshagiri S, Wu P, Wiesmann C, Baker R, Boone DL et al (2004) De-ubiquitination and ubiquitin ligase domains of A20 downregulate NF-kappaB signalling. Nature 430:694–699
Wright A, Reiley WW, Chang M, Jin W, Lee AJ, Zhang M, Sun SC (2007) Regulation of early wave of germ cell apoptosis and spermatogenesis by deubiquitinating enzyme CYLD. Dev Cell 13:705–716
Wu CJ, Conze DB, Li T, Srinivasula SM, Ashwell JD (2006) Sensing of Lys 63-linked polyubiquitination by NEMO is a key event in NF-kappaB activation [corrected]. Nat Cell Biol 8:398–406
Wu H, Tschopp J, Lin SC (2007) Smac mimetics and TNFalpha: a dangerous liaison? Cell 131:655–658
Yang Y, Fang S, Jensen JP, Weissman AM, Ashwell JD (2000) Ubiquitin protein ligase activity of IAPs and their degradation in proteasomes in response to apoptotic stimuli. Science 288:874–877
Yeh WC, Shahinian A, Speiser D, Kraunus J, Billia F, Wakeham A, de la Pompa JL, Ferrick D, Hum B, Iscove N et al (1997) Early lethality, functional NF-kappaB activation, and increased sensitivity to TNF-induced cell death in TRAF2-deficient mice. Immunity 7:715–725
Yeh WC, Itie A, Elia AJ, Ng M, Shu HB, Wakeham A, Mirtsos C, Suzuki N, Bonnard M, Goeddel DV, Mak TW (2000) Requirement for Casper (c-FLIP) in regulation of death receptor-induced apoptosis and embryonic development. Immunity 12:633–642
Yuan J (2006) Divergence from a dedicated cellular suicide mechanism: exploring the evolution of cell death. Mol Cell 23:1–12
Zhang SQ, Kovalenko A, Cantarella G, Wallach D (2000) Recruitment of the IKK signalosome to the p55 TNF receptor: RIP and A20 bind to NEMO (IKKgamma) upon receptor stimulation. Immunity 12:301–311
Zhu G, Wu CJ, Zhao Y, Ashwell JD (2007) Optineurin negatively regulates TNFalpha- induced NF-kappaB activation by competing with NEMO for ubiquitinated RIP. Curr Biol 17:1438–1443

From Biochemical Principles of Apoptosis Induction by TRAIL to Application in Tumour Therapy

Stefanie M. Cordier[1], Kerstin Papenfuss[1], and Henning Walczak

Abstract The tumour necrosis factor-related apoptosis-inducing ligand (TRAIL) is a member of the TNF superfamily which has been shown to selectively kill tumour cells, while sparing normal tissue. This attribute makes TRAIL an attractive drug candidate for cancer therapy. Although most primary tumour cells turned out to be primarily TRAIL-resistant, recent studies evidenced that a variety of cancers can be sensitised to TRAIL-induced apoptosis upon pre-treatment with chemotherapeutic agents or irradiation, while normal cells remain TRAIL-resistant. However, biomarkers that reliably predict which patients may benefit from such combinatorial therapies are required. Thus, it is essential to better understand the mechanisms underlying TRAIL resistance versus sensitivity. In this chapter, we introduce the signalling events which take place during TRAIL-induced apoptosis, describe the physiological function of TRAIL and summarise pre-clinical and clinical results obtained so far with TRAIL-receptor agonists.

1 Introduction

Apoptosis is a form of programmed cell death necessary for tissue remodelling, homeostasis and the development of multi-cellular organisms (Danial and Korsmeyer 2004; Hanahan and Weinberg 2000; Steller 1995; Thompson 1995). Unlike necrosis, apoptosis does not induce an inflammatory response as it specifically eliminates abnormal, infected or aged cells without any damage to the surrounding tissue. Apoptosis can either be induced by intracellular stimuli such as DNA damage or by extracellular engagement of so called death receptors, which are all members of the

H. Walczak(✉), S.M. Cordier and K. Papenfuss
Tumour Immunology Unit, Division of Medicine, Imperial College, London, W12 0NN, UK
e-mail: h.walczak@imperial.ac.uk

[1]Equally Contributing first Authors

Results Probl Cell Differ, doi:10.1007/400_2008_27

tumour necrosis factor receptor superfamily (TNFRSF). So far, six death receptors are described with TNF-R1, CD95 (APO-1, Fas), TRAIL-Receptor 1 (TRAIL-R1, DR4) and TRAIL-R2 (APO-2, DR5, KILLER, TRICK2) being best characterised.

When it was found that TNF-R1 and CD95 stimulation can efficiently kill tumour cells, the hope for generating a new strategy to fight cancer arose. However, TNF-R1 and CD95 agonists not only affect cancer cells, but they also exert fatal unwanted effects when applied systemically, entailing severe side effects such as systemic inflammation and hepatoxicity (Creagan et al. 1988; Creaven et al. 1987; Daniel et al. 2001; Galle et al. 1995; Hersh et al. 1991; Ogasawara et al. 1993); so a broad clinical application of such drugs is rather unlikely. A revival in the field of death-receptor targeting anti-cancer therapy took place when the tumour necrosis factor-related apoptosis inducing-ligand (TRAIL) was identified (Pitti et al. 1996; Wiley et al. 1995). TRAIL selectively eliminates tumour cells without inducing any life-threatening toxicity in vivo (Ashkenazi et al. 1999; Walczak et al. 1999). For reasons that are not yet completely elucidated, tumour cells are more susceptible to the induction of apoptosis by TRAIL than are normal cells. Thus, TRAIL represents a promising tool for the treatment of cancer.

So far, five receptors have been described which are able to bind TRAIL (Fig. 1). Comprising an intracellular death domain (DD), TRAIL-R1 and TRAIL-R2 are capable of transmitting the apoptotic signal to the cell's inside (Pan et al. 1997a, b; Screaton et al. 1997; Sheridan et al. 1997; Walczak et al. 1997; Wu et al. 1997). Both receptors are characterised by the presence of two complete and one partial cysteine rich repeats (CRRs) in their extracellular parts, facilitating TRAIL binding. It is still not completely understood why two apoptosis-inducing TRAIL receptors are expressed in humans though only one receptor is sufficient to induce apoptosis in a variety of tumour cell lines following TRAIL application (Sprick et al. 2002). The functional difference between TRAIL-R1 and TRAIL-R2 remains to be elucidated.

Although TRAIL-R3 (Degli-Esposti et al. 1997b) and TRAIL-R4 (Degli-Esposti et al. 1997a) are highly homologous in their extracellular domains to their apoptosis-inducing counterparts, they are unable to induce apoptosis due to a complete or partial lack of the DD, respectively (Schneider et al. 1997). TRAIL-R3 and -R4 are referred to as decoy receptors. However, a decoy function has so far almost exclusively been demonstrated in overexpression systems; evidence for a decoy function in a more physiological setting is still very rare. In this respect, Merino et al. recently showed for the first time that these two receptors might use different mechanism to inhibit TRAIL-induced apoptosis (Merino et al. 2006). On the one hand, TRAIL-R3 titrates TRAIL within lipid rafts, thereby blocking TRAIL-induced cell death by competition. On the other hand, a TRAIL-dependent interaction of TRAIL-R4 with TRAIL-R2 might result in impaired formation of a death receptor-signalling complex, accompanied by reduced levels of caspase-8 activation, the main executor of apoptosis (Merino et al. 2006). However, as these studies were not performed under physiological expression levels, more studies are required to demonstrate that the role of TRAIL-R3 and TRAIL-R4 is 'regulatory' rather than being a 'decoy'. Accordingly, although all TRAIL-receptors are widely expressed within normal as well as malignant cell types, the expression

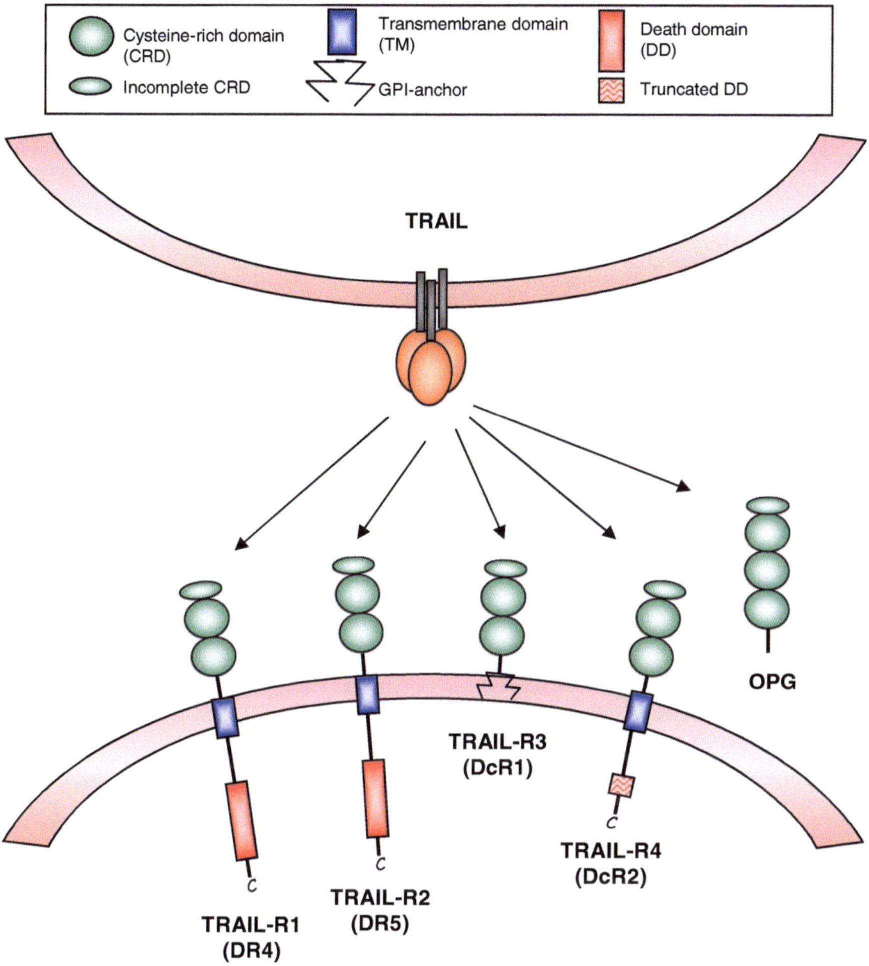

Fig. 1 The human TRAIL/TRAIL-R system. Soluble or membrane-embedded human TRAIL is able to bind to five receptors. Possessing an intracellular death domain, TRAIL-R1 and TRAIL-R2 transmit a signal which results in apoptosis. In contrast, TRAIL-R3 and TRAIL-R4 completely or partially lack a functional death domain, respectively, and are therefore unable to induce apoptosis. OPG is the fifth, rather low-affinity receptor for TRAIL; its function is better defined in osteoclastogenesis

of TRAIL-R3 and TRAIL-R4 does not correlate with the resistance of a given cell towards TRAIL-induced apoptosis. Thus, the mechanism responsible for TRAIL's tumour-selective apoptosis-inducing activity remains elusive.

Osteoprotegerin (OPG) is the fifth, rather low-affinity receptor for TRAIL (Truneh et al. 2000), primary function whose is linked to bone metabolism. Upon binding to RANKL, another member of the TNFSF, OPG competitively inhibits the RANKL–RANK interaction, thereby suppressing osteoclast formation. As TRAIL$^{-/-}$

and TRAIL-R$^{-/-}$ mice do not exhibit any defects in bone architecture, it is rather unlikely that the TRAIL–OPG interaction plays a major role in bone remodelling (Cretney et al. 2002; Diehl et al. 2004; Finnberg et al. 2005).

Surprisingly, not only the expression of TRAIL, but also that of its receptors is widely spread through human tissues, including the spleen, thymus, peripheral blood lymphocytes, prostate, testis, ovary, uterus and multiple tissues along the gastrointestinal tract, as has been shown on mRNA level (Walczak et al. 1997; Wiley et al. 1995). Thus, in contrast to the CD95 system, which is controlled by tight expression of CD95L, the control point for TRAIL-induced apoptosis does not seem to refer to the transcriptional level, but rather regulated by the expression of intracellular pro- and anti-apoptotic molecules involved in TRAIL-R signalling (see chapter TRAIL – Receptor Signalling).

Taken together, TRAIL-R1 and -R2 induce apoptosis in a variety of cells by using similar, but probably not identical pathways that allow for fine-tuned regulation of TRAIL-induced apoptosis. Although TRAIL stimulation correlates with cell death in most cases, a pro-survival function involving NF–κB, JNK and MAPK and a pro-migratory function for TRAIL signalling have been described for some cell types (reviewed in Falschlehner et al. 2007). The exact mechanisms and conditions required to transmit pro-survival signals upon TRAIL-R triggering are currently investigated.

2 TRAIL-Receptor Signalling

Apoptosis is a tightly controlled process regulated by a complex signalling machinery with a variety of check-points at several levels of signalling (Fig. 2). Binding of membrane-embedded or soluble TRAIL (generated by metalloprotease cleavage) to its two apoptosis-inducing receptors TRAIL-R1 and TRAIL-R2 induces receptor oligomerisation, thereby bringing the intracellular DDs of the receptors into close proximity. Protein crystallography experiments suggested that TRAIL binds as a trimer to pre-assembled receptor complexes that are connected via their pre-ligand binding assembly domain (PLAD), which by itself is not capable of transmitting a death signal (Chan et al. 2000). However, once TRAIL is bound to this pre-assembled receptor complex, juxtapositioning of the DDs creates a structure that allows for the recruitment of the Fas-associated death domain (FADD) protein, which binds via its DD to the DD of TRAIL-R1 and -2. Subsequently, pro-caspase-8 and -10 are recruited by the interaction of the death-effector domain (DED) of FADD with the DEDs of these caspases. This complex is known as death-inducing signalling complex (DISC).

Caspases are cysteinyl–aspartate specific proteases which possess a cysteine residue in their active centre that is crucial for cleaving target proteins after aspartic acid residues. Caspases are synthesised as pro-enzymes consisting of a large and a small catalytic subunit and an amino-terminal pro-domain. They can be sub-classified in initiator (i.e. caspase-2, -8, -9, -10) and effector caspases (i.e. caspase-3, -6, -7). While caspase-8 is essential in the induction of apoptosis, the role of caspase-10 in this process remains controversial. Whereas Kischkel et al. reported that caspase-10

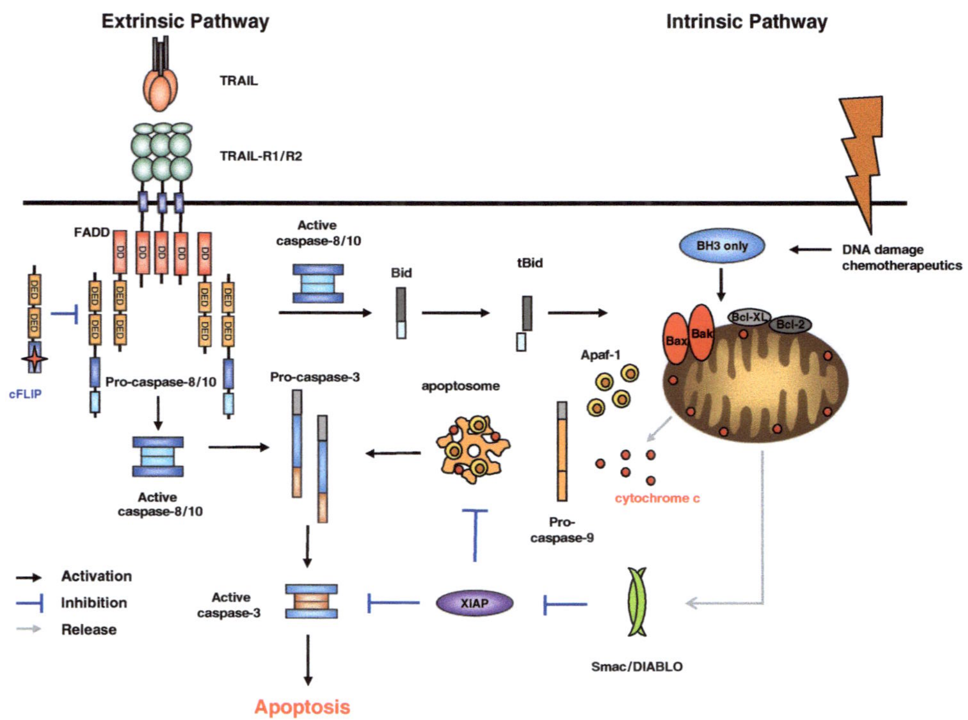

Fig. 2 The extrinsic and intrinsic apoptotic pathways. Upon TRAIL-binding to TRAIL-R1 or -R2, receptors trimerise to form the DISC by recruitment of FADD, pro-caspase-8 and -10. Autocatalytic cleavage of these initiator caspases leads to activation of effector caspases, including caspases 3, and 7, which are responsible for the induction of the typical morphological features associated with apoptosis. Active caspase-8 and -10 can also cleave Bid into truncated Bid (tBid), which subsequently translocates to the mitochondria to trigger cytochrome *c* release. Alternatively, DNA damage induces and/or activates other BH3-only proteins resulting in activation of the pro-apoptotic Bcl-2 family members Bax and Bak. Together with Apaf-1, dATP and pro-caspase-9, cytochrome *c* forms a complex in the cytosol referred to as apoptosome. Apoptosome-activate dcaspase-9 is also able to cleave caspase-3. At the DISC, apoptosis signalling can be inhibited by cFLIP, which structurally resembles caspase-8, but lacks a critical cysteine residue in the active centre (indicated with a star) that is required for enzymatic activity. Additionally, XIAP interferes with full maturation of caspase-3, thereby preventing the transduction of the apoptotic signal. The activity of XIAP is in turn held in check by Smac/DIABLO, which is released from the mitochondria following the action of BH3-only proteins.

can functionally substitute for caspase-8 (Kischkel et al. 2001), Sprick et al. demonstrated that although caspase-10 is recruited to the DISC by FADD, it is not required for apoptosis induction and is unable to functionally substitute for caspase-8 (Sprick et al. 2002). This suggests that caspase-10 may function to alter the apoptotic response by diversifying the apoptotic and/or non-apoptotic signal induced by TRAIL-R1, -R2 and CD95 engagement.

Assembly of the DISC creates a structure allowing for the recruitment of pro-caspases 8 and 10, which results in their activation by auto-catalytic processing. To promote the apoptotic process following pro-caspase-8 (and -10) activation, pro-caspase-3 is activated in a two-step mechanism. Initially, active caspase-8 separates the large from the small subunit. By an auto-catalytic maturation step, caspase-3 then removes its pro-domain, which renders the enzyme fully active. Once activated, caspase-3 cleaves a variety of cellular proteins, including PARP, lamins and cytokeratins. Furthermore, it inactivates ICAD/DFF45, the inhibitor of Caspase activated DNase (CAD/DFF40). Thus, CAD is no longer restrained by ICAD, but is rather able to enter the nucleus and to fragment DNA, thereby producing the 'DNA ladder' characteristic for apoptotic cells (Inohara et al. 1999; Liu et al. 1998).

In contrast to the extrinsic pathway, which requires signals from outside the cell, the intrinsic pathway is triggered autonomously by intracellular activation of caspases, for instance, following DNA damage. A key player in inducing the intrinsic apoptotic pathway is the mitochondrion. Upon genotoxic stress or DNA damage, the permeability of the outer mitochondrial membrane is increased, resulting in the release of cytochrome *c* and other pro-apoptotic molecules from the mitochondrial intermembrane space. Together with Apaf-1 and pro-caspase-9, cytochrome *c* forms a complex referred to as apoptosome (Baliga and Kumar 2003). Like caspase-8, apoptosome-activated caspase-9 is also able to activate pro-caspase-3. Once caspase-3 is activated, it will not only cleave its target proteins, but also new pro-caspase-9 molecules that in turn further activate pro-caspase-3. This positive feedback loop ensures apoptosis to be inevitably carried out.

Mitochondrial integrity is tightly regulated by a functionally diverse class of proteins that belong to the Bcl-2 family. Because of their domain structure, members of the Bcl-2 family can be subdivided into three groups (Borner 2003). Anti-apoptotic proteins like Bcl-2, Bcl-X_L and Mcl-1 are associated with the outer membrane of the mitochondria in order to maintain mitochondrial integrity, to prevent cytochrome *c* release and to promote cell survival. In contrast, members of the pro-apoptotic Bcl-2 family (i.e. Bax, Bak, Bok) destabilise mitochondrial integrity possibly by generating a transmembrane pore through the outer mitochondrial membrane during apoptosis, thereby inducing loss of the mitochondrial transmembrane potential and release of pro-apoptotic factors. Pro-apoptotic Bcl-2 family members are activated by molecules referred to as BH-3 only proteins (e.g. Bid, Bim, Bad, Bmf, Puma, Noxa). It remains controversial as to whether they only do so indirectly via neutralisation of Bcl-2-like anti-apoptotic proteins or also directly by activating the pro-apoptotic Bcl-2 proteins Bax and Bak (Certo et al. 2006; Strasser 2005; Youle and Strasser 2008).

BH3-only proteins are activated by a variety of intracellular stimuli including DNA damage and cellular stress. TRAIL-receptor cross-linking also activates a BH3-only protein. This protein is Bid. Bid is cleaved by caspase-8 (and -10) to truncated Bid (tBid). tBid then translocates to mitochondria to induce the release of pro-apoptotic factors by the not yet completely understood mechanism referred to above. Thus, Bid forms a bridge connecting the extracellular and intracellular pathways.

Depending on the need of the intrinsic apoptotic pathway to undergo death receptor induced apoptosis, cells can be classified as type I and type II cells, respectively. Type I cells are characterised by strong DISC formation and therefore strong caspase-8 activation, which is sufficient for subsequent caspase-3 activation. In contrast, DISC formation in type II cells is rather weak. Thus, these cells additionally require the mitochondrial amplification loop to efficiently activate effector caspases to undergo apoptosis (Barnhart et al. 2003; Ozoren and El-Deiry 2002).

The apoptotic signal is regulated at several stages. Due to the presence of two DEDs, the cellular FLICE-like inhibitory protein (cFLIP) competes with caspase-8 for binding to FADD (Krueger et al. 2001). Displacement of caspase-8 from the DISC prevents the initiation of the caspase cascade responsible for apoptosis transmission. The amount of cFLIP within a cell inversely correlates with the amount of caspases that are activated at the DISC and therefore with the decision whether apoptosis is induced. Three splice variants of cFLIP are reported, i.e. $cFLIP_L$, $cFLIP_s$ and $cFLIP_R$ (Golks et al. 2005). Comprising two DEDs and an additional C-terminal caspase domain, $cFLIP_L$ closely resembles caspase-8 in its overall structural organisation. However, due to a lack of the critical cysteine residue within the active centre, $cFLIP_L$ does not possess proteolytic activity. Whereas $cFLIP_S$ and $cFLIP_R$ can inhibit cleavage of pro-caspase-8 at the DISC altogether, $cFLIP_L$ rather interferes with the full maturation of DISC-recruited pro-caspase-8. However interestingly, a heterodimer of caspase-8 and $cFLIP_L$ has been shown to possess stronger caspase-8 activity than a monodimer of solely caspase-8 in a cell free system (Micheau et al. 2002). Also possessing a pro-apoptotic activity in this context, the role of $cFLIP_L$ might be more complex than initially assumed.

Other proteins inhibiting the apoptotic process are the cellular inhibitor of apoptosis proteins (IAPs) (Salvesen and Duckett 2002). Among these, XIAP is the most prominent member that is known to prevent the activation of caspase-3 and -9 by direct interaction (Riedl et al. 2001). XIAP blocks the removal of the inhibitory pro-domain of caspase-3, thereby inhibiting the before-mentioned maturation step. However, XIAP has also been shown to catalyse the ubiquitination of caspase-3, leading to its proteasomal degradation (Vaux and Silke 2005). Consequently, the activation of the initiator caspases such as caspase-8 and -10 are not necessarily associated with apoptosis. Type II cells are often hallmarked by high expression levels of XIAP, providing an additional if not alternative explanation for the need of the mitochondrial pathway in type II cells.

The activity of IAPs themselves is in turn controlled by another set of proteins that antagonises their function. Once released from the mitochondrial intermembrane space, the pro-apoptotic Smac/DIABLO protein interacts with XIAP, which removes it from caspase-3 and -9. Caspase-3 is then auto-catalytically activated, thereby allowing apoptosis to proceed.

Taken together, TRAIL-induced apoptosis is a complex, highly regulated process that is influenced by a variety of pro- as well as anti-apoptotic proteins. While caspases are the main executors of apoptosis, intracellular factors like anti-apoptotic Bcl-2 family members, cFLIP and IAPs are able to reduce the sensitivity of a given cell towards apoptosis. It is therefore not surprising that many tumour cells overexpress these inhibitory molecules or downregulate pro-apoptotic Bcl-2 proteins. In many cancers, the balance of anti- and pro-apoptotic effectors is shifted in favour of the former, indicating that cells continue to replicate in spite of being damaged.

Although the molecules so far detected in the TRAIL–R DISC are the same ones as those of the CD95 DISC, the biological outcome of activation by these two receptor–ligand systems is extremely diverse. While systemic CD95 stimulation also kills normal tissue including hepatocytes, TRAIL can specifically eliminate malignantly transformed cells without damaging healthy normal tissue. Therefore, it is highly likely that the composition of the receptor-associated signalling complex in these two systems will in fact be different. New studies are required which compare both receptor complexes following stimulation in order to detect novel factors that are only present in one of the two systems and might therefore explain the difference in functional outcome.

3 Physiological Role of the TRAIL/TRAIL-Receptor System

The physiological role of TRAIL has so far mainly been studied in the immune system. Although TRAIL is not expressed on the surface of freshly isolated T cells, NK cells, NKT cells, B cells, dendritic cells (DCs) and monocytes, high levels of TRAIL can be induced by stimulation. For instance, IL-2, interferons (IFNs) and IL-15 can induce the expression of TRAIL on NK cells (Johnsen et al. 1999; Kashii et al. 1999; Kayagaki et al. 1999). Following infection and subsequent DC activation, type I and type II IFNs are secreted facilitating the elimination of infected cells via TRAIL by two means. On the one hand, they induce the expression of TRAIL on the surface of NK cells, monocytes and DCs, thereby allowing for direct killing of target cells. At the same time, IFNs bind to their respective receptors on infected cells, which sensitises these cells to TRAIL-induced apoptosis. Due to the autocrine production of IFN-γ, a subset of freshly isolated liver NK cells constitutively expresses TRAIL (Takeda et al. 2001). Besides CD95L and perforin, TRAIL contributes to most of the cytotoxic activity of NK cells against virus-infected target cells or cancer cells. Accordingly, Smyth et al. demonstrated that NK cell-derived TRAIL induced by IFN-γ treatment prevents formation of primary tumours and experimental metastases (Smyth et al. 2001).

The physiological role of TRAIL is not constrained to innate immune cells. Recent studies suggest a role for TRAIL in T cell homeostasis. So called ‘helpless’ $CD8^+$ T cells that were primed in the absence of ‘help’ $CD4^+$ T cells are able to mediate effector functions upon first antigen encounter. However, following re-stimulation, they do not undergo a second round of clonal expansion but, in contrast to ‘helped’ $CD8^+$ T cells, they express TRAIL (Janssen et al. 2005). TRAIL-deficient ‘helpless’ T cells are able to expand a second time upon re-stimulation, indicating that TRAIL expressed by wild-type 'helpless' T cells is responsible for their lack in expansion, possibly due to activation-induced cell death of these 'helpless' T cells. The repression of TRAIL expression imposed by $CD4^+$ T cell help might therefore contribute to the generation of memory $CD8^+$ T cells and represent a novel mechanism for controlling adaptive immune responses.

In addition, TRAIL has recently been shown to contribute to the regulation of infectious immunological tolerance following presentation of apoptotic cells (Griffith et al. 2007). This phenomenon, mediated by $CD8^+$ regulatory T cells, had already been described over 30 years ago as being executed by T suppressor cells (Gershon and Kondo 1971; Greene and Benacerraf 1980; Sy et al. 1977). Griffith et al. could now show that apoptotic cells induce the expression of TRAIL on $CD8^+$ T cells. These thereby generated $CD8^+$ regulatory T cells transfer unresponsiveness to naïve nontolerant recipient mice. In contrast, $CD8^+$ regulatory T cells of TRAIL-deficient animals were unable to transfer tolerance, underlining that TRAIL acts as important effector molecule in this tolerance pathway. Opposed to the 'helpless' T cells described above, these regulatory $CD8^+$ T cells were not themselves subjected to activation-induced cell death, but rather eliminated (self-reactive) T cells specific for apoptotic antigens, thereby conferring tolerance. Using appropriate mouse models, it will be interesting to determine under what conditions $CD8^+$ T cells possessing a 'helpless' phenotype induce immunological tolerance or are themselves committed to activation-induced cell death.

To further analyse the physiological role of the TRAIL/TRAIL-R system, $TRAIL^{-/-}$ and $TRAIL\text{-}R^{-/-}$ mice were generated (N.B. mice only express one apoptosis-inducing receptor that is equally homologous to human TRAIL-R1 and TRAIL-R2). Being viable, fertile and not displaying any developmental defects, absence of a functional TRAIL/TRAIL-R system does not result in an overt phenotype (Cretney et al. 2002; Sedger et al. 2002). However, Syngeneic tumour models indicate that the TRAIL/TRAIL-R system is involved in tumourigenesis. Accordingly, growth and incidence of tumours are increased in the absence of TRAIL following transplantation of a variety of TRAIL sensitive tumour lines. Autochthonous tumour models point to an ambiguous role of the TRAIL/TRAIL-R system in tumourigenesis. Depending on the experimental setting and mouse model applied, TRAIL or TRAIL-R deficiency did or did not affect tumour development. Whereas the incidence of spontaneous tumour development at early stages was independent of TRAIL and TRAIL-R (Cretney et al. 2002; Diehl et al. 2004; Grosse-Wilde et al. 2008; Sedger et al. 2002), aged mice displayed more lymphomas in the absence of TRAIL (Zerafa et al. 2005). Controversial results were reported for tumour formation in absence of p53 in an inhibited TRAIL/TRAIL-R system. While the frequency of sarcomas and lymphomas in $TRAIL^{-/-}$ $p53^{+/-}$ mice was increased in one study (Zerafa et al. 2005), others were unable to detect a change in tumours in $TRAIL\text{-}R^{-/-}$ mice on a $p53^{-/-}$ background (Yue et al. 2005).

Intriguingly, we recently determined that the Trail-r gene is a metastasis suppressor gene in autochthonous skin tumourigenesis (Grosse-Wilde et al. 2008). While the development and frequency of primary tumours was not affected by the presence of TRAIL-R in an autochthonous DMBA/TPA-induced skin carcinoma model, significantly fewer metastases developed in wild-type mice compared to TRAIL-R$^{-/-}$ mice. Detachment of tumour cells from the primary tumour is an obligatory step for metastasis formation. Cell detachment sensitises TRAIL-resistant tumour cells to TRAIL-induced apoptosis by inactivation of the ERK signalling pathway in vitro. Thus, TRAIL-R expressing tumour cells might become susceptible to TRAIL-mediated cell death during the process of metastasis formation. Further studies are required to determine which cells are responsible for the killing of detached tumour cells in vivo.

Taken together, the TRAIL/TRAIL-R system does not seem to play a general role early in tumourigenesis as it fails to suppress tumour growth, especially in autochthonous tumour models. However, TRAIL-expressing immune cells, most probably NK cells and lymphocytes, are able to kill tumour cells when detached. Being an essential step in metastasis formation, detachment of tumour cells from the primary tumour origin sensitises these cells to TRAIL-induced apoptosis. Thus, patients with tumours that express TRAIL-R1 and/or TRAIL-R2 might benefit from a treatment with TRAIL-R agonists to inhibit metastasis formation following the removal of the primary tumour.

4 TRAIL as a Therapeutic Agent

In contrast to systemic treatment with CD95L or TNF, TRAIL selectively induces apoptosis in about 50% of tumour cell lines while leaving normal cells unharmed (Ashkenazi et al. 1999; Walczak et al. 1999). Thus, TRAIL application might display a promising strategy in the fight against cancer.

Unfortunately, recent studies revealed that most primary tumour cells are TRAIL-resistant in the first place (Todaro et al. 2008). However, many of these primary cancer cells can be sensitised to TRAIL-induced apoptosis by chemotherapeutic agents or irradiation. Currently, several companies pursue TRAIL-R-targeted therapies in clinical trials using TRAIL-R agonists alone or in combination with other cancer therapeutics. This chapter will introduce a variety of TRAIL-R agonists developed so far, discuss new bio-engineering approaches invented to improve the targeting of TRAIL-R agonists to the tumour site and will summarise the available data about their effects on primary tumours in vitro and in clinical trials.

4.1 TRAIL-R Agonists and Their Toxicities

In order to trigger the TRAIL-mediated apoptotic pathway, soluble recombinant versions of TRAIL as well as agonistic antibodies targeting TRAIL-R1 or TRAIL-R2 can be applied. Ideally, these agonists should display high anti-tumour activity and at the same time low toxicity to normal cells, thereby ensuring safe and efficient application as anti-cancer drugs in the clinics.

4.1.1 Agonistic TRAIL-R-Specific Monoclonal Antibodies

It is still a matter of debate whether TRAIL-R3 and TRAIL-R4 truly act as decoy receptors and whether their overexpression protects cancer cells from TRAIL-induced apoptosis (reviewed in Buchsbaum et al. 2006). However, also based on the rationale of overcoming a potential safeguarding effect of TRAIL-R3 and TRAIL-R4, agonistic monoclonal antibodies specifically targeting apoptosis-inducing TRAIL-R1 or TRAIL-R2 have been developed. Additionally, the in vivo half-life of these monoclonal antibodies is more than 100 times that of recombinant TRAIL (14–21 days as compared to 30–60 min). These properties which at first sight seem advantageous may, however, also be the source of potential toxicity to normal cells. Yet, the TRAIL-R2-specific antibody TRA-8 for instance has been reported to kill leukemia cells, astrocytoma and engrafted breast cells while sparing normal human astrocytes, B and T cells as wells as primary human hepatocytes (Buchsbaum et al. 2003; Ichikawa et al. 2001).

A third difference is that monoclonal antibodies are bifunctional molecules, i.e. apart from their two antigen-combining moieties, they also contain an Fc-domain, which is capable of inducing a number of effector functions. Due to the formation of higher order complexes and the recruitment and activation of innate immune cells, additional cross-linking of TRAIL-R-specific antibodies by Fc-receptor-expressing immune cells can lead to a higher efficiency in the anti-tumour response (Takeda et al. 2004). A combination of a TRAIL-R antibody with CD40- and 4-1BB-targeting antibodies was even able to completely eradicate syngeneic tumours without any observed toxicity in mice (Uno et al. 2006). In this model, anti-TRAIL-R antibodies did not only kill TRAIL-sensitive tumour cells, but also recruited Fc-receptor expressing cells such as DCs and macrophages via the constant region of the antibody. These antigen-presenting cells (APCs) subsequently engulf the apoptotic tumour cells, process tumour antigens and present them to surrounding T cells. Concomitant stimulation with anti-CD40 and anti-4-1BB antibodies induces further APC and T cell activation in order to efficiently stimulate surrounding cytotoxic T cells (CTLs). Being properly activated, CTLs are then able to kill the TRAIL-resistant tumour burden expressing tumour-associated antigens. It will be interesting to determine the molecular signatures of such ‘properly’ activated CTLs.

Yet again, besides leading to increased anti-tumour responses, cross-linking of TRAIL-R-specific antibodies may also result in higher toxicity for normal cells, including primary human hepatocytes (Mori et al. 2004) and cholangiocytes (Takeda et al. 2008).

4.1.2 Recombinant TRAIL

In contrast to TRAIL-R1- or TRAIL-R2-specific antibodies, recombinant forms of TRAIL allow for the activation of TRAIL-R1 and TRAIL-R2 at the same time. This might be a promising strategy as the expression profile of TRAIL-Rs on tumours seems to vary substantially. So far, a variety of soluble TRAIL versions has been generated; many of these versions encode the extracellular domain of human TRAIL amino-terminally fused to an oligomerisation motif, e.g. a poly-histidine tag (His-TRAIL) (Pitti et al. 1996), a FLAG-epitope (Schneider 2000), a leucine zipper (Walczak et al. 1999) or an isoleucine zipper motif (Ganten et al. 2006). These additional tags improve receptor oligomerisation which is necessary to successfully transmit the death signal. Yet again, as has been discussed for TRAIL-R- specific antibodies, the ability of at least some of these recombinant, tagged versions of TRAIL to form higher-order complexes might coincide with increased toxicity to normal cells (Koschny et al. 2007c; Lawrence et al. 2001).

It seems that two main factors determine TRAIL sensitivity of normal human cells, i.e. the form of the recombinant TRAIL used and the model system chosen. Highly oligomerised forms of TRAIL, e.g. cross-linked FLAG–TRAIL were reported to induce killing of primary human hepatocytes, keratinocytes and astrocytes in some model systems (reviewed in Koschny et al. 2007c). However, it is still a matter of debate which of the model systems most reliably resembles the physiological conditions. studies by Ganten et al. in primary human hepatocytes shed new light on this matter (Ganten et al. 2005). Here, freshly isolated primary human hepatocytes at day one of in vitro culture were efficiently killed by highly aggregated forms of TRAIL. However, on day four of in vitro culture, on which the phenotype of primary human hepatocytes resembles normal liver tissue, the primary human hepatocytes turned out to be TRAIL-resistant. These results correspond to the ones obtained in an elegant in vivo study by Hao et al. in which orthotopically xenotransplanted human liver cells in mice were not killed by systemic treatment with non-tagged TRAIL (Hao et al. 2004). Furthermore, application of TRAIL alone or in combination with chemotherapeutics in vivo, as has been shown in mice, cynomolgous monkeys and chimpanzees, did not lead to any signs of toxicity (Ashkenazi et al. 1999). However, one has to bear in mind that toxicity could potentially occur under certain sensitising conditions like viral or non-viral hepatitis or in a pro-inflammatory milieu (Liang et al. 2007; Mundt et al. 2005).

Non-tagged TRAIL seemed to exhibit the lowest potential for toxicity to normal cells in vitro when compared to highly oligomerised forms of TRAIL, e.g. His- or FLAG–TRAIL, but nevertheless showed substantial killing activity, especially when combined with chemotherapeutics. Thus, a non-tagged form of human soluble TRAIL called Apo2L/TRAIL has been chosen as the first recombinant TRAIL drug candidate to enter clinical development (see below). Studies comparing recombinant versions of TRAIL to TRAIL-R-specific antibodies are still missing today. However, for the CD95 and TNF receptors–ligand systems it is known that the killing potential of the recombinant cognate ligand is superior to the respective antibody (Schlosser et al. 2000). The same phenomenon is likely to apply to the TRAIL system. Thus, despite having a much lower half-life than TRAIL-R-specific antibodies, recombinant forms of TRAIL may turn out to be powerful novel cancer therapeutics. Accordingly, Apo2L/TRAIL was reported to possess high anti-tumour activity in vivo, which may, at least in part, be attributable to its excellent tumour penetration (Kelley and Ashkenazi 2004; Koschny et al. 2007a).

Taking all this into consideration, the data obtained so far suggest TRAIL-R agonists, including TRAIL-R1 and TRAIL-R2-specific antibodies as well as soluble recombinant forms of TRAIL to be promising novel tools for future treatment of cancers. Hence, a novel class of cancer drugs is emerging.

4.2 Targeted TRAIL Delivery

Large amounts of recombinant TRAIL are required to inhibit tumour growth in vivo as most protein is already cleared within 5 h following injection (Walczak et al. 1999). In line with this, the anti-tumour activity of TRAIL is most pronounced when TRAIL is administered shortly after tumour implantation (Ashkenazi et al. 1999; Chinnaiyan et al. 2000; Gliniak and Le 1999; Walczak et al. 1999). However, as this is an unrealistic scenario with respect to cancer therapy in humans, new strategies need to be devised. For this purpose, a recombinant, replication-deficient adenoviral vector encoding human TRAIL (Ad5-TRAIL) was engineered (Griffith and Broghammer 2001). The use of Ad5-TRAIL possesses many advantages over a recombinant TRAIL-based therapy: First of all, it kills TRAIL-sensitive tumour cells as efficiently as a soluble ligand, but does not require consecutive applications. Moreover, tumour growth is still inhibited when Ad5-TRAIL is not applied directly following tumour implantation, but rather 10 days later. In contrast to the short half-life of recombinant TRAIL, the expression of TRAIL following Ad5-TRAIL intratumoural injection is maintained over a long period, with sustained, regional high TRAIL expression levels. Even though the expression of TRAIL is not restricted to tumour cells, at least in prostate cancer, Ad5-TRAIL treatment is nevertheless safe and specifically kills tumour cells while sparing normal tissue. Infection of normal epithelial prostate cells is even advantageous as TRAIL expressed by these resistant cells also contributes to the killing of adjacent tumour tissue. Interim results from a phase I clinical study applying Ad5-TRAIL gene therapy so far looked promising as the treatment was well tolerated in three patients suffering from organ-confined prostate cancer.

As Ad5-TRAIL is induced only locally at the tumour site, it is unlikely that it travels to other locations within the body. Therefore, Ad5-TRAIL is less efficient in the elimination of metastasis. For this purpose, it might be beneficial to combine Ad5-therapy with systemic soluble TRAIL treatment.

The main problem with regard to adenoviral therapy is that many tumour cells (especially in the prostate) exhibit reduced expression of the Coxsackie-adenovirus receptor (CAR), which is responsible for recognition and subsequent internalisation of adenoviruses (Okegawa et al. 2000; Rauen et al. 2002). If only a few vectors were taken up, the expression of TRAIL at the site of the tumour would be reduced and consequently the efficiency of the therapy decreased. To overcome this limitation, Ad5-TRAIL therapy can be combined with the administration of histone deacetylase inhibitors (HDACi), which were reported to increase CAR expression (Hemminki et al. 2003; Sachs et al. 2004). As HDACi are also reported to sensitise tumour cells to TRAIL-induced apoptosis (Inoue et al. 2004; Lindemann et al. 2007), such a combinatorial treatment presents a feasible alternative for the killing of tumour cells (VanOosten et al. 2007). However, more studies are required to rule out possible side effects associated with adenoviral therapy.

The widespread expression of TRAIL-Rs on normal tissues limits the success of TRAIL-based cancer therapies as soluble TRAIL (sTRAIL) does not specifically accumulate at the tumour site. Additionally, untagged soluble TRAIL is characterised by a fast-on/fast-off binding rate. To overcome these problems, soluble TRAIL can be genetically linked to a single chain variable antibody fragment (scFv), which retains fast-on/slow-off rates (Bremer et al. 2004). The application of scFv:sTRAIL fusion proteins that bind via the scFv portion to a pre-selected tumour-specific target antigen induce accumulation of cross-linked TRAIL at the tumour site.

A variety of TRAIL fusion proteins were generated lately, including scFvEGP:sTRAIL (Bremer et al. 2004), in which TRAIL is genetically linked to the pan-carcinoma associated cell surface antigen EGP (EpCAM) that is highly expressed on a variety of human tumours including breast, colorectal carcinoma (CRC) and small cell lung cancer (SCLC) (Balzar et al. 1999; Moldenhauer et al. 1987).

Other tumour entities that can be targeted by scFv:sTRAIL fusion proteins are T cell leukemias and lymphomas. In these malignancies, current therapeutic protocols often result in severe side effects and limited response. To address this medical need, a scFvCD7:sTRAIL fusion protein was generated, in which soluble TRAIL was genetically linked to scFv specific for the T cell surface antigen CD7. Although CD7 is also expressed on resting leukocytes, activated T cells and vascular endothelial cells, scFvCD7:sTRAIL specifically induces apoptosis in malignantly transformed T cell lines and cells freshly isolated from T cell acute lymphoblastic leukaemia patients (Bremer et al. 2005a). Similar results were obtained for B cell malignancies upon application of scFvCD19:sTRAIL (Stieglmaier et al. 2008). CD19 is a glycoprotein of the immunoglobulin superfamily expressed during all stages of B cell development (Stamenkovic and Seed 1988) and on the majority of chronic B lymphocytic leukaemia, non-Hodgin lymphoma and acute B lymphoblastic leukaemia cells (Uckun et al. 1988).

The scFv425:sTRAIL fusion protein is another example of tumour-restricted TRAIL accumulation. Due to binding of the EGFR-blocking antibody fragment scFv425 to EGF receptors (EGFR), which are known to be overexpressed or dysregulated in a variety of human tumours, large amounts of TRAIL accumulate at the tumour site (Bremer et al. 2005b). In addition to tumour-restricted TRAIL aggregation, the EGFR-blocking antibody fragment also causes dephosphorylation of EGFR, thereby inhibiting downstream mitogenic signalling. Inhibition of the PI3K and MAPK pathways is accompanied by downregulation of $cFLIP_L$ and dephosphorylation of Bad, consequently sensitising tumour cells to TRAIL-induced apoptosis at the DISC and mitochondrial levels, respectively.

While all the described fusion proteins are marked by potent anti-tumour efficiency, no severe side effects have been reported so far, making fusion proteins a safe alternative to conventional cytotoxic therapy. As reported for the anti-CD20 antibody rituximab, some tumours might lose target antigen expression during tumour progression, thereby becoming resistant to antibody therapy (Kennedy et al. 2002). A surplus of the sTRAIL domain on the target cell surface does not only induce apoptosis of the cell expressing the target antigen, but is also capable of killing neighbouring tumour cells that have lost the expression of the target antigen. This bystander anti-tumour effect makes TRAIL fusion proteins potentially valuable tools in the fight against cancer (Fig. 3). However, the success of this kind of therapeutics strongly depends on the expression of the target antigen by at least a considerable fraction of the tumour cells.

4.3 Efficiency of TRAIL in Primary Tumours

A variety of studies which investigated the effect of TRAIL on tumour cell lines so far have yielded very promising results. In contrast to this, the effect in primary tumour cells seems to be more diverse. Pre-clinical studies applying TRAIL to freshly isolated human myeloma cells showed that TRAIL can efficiently induce apoptosis in these otherwise chemotherapy-resistant cells (Gazitt 1999; Mitsiades et al. 2001). However, TRAIL could not do so in acute lymphoblastic leukaemia, acute myelogenous leukaemia, acute promyelocytic leukaemia and in primary B cell acute or chronic lymphocytic leukaemia (Clodi et al. 2000; MacFarlane et al. 2002). The factors that may determine TRAIL resistance of primary tumour cells have only been revealed or suggested for a few types of cancer. For example, Riccioni et al. (Riccioni et al. 2005) reported a correlation between TRAIL resistance and the expression of decoy receptors in myeloid leukaemia. Furthermore, it could be shown that TRAIL resistance in primary glioblastoma is dependent on the expression of the tumour suppressor PTEN (phosphatase and tensin homologue deleted on chromosome TEN) and cFLIP (Panner et al. 2005). Expression of wildtype PTEN and low levels of cFLIP rendered the cells TRAIL-sensitive, whereas the expression of mutated PTEN together with high levels cFLIP confers TRAIL resistance. However, the expression levels of cFLIP seem to be irrelevant for (oligo-) astrocytoma specimens of all four WHO grades of malignancy (Koschny et al. 2007a) as well as in isolated tumour cells form medullablastoma, meningeoma, esthesioneuroblastoma and soft tissue sarcoma, all of which are TRAIL-resistant (Clayer et al. 2001). Intriguingly, for pancreatic cancer and cholangiocarcinoma cells TRAIL treatment has been observed to enhance migration and metastatic spread in vitro and in vivo (Ishimura et al. 2006; Trauzold et al. 2006).

Taken together, as most primary tumour cells – unlike cancer cell lines – are TRAIL-resistant and TRAIL treatment was even counterproductive in some cases, a broad application of TRAIL as a single agent is unlikely and attempts to do so should be questioned. It is of major importance to carefully characterise the tumour specimen with regard to its TRAIL sensitivity prior to treatment in order to be able to administer a tailored therapy specific to the patient's sensitivity profile. For this purpose, it is necessary to develop biomarkers and appropriate sensitivity tests (McCarthy et al. 2005). As the expression of O-glycosylating enzymes seems to

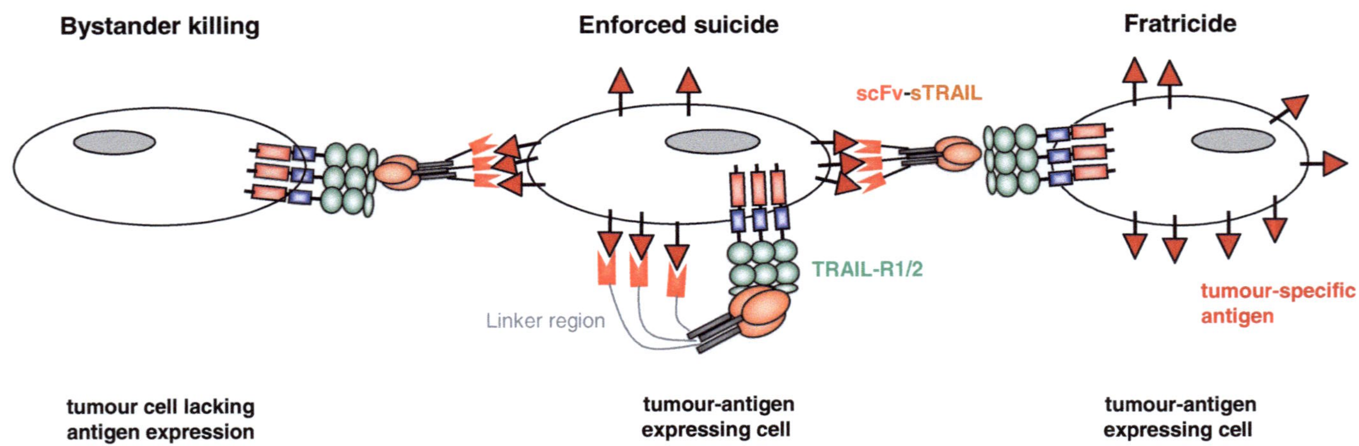

Fig. 3 Model of death induction by scFv–sTRAIL fusion proteins. The fusion protein binds via the scFv fragment to its target antigen (red) which is specifically expressed on tumour cells. These tumour cells are subsequently killed by sTRAIL-binding to TRAIL-R1 and/or TRAIL-R2, respectively. Adjacent TRAIL-R-positive tumour cells having lost tumour antigen expression are also killed by scFv–sTRAIL (bystander effect)

correlate with TRAIL sensitivity, these enzymes might be valuable markers to predict the prospect of success of a TRAIL-based therapy (Wagner et al. 2007). The expression of the O-glycosylating enzyme GALNT3 for instance correlates with TRAIL sensitivity in CRC and the expression of GALNT14 with TRAIL sensitivity in non-small cell lung cancer (NSCLC), pancreatic cancer and melanoma cell lines. Thus, specific O-glycosylating enzymes could potentially be used as predictive biomarkers for responsiveness to TRAIL-based cancer therapy. More of such biomarkers are needed for different tumour entities.

4.4 Sensitisation to TRAIL-Induced Apoptosis

As discussed above, TRAIL as a single agent is not able to induce apoptosis in most primary tumour cells. Fortunately, encouraging results have been obtained showing that the additional use of other cancer drugs sensitises tumour cells to the effects of TRAIL. An overview of the cytotoxic agents, the potential mechanism and the model system used is given in Table 1. Table 2 summarises related

Table 1 Combinatorial treatment

Primary Tumour	TRAIL in combination with	Proposed mechanism	Reference
ALL	Vincristine (microtubule inhibitor)		(Bremer et al. 2005a)
AML	HDAC inhibitor		(Nebbioso et al. 2005)
B-CLL	Cyclohexamide	Down-regulation of cFlipL	(Olsson et al. 2001)
CLL	HDAC inhibitor	Signal via TRAIL-R1	(Inoue et al. 2004; MacFarlane et al. 2005)
Colon cancer	Irinotecan, 5-FU	Up-regulation of TRAIL-R2	(Naka et al. 2002)
Erytholeukemic cells	Irradiation	Up-regulation of TRAIL-R1	(Di Pietro et al. 2001)
Multiple myeloma	NFκB inhibitor SN50	Down-regulation of Bcl-2, Bfl-1, IAPs, up-regulation of Bax	(Mitsiades et al., 2002)
(oligo-) astrocytoma	Bortezomib	Up-regulation of TRAIL-R1/R2, Bax/Bak Down-regulation of cFlipL	(Koschny et al. 2007b)
Pancreatic cancer	Gemcitabine, Doxorubicine, Cisplatin, Etoposide, Methotrexat		(Hylander et al. 2005)
Soft tissue sarcoma	Cyclophosphamide		(Clayer et al. 2001)

Table 2 Potential toxicities of TRAIL-comprising combinatorial treatments on several normal tissues as an extrapolation of effects on normal cells cultured in vitro

Cultured normal cell	TRAIL in combination with	Toxicity	Reference
Erythoblasts	Irradiation	No	(Di Pietro et al. 2001)
Hepatocytes	5-FU, Gemcitabine, Irinotecan, Oxaliplatin, Bortezomib (low dose)	No	(Ganten et al. 2005; Ganten et al. 2006)
Hepatocytes	Cisplatin (high dose: 240 μM), Bortezomib (high dose: 3 μM)	Yes	(Ganten et al. 2006; Koschny et al. 2007a)
Hepatocytes	HDAC inhibitor	No	(Pathil et al. 2006)
Keratinocytes	MG115 (Proteasome inhibitor)	Yes	(Leverkus et al. 2003)
Myeloid Progenitors	HDAC inhibitor	No	(Nebbioso et al. 2005)
Osteoblasts	Etoposide, Cisplatin; Doxorubicin, Methotrexate, Cyclophosphamide	No	(Atkins et al. 2002; Evdokiou et al. 2002)
Osteoblasts	Etoposide	No	(Van Valen et al. 2003)
Osteoblasts	Cisplatin, Doxorubicine	Yes	(Van Valen et al. 2003)
PBMC	HDAC inhibitor	No	(Inoue et al. 2004; Nakata et al. 2004)
Prostate stromal cells	Doxorubicin	No	(Wu et al. 2002)

toxicities of combinatorial treatment in normal cells. The synergistic effect of cytotoxic agents and TRAIL is believed to be mainly due to changes on the transcriptional levels of proteins important for the TRAIL pathway (Cretney et al. 2006; Wajant et al. 2002). Many studies suggest that changes on the receptor level, e.g. upregulation of TRAIL-R1 and TRAIL-R2, are already sufficient for the observed sensitising effect. However, although upregulation may correlate with the sensitising effect observed, in most cases it does not seem to be the cause of sensitisation. In our own studies, we could show that combinatorial treatment with 5-FU or bortezomib led to an upregulation of TRAIL-R1 and TRAIL-R2 (Koschny et al. 2007a). However, this change on the receptor level was not responsible for sensitisation. Instead, sensitising agents improve the cells' capability to form a TRAIL DISC, thereby shifting the threshold for apoptosis induction by TRAIL in tumour cells (Ganten et al. 2005). In addition, many drugs have been shown to downregulate anti-apoptotic molecules like IAPs, cFLIP, Bcl-2, Bcl-X_L and Mcl-1 and to upregulate pro-apoptotic factors including death receptors, caspase-8, FADD, Bak or Bax (Held and Schulze-Osthoff 2001; Kelley and Ashkenazi 2004; Mitsiades et al. 2002).

For a safe use of a combinatorial therapy, it is important that tumour cells are preferentially sensitised to TRAIL-induced apoptosis while normal cells remain resistant. So far, only very high doses of the frequently applied chemotherapeutic agent cisplatin or the proteasome inhibitor bortezomib were shown to induce toxicity in primary human hepatocytes at day 4 of in vitro culture when combined with TRAIL (Ganten et al. 2005). However, the concentration of bortezomib was about 40 times higher than actually needed for TRAIL sensitisation of tumour cells. Thus, combining TRAIL with bortezomib may open a therapeutic window for treatment of tumour patients without severe toxicity. It is noteworthy that, data obtained with different proteasome inhibitors show that each combination has to be assessed carefully, even though the sensitisers belong to the same class of cytotoxic agents. In this respect, normal primary human keratinocytes were sensitised to TRAIL even with low concentrations of the proteasome inhibitor MG-115 (Sudarshan et al. 2005). Most probably, MG-115 activates the mitochondrial pathway in these cells. Thereby, XIAP, which is expressed at high levels in primary keratinocytes, is inhibited by Smac/DIABLO, allowing for caspases-3 activation and apoptosis to occur (Leverkus et al. 2003). Interestingly, a study employing a combination of a TRAIL-R-specific antibody and bortezomib in mice in vivo did not lead to alterations in the skin and did not report toxicity (Shanker et al. 2008).

As mentioned earlier, HDAC is constitute another class of TRAIL-sensitising agents. Although some studies hint towards an HDACi-mediated activation of caspase-2 and subsequent caspase-8 activation (VanOosten et al. 2007), the exact mechanism of HDACi-dependent sensitisation to TRAIL-induced apoptosis is not yet completely clear. However, it efficiently induces apoptosis in hepatoma cell lines (Lindemann et al. 2007; Schuchmann et al. 2006), B cell lymphomas (Lindemann et al. 2007), primary AML and CCL cells (Inoue et al. 2004; MacFarlane et al. 2005; Nebbioso et al. 2005), while primary human hepatocytes, normal peripheral mononuclear blood cells and myeloid progenitors remain unharmed.

Taken together, these pre-clinical data point towards a great potential of combinatorial treatments that comprise TRAIL-R agonists in the therapy of cancer. However, with many of such drugs now in clinical trials and with the described diversity of outcomes of TRAIL-R stimulation in cancerous and non-cancerous cells, there is a clear need for a better understanding of the principles of TRAIL apoptosis sensitivity versus resistance. Thereby, it should be possible to further refine and tailor TRAIL-based approaches in cancer therapy.

4.5 Clinical Development of Apoptosis-Inducing TRAIL Receptor Agonists (TRAs)

Based on the promising pre-clinical findings of TRAIL-R-targeting approaches, several companies have begun to develop TRAIL receptor agonists (TRAs), which define a sub-class of the pro-apoptotic receptor agonists (PARAs) that have recently been described by Ashkenazi and Herbst (Ashkenazi and Herbst 2008). The progress of one recombinant ligand, one anti-TRAIL-R1 and five anti-TRAIL-R2 antibodies in clinical trials will be summarised.

Human Genomic Sciences (HGS) is developing two fully humanised monoclonal antibodies Mapatumumab (HGS-ETR1) that activates TRAIL-R1 and Lexatumumab

(HGS-ETR2) which is specific for TRAIL-R2, respectively. These two antibodies have been very successful in pre-clinical studies and induced apoptosis across a wide range of human tumour cell lines as well as in primary cells isolated from solid as well as haematological malignancies. In all studies conducted so far, Mapatumumab was generally well tolerated, with the maximum tolerated dose yet to be reached. It has yielded stable disease as best clinical response in a phase Ia setting (Tolcher et al. 2007). In contrast, phase Ib studies in which Mapatumumab was tested in combination with either gemcitabine-cisplatin or paclitaxel-cisplatin have yielded partial responses (28 and 23%, respectively) (Chow et al. 2006; Hotte et al. 2005). In this case, a dose-limiting toxicity could be observed for one patient. Mapatumumab's activity could also be validated in three Phase II studies with patients suffering from Non-Hodgkin's lymphoma (NHL), CRC and NSCLC. For NHL, Mapatumumab as a single agent has yielded three objective clinical responses in patients suffering from NHL. However, phase II studies in CRC and NSCLC have produced stable disease as the best response in 34 and 29% of the cases, respectively. The mono-therapy was well tolerated with only one drug-related serious adverse event recorded. Another phase II study is currently investigating the efficiency and safety of Mapatumumab in combination with bortezomib in patients suffering from advanced multiple myeloma (study number: HGS 1012-C1055).

The results for Lexatumumab resemble those obtained for Mapatumumab. In a phase Ia clinical study several patients have reached stable disease with Lexatumumab as a monotherapeutic agent, but no tumour response has yet been recorded (Patnaik et al. 2006). In contrast, combinations of Lexatumumab with FOLFIRI (a chemotherapy cocktail made up of the drugs folic acid, 5-fluorouracil and irinotecan) or doxorubicin were well tolerated and induced tumour shrinkage and partial response in a wide range of cancer types. Several grade 3 toxicities, among them elevated liver enzymes, were related to Lexatumumab treatment and the maximum tolerated dose was reached at 20 mg kg^{-1}. Nevertheless, Lexatumumab could safely be administered, making further evaluations with regard to combinatorial therapy warranted. It is noteworthy that a pre-clinical study showed a complete regression of various tumour cell line xenografts in vivo upon treatment with Lexatumumab and the Smac-mimetic SM-164.

The humanised anti-TRAIL-R2 antibody TRA-8 is developed for treatment of solid tumours and lymphoma by Daiichi Sankyo. It exhibits high anti-tumour activity against astrocytoma and leukemia cells in vitro, against engrafted breast cancer cells in vivo and is currently in phase I clinical development.

Novartis is producing the TRAIL-R2-specific antibody LBY135, which is able to induce apoptosis in 50% of a panel of 40 human colon cancer cell lines with an IC_{50} of <10 nM. Currently, Novartis is recruiting patients for a phase I/II study of LBY135 alone or in combination with capecitabine in advanced solid tumours (Nevada Cancer Institute). So far, the anti-tumour activity of LBY135 has been proven in human CRC xenograft models in mice (Buchsbaum et al. 2003; Ichikawa et al. 2001).

The fully humanised TRAIL-R2-targeting antibody Apomab has been developed by Genentech. Currently, Apomab is in phase I and phase II clinical trials for solid tumours. Preliminary results of the phase Ia study revealed that Apomab was safe and well tolerated and yielded 52% stable disease. Two dose-limiting toxicities occurred comprising asymptomatic transaminitis and pulmonary embolism in one patient each (Camidge et al. 2007). In 2007, a phase II study was initiated, evaluating Apomab as a mono-therapeutic agent for the treatment of sarcoma and in combination with the VEGF-blocking anti-angiogenic antibody vastin to treat NSCLC. An additional study evaluating the effect of Apomab in combination with the CD20-targeting antibody Rituximab as a first line treatment for NSCLC is planned.

A fully humanised monoclonal antibody against TRAIL-R2 referred to as AMG 655 has been developed by Amgen. In phase Ib clinical trials, it showed anti-tumour effects against CRC and NSCLC, in which it led to metabolic partial responses or partial responses, respectively. So far, neither dose limiting toxicities nor severe side effects were recorded when AMG 655 was applied at doses of 20 mg kg^{-1} every 2 weeks. However, 9 of 11 patients showed adverse effects including hypomagnesaemia, fever and fatigue (LoRusso et al. 2007).

The only recombinant form of TRAIL so far tested in clinical trials is an untagged version of human TRAIL, referred to as Apo2L/TRAIL that is developed by Genentech in cooperation with Amgen. Pharmacokinetics and safety studies (phase Ib/II) were carried out in patients suffering from low-grade NHL. Preliminary results have proven Apo2L/TRAIL to be safe and active either alone or in combination with Rituximab. To date no dose-limiting toxcities have been reported; of the five patients investigated, two showed complete response, one partial response and two stable disease. More NHL patients are being recruited for further dose optimisation (Herbst et al. 2006) (Table 3).

Table 3 Development status of TARAs

TRA	Combination	Development stage
Mapatumumab		*Phase II completed:* NHL, CRC, NSCLC
	+ Paclitaxel + Carboplatin	*Phase Ib:* advanced solid tumours
	+ Gemcitabine	*Phase Ib:* advanced solid tumours
	+ Cisplatin + Bortezomib	*Phase II:* advanced multiple myeloma
Lexatumumab		*Phase I:* advanced solid tumours
	+ Chemotherapy	*Phase Ib:* advanced solid tumours
TRA-8		*Phase I:* advanced solid tumours and lymphomas (not yet recruiting)
LBY135		*Phase I/II:* advanced solid tumours
	+ Capecitabine	*Phase I/II:* advanced solid tumours (recruiting since 2006)
Apomab		*Phase I/II:* advanced solid tumours
	+ Avastin	*Phase II:* advanced solid tumours (initiated in 2007)
AMG655		*Phase II:* NSCLC, colorectal cancer (initiated in 2005)
Apo2L/TRAIL		*Phase Ib*
	+ Rituximab	*Phase Ib/II:* NHL (recruiting since 2006)
Ad5-TRAIL		*Phase I*

5 Future Perspectives

The pre-clinical and clinical data summarised in this chapter warrant the targeting of TRAIL's apoptosis-inducing receptors with TRAIL-receptor agonists as a promising novel approach in cancer therapy. However, to appreciate the full potential of this new approach, it is essential to deepen our understanding of the biochemical mechanisms conferring TRAIL resistance to tumour cells. The current limitations in cancer treatment can only be overcome by knowledge-based decisions for a particular therapeutic combination in a given tumour patient.

References

Ashkenazi A, Herbst RS (2008) To kill a tumor cell: the potential of proapoptotic receptor agonists. J Clin Invest 118:1979–1990

Ashkenazi A, Pai RC, Fong S, Leung S, Lawrence DA, Marsters SA, Blackie C, Chang L, McMurtrey AE, Hebert A, DeForge L, Koumenis IL, Lewis D, Harris L, Bussiere J, Koeppen H, Shahrokh Z, Schwall RH (1999) Safety and antitumor activity of recombinant soluble Apo2 ligand. J Clin Invest 104:155–162

Atkins GJ, Bouralexis S, Evdokiou A, Hay S, Labrinidis A, Zannettino AC, Haynes DR, Findlay DM (2002) Human osteoblasts are resistant to Apo2L/TRAIL-mediated apoptosis. Bone 31:448–456

Baliga B, Kumar S (2003) Apaf-1/cytochrome c apoptosome: an essential initiator of caspase activation or just a sideshow? Cell Death Differ 10:16–18

Balzar M, Winter MJ, de Boer CJ, Litvinov SV (1999) The biology of the 17-1A antigen (Ep-CAM). J Mol Med 77:699–712

Barnhart BC, Alappat EC, Peter ME (2003) The CD95 type I/type II model. Semin Immunol 15:185–193

Borner C (2003) The Bcl-2 protein family: sensors and checkpoints for life-or-death decisions. Mol Immunol 39:615–647

Bremer E, Kuijlen J, Samplonius D, Walczak H, de Leij L, Helfrich W (2004) Target cell-restricted and -enhanced apoptosis induction by a scFv:sTRAIL fusion protein with specificity for the pancarcinoma-associated antigen EGP2. Int J Cancer 109:281–290

Bremer E, Samplonius DF, Peipp M, van Genne L, Kroesen BJ, Fey GH, Gramatzki M, de Leij LF, Helfrich W (2005a) Target cell-restricted apoptosis induction of acute leukemic T cells by a recombinant tumor necrosis factor-related apoptosis-inducing ligand fusion protein with specificity for human CD7. Cancer Res 65:3380–3388

Bremer E, Samplonius DF, van Genne L, Dijkstra MH, Kroesen BJ, de Leij LF, Helfrich W (2005b) Simultaneous inhibition of epidermal growth factor receptor (EGFR) signaling and enhanced activation of tumor necrosis factor-related apoptosis-inducing ligand (TRAIL) receptor-mediated apoptosis induction by an scFv:sTRAIL fusion protein with specificity for human EGFR. J Biol Chem 280:10025–10033

Buchsbaum DJ, Zhou T, Grizzle WE, Oliver PG, Hammond CJ, Zhang S, Carpenter M, LoBuglio AF (2003) Antitumor efficacy of TRA-8 anti-DR5 monoclonal antibody alone or in combination with chemotherapy and/or radiation therapy in a human breast cancer model. Clin Cancer Res 9:3731–3741

Buchsbaum DJ, Zhou T, Lobuglio AF (2006) TRAIL receptor-targeted therapy. Future Oncol 2:493–508

Camidge D, Gordon M, Eckhardt S, Kurzroc R, Durbin B, Ing J, Ling J, Sager J, Mendelsohn D (2007) ASCO Annual Meeting Proceedings Part I. J Clin Oncol 25:3582

Certo M, Del Gaizo Moore V, Nishino M, Wei G, Korsmeyer S, Armstrong SA, Letai A (2006) Mitochondria primed by death signals determine cellular addiction to antiapoptotic BCL-2 family members. Cancer Cell 9:351–365
Chan FK, Chun HJ, Zheng L, Siegel RM, Bui KL, Lenardo MJ (2000) A domain in TNF receptors that mediates ligand-independent receptor assembly and signaling. Science 288:2351–2354
Chinnaiyan AM, Prasad U, Shankar S, Hamstra DA, Shanaiah M, Chenevert TL, Ross BD, Rehemtulla A (2000) Combined effect of tumor necrosis factor-related apoptosis-inducing ligand and ionizing radiation in breast cancer therapy. Proc Natl Acad Sci U S A 97:1754–1759
Chow LQ, Eckhardt SG, Gustafson DL, O'Bryant C, Hariharan S, Diab S, Fox NL, Corey A, Padavic K, Brown M, Cohen RB (2006) HGS-ETR1, an antibody targeting TRAIL-R1, in combination with paclitaxel and carboplatin in patients with advanced solid malignancies: results of a phase 1 and PK study. J Clin Oncol ASCO Annual Meeting Proceedings Part I. 24:2515
Clayer M, Bouralexis S, Evdokiou A, Hay S, Atkins GJ, Findlay DM (2001) Enhanced apoptosis of soft tissue sarcoma cells with chemotherapy: a potential new approach using TRAIL. J Orthop Surg 9:19–22
Clodi K, Wimmer D, Li Y, Goodwin R, Jaeger U, Mann G, Gadner H, Younes A (2000) Expression of tumour necrosis factor (TNF)-related apoptosis-inducing ligand (TRAIL) receptors and sensitivity to TRAIL-induced apoptosis in primary B-cell acute lymphoblastic leukaemia cells. Br J Haematol 111:580–586
Creagan ET, Kovach JS, Moertel CG, Frytak S, Kvols LK (1988) A phase I clinical trial of recombinant human tumor necrosis factor. Cancer 62:2467–2471
Creaven PJ, Plager JE, Dupere S, Huben RP, Takita H, Mittelman A, Proefrock A (1987) Phase I clinical trial of recombinant human tumor necrosis factor. Cancer Chemother Pharmacol 20:137–144
Cretney E, Takeda K, Yagita H, Glaccum M, Peschon JJ, Smyth MJ (2002) Increased susceptibility to tumor initiation and metastasis in TNF-related apoptosis-inducing ligand-deficient mice. J Immunol 168:1356–1361
Cretney E, Shanker A, Yagita H, Smyth MJ, Sayers TJ (2006) TNF-related apoptosis-inducing ligand as a therapeutic agent in autoimmunity and cancer. Immunol Cell Biol 84:87–98
Danial NN, Korsmeyer SJ (2004) Cell death: critical control points. Cell 116:205–219
Daniel PT, Wieder T, Sturm I, Schulze-Osthoff K (2001) The kiss of death: promises and failures of death receptors and ligands in cancer therapy. Leukemia 15:1022–1032
Degli-Esposti MA, Dougall WC, Smolak PJ, Waugh JY, Smith CA, Goodwin RG (1997a) The novel receptor TRAIL-R4 induces NF-kappaB and protects against TRAIL-mediated apoptosis, yet retains an incomplete death domain. Immunity 7:813–820
Degli-Esposti MA, Smolak PJ, Walczak H, Waugh J, Huang CP, DuBose RF, Goodwin RG, Smith CA (1997b) Cloning and characterization of TRAIL-R3, a novel member of the emerging TRAIL receptor family. J Exp Med 186:1165–1170
Di Pietro R, Secchiero P, Rana R, Gibellini D, Visani G, Bemis K, Zamai L, Miscia S, Zauli G (2001) Ionizing radiation sensitizes erythroleukemic cells but not normal erythroblasts to tumor necrosis factor-related apoptosis-inducing ligand (TRAIL) – mediated cytotoxicity by selective up-regulation of TRAIL-R1. Blood 97:2596–2603
Diehl GE, Yue HH, Hsieh K, Kuang AA, Ho M, Morici LA, Lenz LL, Cado D, Riley LW, Winoto A (2004) TRAIL-R as a negative regulator of innate immune cell responses. Immunity 21:877–889
Evdokiou A, Bouralexis S, Atkins GJ, Chai F, Hay S, Clayer M, Findlay DM (2002) Chemotherapeutic agents sensitize osteogenic sarcoma cells, but not normal human bone cells, to Apo2L/TRAIL-induced apoptosis. Int J Cancer 99:491–504
Falschlehner C, Emmerich CH, Gerlach B, Walczak H (2007) TRAIL signalling: decisions between life and death. Int J Biochem Cell Biol 39:1462–1475
Finnberg N, Gruber JJ, Fei P, Rudolph D, Bric A, Kim SH, Burns TF, Ajuha H, Page R, Wu GS, Chen Y, McKenna WG, Bernhard E, Lowe S, Mak T, El-Deiry WS (2005) DR5 knockout mice are compromised in radiation-induced apoptosis. Mol Cell Biol 25:2000–2013
Galle PR, Hofmann WJ, Walczak H, Schaller H, Otto G, Stremmel W, Krammer PH, Runkel L (1995) Involvement of the CD95 (APO-1/Fas) receptor and ligand in liver damage. J Exp Med 182:1223–1230

Ganten TM, Koschny R, Haas TL, Sykora J, Li-Weber M, Herzer K, Walczak H (2005) Proteasome inhibition sensitizes hepatocellular carcinoma cells, but not human hepatocytes, to TRAIL. Hepatology 42:588–597

Ganten TM, Koschny R, Sykora J, Schulze-Bergkamen H, Buchler P, Haas TL, Schader MB, Untergasser A, Stremmel W, Walczak H (2006) Preclinical differentiation between apparently safe and potentially hepatotoxic applications of TRAIL either alone or in combination with chemotherapeutic drugs. Clin Cancer Res 12:2640–2646

Gazitt Y (1999) TRAIL is a potent inducer of apoptosis in myeloma cells derived from multiple myeloma patients and is not cytotoxic to hematopoietic stem cells. Leukemia 13:1817–1824

Gershon RK, Kondo K (1971) Infectious immunological tolerance. Immunology 21:903–914

Gliniak B, Le T (1999) Tumor necrosis factor-related apoptosis-inducing ligand's antitumor activity in vivo is enhanced by the chemotherapeutic agent CPT-11. Cancer Res 59:6153–6158

Golks A, Brenner D, Fritsch C, Krammer PH, Lavrik IN (2005) c-FLIPR, a new regulator of death receptor-induced apoptosis. J Biol Chem 280:14507–14513

Greene MI, Benacerraf B (1980) Studies on hapten specific T cell immunity and suppression. Immunol Rev 50:163–186

Griffith TS, Broghammer EL (2001) Suppression of tumor growth following intralesional therapy with TRAIL recombinant adenovirus. Mol Ther 4:257–266

Griffith TS, Kazama H, VanOosten RL, Earle JK Jr, Herndon JM, Green DR, Ferguson TA (2007) Apoptotic cells induce tolerance by generating helpless CD8 + T cells that produce TRAIL. J Immunol 178:2679–2687

Grosse-Wilde A, Voloshanenko O, Bailey SL, Longton GM, Schaefer U, Csernok AI, Schutz G, Greiner EF, Kemp CJ, Walczak H (2008) TRAIL-R deficiency in mice enhances lymph node metastasis without affecting primary tumor development. J Clin Invest 118:100–110

Hanahan D, Weinberg RA (2000) The hallmarks of cancer. Cell 100:57–70

Hao C, Song JH, Hsi B, Lewis J, Song DK, Petruk KC, Tyrrell DL, Kneteman NM (2004) TRAIL inhibits tumor growth but is nontoxic to human hepatocytes in chimeric mice. Cancer Res 64:8502–8506

Held J, Schulze-Osthoff K (2001) Potential and caveats of TRAIL in cancer therapy. Drug Resist Updat 4:243–252

Hemminki A, Kanerva A, Liu B, Wang M, Alvarez RD, Siegal GP, Curiel DT (2003) Modulation of coxsackie-adenovirus receptor expression for increased adenoviral transgene expression. Cancer Res 63:847–853

Herbst RS, Mendolson DS, Ebbinghaus S, Gordon MS, O'Dwyer P, Lieberman G, Ing J, Kurzrock R, Novotny W, Eckhardt G (2006) A phase I safety and pharmacokinetic (PK) study of recombinant Apo2L/TRAIL, an apoptosis-inducing protein in patients with advanced cancer. ASCO Annual Meeting Proceedings Part I. J Clin Oncol 24(Suppl):3013

Hersh EM, Metch BS, Muggia FM, Brown TD, Whitehead RP, Budd GT, Rinehart JJ, Crawford ED, Bonnet JD, Behrens BC (1991) Phase II studies of recombinant human tumor necrosis factor alpha in patients with malignant disease: a summary of the Southwest Oncology Group experience. J Immunother 10:426–431

Hotte SJ, Hirte HW, Chen EX, Le LH, Corey A, Maclean M, Iacobucci A, Fox NL, Oza AM (2005) HGS-ETR1, a Fully Human Monoclonal Antibody to the Tumor Necrosis Factor-Related Apoptosis-Inducing Ligand Death Receptor 1 (TRAIL-R1) in Patients with Advanced solid cancer: results of a Phase 1 Trial. J Clin Oncol ASCO Annual Meeting Proceedings 23:3052

Hylander BL, Pitoniak R, Penetrante RB, Gibbs JF, Oktay D, Cheng J, Repasky EA (2005) The anti-tumor effect of Apo2L/TRAIL on patient pancreatic adenocarcinomas grown as xenografts in SCID mice. J Transl Med 3:22

Ichikawa K, Liu W, Zhao L, Wang Z, Liu D, Ohtsuka T, Zhang H, Mountz JD, Koopman WJ, Kimberly RP, Zhou T (2001) Tumoricidal activity of a novel anti-human DR5 monoclonal antibody without hepatocyte cytotoxicity. Nat Med 7:954–960

Inohara N, Koseki T, Chen S, Benedict MA, Nunez G (1999) Identification of regulatory and catalytic domains in the apoptosis nuclease DFF40/CAD. J Biol Chem 274:270–274

Inoue S, MacFarlane M, Harper N, Wheat LM, Dyer MJ, Cohen GM (2004) Histone deacetylase inhibitors potentiate TNF-related apoptosis-inducing ligand (TRAIL)-induced apoptosis in lymphoid malignancies. Cell Death Differ 11(Suppl 2):S193–S206

Ishimura N, Isomoto H, Bronk SF, Gores GJ (2006) Trail induces cell migration and invasion in apoptosis-resistant cholangiocarcinoma cells. Am J Physiol Gastrointest Liver Physiol 290:G129–136

Janssen EM, Droin NM, Lemmens EE, Pinkoski MJ, Bensinger SJ, Ehst BD, Griffith TS, Green DR, Schoenberger SP (2005) CD4 + T-cell help controls CD8 + T-cell memory via TRAIL-mediated activation-induced cell death. Nature 434:88–93

Johnsen AC, Haux J, Steinkjer B, Nonstad U, Egeberg K, Sundan A, Ashkenazi A, Espevik T (1999) Regulation of APO-2 ligand/trail expression in NK cells-involvement in NK cell-mediated cytotoxicity. Cytokine 11:664–672

Kashii Y, Giorda R, Herberman RB, Whiteside TL, Vujanovic NL (1999) Constitutive expression and role of the TNF family ligands in apoptotic killing of tumor cells by human NK cells. J Immunol 163:5358–5366

Kayagaki N, Yamaguchi N, Nakayama M, Takeda K, Akiba H, Tsutsui H, Okamura H, Nakanishi K, Okumura K, Yagita H (1999) Expression and function of TNF-related apoptosis-inducing ligand on murine activated NK cells. J Immunol 163:1906–1913

Kelley SK, Ashkenazi A (2004) Targeting death receptors in cancer with Apo2L/TRAIL. Curr Opin Pharmacol 4:333–339

Kennedy GA, Tey SK, Cobcroft R, Marlton P, Cull G, Grimmett K, Thomson D, Gill D (2002) Incidence and nature of CD20-negative relapses following rituximab therapy in aggressive B-cell non-Hodgkin's lymphoma: a retrospective review. Br J Haematol 119:412–416

Kischkel FC, Lawrence DA, Tinel A, LeBlanc H, Virmani A, Schow P, Gazdar A, Blenis J, Arnott D, Ashkenazi A (2001) Death receptor recruitment of endogenous caspase-10 and apoptosis initiation in the absence of caspase-8. J Biol Chem 276:46639–46646

Koschny R, Ganten TM, Sykora J, Haas TL, Sprick MR, Kolb A, Stremmel W, Walczak H (2007a) TRAIL/bortezomib cotreatment is potentially hepatotoxic but induces cancer-specific apoptosis within a therapeutic window. Hepatology 45:649–658

Koschny R, Holland H, Sykora J, Haas TL, Sprick MR, Ganten TM, Krupp W, Bauer M, Ahnert P, Meixensberger J, Walczak H (2007b) Bortezomib sensitizes primary human astrocytoma cells of WHO grades I to IV for tumor necrosis factor-related apoptosis-inducing ligand-induced apoptosis. Clin Cancer Res 13:3403–3412

Koschny R, Walczak H, Ganten TM (2007c) The promise of TRAIL – potential and risks of a novel anticancer therapy. J Mol Med 85:923–935

Krueger A, Schmitz I, Baumann S, Krammer PH, Kirchhoff S (2001) Cellular FLICE-inhibitory protein splice variants inhibit different steps of caspase-8 activation at the CD95 death-inducing signaling complex. J Biol Chem 276:20633–20640

Lawrence D, Shahrokh Z, Marsters S, Achilles K, Shih D, Mounho B, Hillan K, Totpal K, DeForge L, Schow P, Hooley J, Sherwood S, Pai R, Leung S, Khan L, Gliniak B, Bussiere J, Smith CA, Strom SS, Kelley S, Fox JA, Thomas D, Ashkenazi A (2001) Differential hepatocyte toxicity of recombinant Apo2L/TRAIL versions. Nat Med 7:383–385

Leverkus M, Sprick MR, Wachter T, Mengling T, Baumann B, Serfling E, Brocker EB, Goebeler M, Neumann M, Walczak H (2003) Proteasome inhibition results in TRAIL sensitization of primary keratinocytes by removing the resistance-mediating block of effector caspase maturation. Mol Cell Biol 23:777–790

Liang X, Liu Y, Zhang Q, Gao L, Han L, Ma C, Zhang L, Chen YH, Sun W (2007) Hepatitis B virus sensitizes hepatocytes to TRAIL-induced apoptosis through Bax. J Immunol 178:503–510

Lindemann RK, Newbold A, Whitecross KF, Cluse LA, Frew AJ, Ellis L, Williams S, Wiegmans AP, Dear AE, Scott CL, Pellegrini M, Wei A, Richon VM, Marks PA, Lowe SW, Smyth MJ, Johnstone RW (2007) Analysis of the apoptotic and therapeutic activities of histone deacetylase inhibitors by using a mouse model of B cell lymphoma. Proc Natl Acad Sci U S A 104:8071–8076

Liu X, Li P, Widlak P, Zou H, Luo X, Garrard WT, Wang X (1998) The 40-kDa subunit of DNA fragmentation factor induces DNA fragmentation and chromatin condensation during apoptosis. Proc Natl Acad Sci U S A 95:8461–8466

LoRusso P, Hong D, Heath E, Kurzrock R, Wang D, Hsu M, Goyal L, Wiezorek J, Storgard C, Herbst R (2007) First-in-human study of AMG 655, a pro-apoptotic TRAIL receptor-2 agonist, in adult patients with advanced solid tumors. J Clin Oncol ASCO Annual Meeting Proceedings Part I 25:3534

MacFarlane M, Harper N, Snowden RT, Dyer MJ, Barnett GA, Pringle JH, Cohen GM (2002) Mechanisms of resistance to TRAIL-induced apoptosis in primary B cell chronic lymphocytic leukaemia. Oncogene 21:6809–6818

MacFarlane M, Inoue S, Kohlhaas SL, Majid A, Harper N, Kennedy DB, Dyer MJ, Cohen GM (2005) Chronic lymphocytic leukemic cells exhibit apoptotic signaling via TRAIL-R1. Cell Death Differ 12:773–782

McCarthy MM, Sznol M, DiVito KA, Camp RL, Rimm DL, Kluger HM (2005) Evaluating the expression and prognostic value of TRAIL-R1 and TRAIL-R2 in breast cancer. Clin Cancer Res 11:5188–5194

Merino D, Lalaoui N, Morizot A, Schneider P, Solary E, Micheau O (2006) Differential inhibition of TRAIL-mediated DR5-DISC formation by decoy receptors 1 and 2. Mol Cell Biol 26:7046–7055

Micheau O, Thome M, Schneider P, Holler N, Tschopp J, Nicholson DW, Briand C, Grutter MG (2002) The long form of FLIP is an activator of caspase-8 at the Fas death-inducing signaling complex. J Biol Chem 277:45162–45171

Mitsiades CS, Treon SP, Mitsiades N, Shima Y, Richardson P, Schlossman R, Hideshima T, Anderson KC (2001) TRAIL/Apo2L ligand selectively induces apoptosis and overcomes drug resistance in multiple myeloma: therapeutic applications. Blood 98:795–804

Mitsiades N, Mitsiades CS, Poulaki V, Chauhan D, Richardson PG, Hideshima T, Munshi N, Treon SP, Anderson KC (2002) Biologic sequelae of nuclear factor-kappaB blockade in multiple myeloma: therapeutic applications. Blood 99:4079–4086

Moldenhauer G, Momburg F, Moller P, Schwartz R, Hammerling GJ (1987) Epithelium-specific surface glycoprotein of Mr 34,000 is a widely distributed human carcinoma marker. Br J Cancer 56:714–721

Mori E, Thomas M, Motoki K, Nakazawa K, Tahara T, Tomizuka K, Ishida I, Kataoka S (2004) Human normal hepatocytes are susceptible to apoptosis signal mediated by both TRAIL-R1 and TRAIL-R2. Cell Death Differ 11:203–207

Mundt B, Wirth T, Zender L, Waltemathe M, Trautwein C, Manns MP, Kuhnel F, Kubicka S (2005) Tumour necrosis factor related apoptosis inducing ligand (TRAIL) induces hepatic steatosis in viral hepatitis and after alcohol intake. Gut 54:1590–1596

Naka T, Sugamura K, Hylander BL, Widmer MB, Rustum YM, Repasky EA (2002) Effects of tumor necrosis factor-related apoptosis-inducing ligand alone and in combination with chemotherapeutic agents on patients' colon tumors grown in SCID mice. Cancer Res 62:5800–5806

Nakata S, Yoshida T, Horinaka M, Shiraishi T, Wakada M, Sakai T (2004) Histone deacetylase inhibitors upregulate death receptor 5/TRAIL-R2 and sensitize apoptosis induced by TRAIL/APO2-L in human malignant tumor cells. Oncogene 23:6261–6271

Nebbioso A, Clarke N, Voltz E, Germain E, Ambrosino C, Bontempo P, Alvarez R, Schiavone EM, Ferrara F, Bresciani F, Weisz A, de Lera AR, Gronemeyer H, Altucci L (2005) Tumor-selective action of HDAC inhibitors involves TRAIL induction in acute myeloid leukemia cells. Nat Med 11:77–84

Ogasawara J, Watanabe-Fukunaga R, Adachi M, Matsuzawa A, Kasugai T, Kitamura Y, Itoh N, Suda T, Nagata S (1993) Lethal effect of the anti-Fas antibody in mice. Nature 364:806–809

Okegawa T, Li Y, Pong RC, Bergelson JM, Zhou J, Hsieh JT (2000) The dual impact of coxsackie and adenovirus receptor expression on human prostate cancer gene therapy. Cancer Res 60:5031–5036

Olsson A, Diaz T, Aguilar-Santelises M, Osterborg A, Celsing F, Jondal M, Osorio LM (2001) Sensitization to TRAIL-induced apoptosis and modulation of FLICE-inhibitory protein in B chronic lymphocytic leukemia by actinomycin D. Leukemia 15:1868–1877

Ozoren N, El-Deiry WS (2002) Defining characteristics of Types I and II apoptotic cells in response to TRAIL. Neoplasia 4:551–557

Pan G, Ni J, Wei YF, Yu G, Gentz R, Dixit VM (1997a) An antagonist decoy receptor and a death domain-containing receptor for TRAIL. Science 277:815–818

Pan G, O'Rourke K, Chinnaiyan AM, Gentz R, Ebner R, Ni J, Dixit VM (1997b) The receptor for the cytotoxic ligand TRAIL. Science 276:111–113

Panner A, James CD, Berger MS, Pieper RO (2005) mTOR controls FLIPS translation and TRAIL sensitivity in glioblastoma multiforme cells. Mol Cell Biol 25:8809–8823

Pathil A, Armeanu S, Venturelli S, Mascagni P, Weiss TS, Gregor M, Lauer UM, Bitzer M (2006) HDAC inhibitor treatment of hepatoma cells induces both TRAIL-independent apoptosis and restoration of sensitivity to TRAIL. Hepatology 43:425–434

Patnaik A, Wakelee H, Mita M, Fitzgerald A, Hill M, Fox N, Howard T, Ullrich S, Tolcher A, Sikic B (2006) HGS-ETR2 – A fully human monoclonal antibody to TRAIL-R2: results of a phase I trial in patients with advanced solid tumors. J Clin Oncol ASCO Annual Meeting Proceedings Part I. Vol 24 24:3012

Pitti RM, Marsters SA, Ruppert S, Donahue CJ, Moore A, Ashkenazi A (1996) Induction of apoptosis by Apo-2 ligand, a new member of the tumor necrosis factor cytokine family. J Biol Chem 271:12687–12690

Rauen KA, Sudilovsky D, Le JL, Chew KL, Hann B, Weinberg V, Schmitt LD, McCormick F (2002) Expression of the coxsackie adenovirus receptor in normal prostate and in primary and metastatic prostate carcinoma: potential relevance to gene therapy. Cancer Res 62:3812–3818

Riccioni R, Pasquini L, Mariani G, Saulle E, Rossini A, Diverio D, Pelosi E, Vitale A, Chierichini A, Cedrone M, Foa R, Lo Coco F, Peschle C, Testa U (2005) TRAIL decoy receptors mediate resistance of acute myeloid leukemia cells to TRAIL. Haematologica 90:612–624

Riedl SJ, Renatus M, Schwarzenbacher R, Zhou Q, Sun C, Fesik SW, Liddington RC, Salvesen GS (2001) Structural basis for the inhibition of caspase-3 by XIAP. Cell 104:791–800

Sachs MD, Ramamurthy M, Poel H, Wickham TJ, Lamfers M, Gerritsen W, Chowdhury W, Li Y, Schoenberg MP, Rodriguez R (2004) Histone deacetylase inhibitors upregulate expression of the coxsackie adenovirus receptor (CAR) preferentially in bladder cancer cells. Cancer Gene Ther 11:477–486

Salvesen GS, Duckett CS (2002) IAP proteins: blocking the road to death's door. Nat Rev Mol Cell Biol 3:401–410

Schlosser SF, Azzaroli F, Dao T, Hingorani R, Nicholas Crispe I, Boyer JL (2000) Induction of murine hepatocyte death by membrane-bound CD95 (Fas/APO-1)-ligand: characterization of an in vitro system. Hepatology 32:779–785

Schneider P (2000) Production of recombinant TRAIL and TRAIL receptor: Fc chimeric proteins. Methods Enzymol 322:325–345

Schneider P, Bodmer JL, Thome M, Hofmann K, Holler N, Tschopp J (1997) Characterization of two receptors for TRAIL. FEBS Lett 416:329–334

Schuchmann M, Schulze-Bergkamen H, Fleischer B, Schattenberg JM, Siebler J, Weinmann A, Teufel A, Worns M, Fischer T, Strand S, Lohse AW, Galle PR (2006) Histone deacetylase inhibition by valproic acid down-regulates c-FLIP/CASH and sensitizes hepatoma cells towards CD95- and TRAIL receptor-mediated apoptosis and chemotherapy. Oncol Rep 15:227–230

Screaton GR, Mongkolsapaya J, Xu XN, Cowper AE, McMichael AJ, Bell JI (1997) TRICK2, a new alternatively spliced receptor that transduces the cytotoxic signal from TRAIL. Curr Biol 7:693–696

Sedger LM, Glaccum MB, Schuh JC, Kanaly ST, Williamson E, Kayagaki N, Yun T, Smolak P, Le T, Goodwin R, Gliniak B (2002) Characterization of the in vivo function of TNF-alpha-related apoptosis-inducing ligand, TRAIL/Apo2L, using TRAIL/Apo2L gene-deficient mice. Eur J Immunol 32:2246–2254

Shanker A, Brooks AD, Tristan CA, Wine JW, Elliott PJ, Yagita H, Takeda K, Smyth MJ, Murphy WJ, Sayers TJ (2008) Treating metastatic solid tumors with bortezomib and a tumor necrosis factor-related apoptosis-inducing ligand receptor agonist antibody. J Natl Cancer Inst 100:649–662

Sheridan JP, Marsters SA, Pitti RM, Gurney A, Skubatch M, Baldwin D, Ramakrishnan L, Gray CL, Baker K, Wood WI, Goddard AD, Godowski P, Ashkenazi A (1997) Control of TRAIL-induced apoptosis by a family of signaling and decoy receptors. Science 277:818–821

Smyth MJ, Cretney E, Takeda K, Wiltrout RH, Sedger LM, Kayagaki N, Yagita H, Okumura K (2001) Tumor necrosis factor-related apoptosis-inducing ligand (TRAIL) contributes to interferon gamma-dependent natural killer cell protection from tumor metastasis. J Exp Med 193:661–670

Sprick MR, Rieser E, Stahl H, Grosse-Wilde A, Weigand MA, Walczak H (2002) Caspase-10 is recruited to and activated at the native TRAIL and CD95 death-inducing signalling complexes in a FADD-dependent manner but can not functionally substitute caspase-8. EMBO J 21:4520–4530

Stamenkovic I, Seed B (1988) CD19, the earliest differentiation antigen of the B cell lineage, bears three extracellular immunoglobulin-like domains and an Epstein-Barr virus-related cytoplasmic tail. J Exp Med 168:1205–1210

Steller H (1995) Mechanisms and genes of cellular suicide. Science 267:1445–1449

Stieglmaier J, Bremer E, Kellner C, Liebig TM, ten Cate B, Peipp M, Schulze-Koops H, Pfeiffer M, Buhring HJ, Greil J, Oduncu F, Emmerich B, Fey GH, Helfrich W (2008) Selective induction of apoptosis in leukemic B-lymphoid cells by a CD19-specific TRAIL fusion protein. Cancer Immunol Immunother 57:233–246

Strasser A (2005) The role of BH3-only proteins in the immune system. Nat Rev Immunol 5:189–200

Sudarshan S, Holman DH, Hyer ML, Voelkel-Johnson C, Dong JY, Norris JS (2005) In vitro efficacy of Fas ligand gene therapy for the treatment of bladder cancer. Cancer Gene Ther 12:12–18

Sy MS, Miller SD, Kowach HB, Claman HN (1977) A splenic requirement for the generation of suppressor T cells. J Immunol 119:2095–2099

Takeda K, Smyth MJ, Cretney E, Hayakawa Y, Yamaguchi N, Yagita H, Okumura K (2001) Involvement of tumor necrosis factor-related apoptosis-inducing ligand in NK cell-mediated and IFN-gamma-dependent suppression of subcutaneous tumor growth. Cell Immunol 214:194–200

Takeda K, Yamaguchi N, Akiba H, Kojima Y, Hayakawa Y, Tanner JE, Sayers TJ, Seki N, Okumura K, Yagita H, Smyth MJ (2004) Induction of tumor-specific T cell immunity by anti-DR5 antibody therapy. J Exp Med 199:437–448

Takeda K. Kojima Y, Ikejima K, HAradaK, Yamashina S, Okumura K, Aoyama T, Frese S, Ikeda h, Haynes NM, Cretney E, Yagita H, Sueyoshi N, Nakanuma Y, Smyth MJ(2008) Death racepter 5 mediated-apoptosis contributes to cholestatic liver disease. Proc Matl Acad Sci U S A 105(31):10895–500

Thompson CB (1995) Apoptosis in the pathogenesis and treatment of disease. Science 267:1456–1462

Todaro M, Lombardo Y, Francipane MG, Alea MP, Cammareri P, Iovino F, Di Stefano AB, Di Bernardo C, Agrusa A, Condorelli G, Walczak H, Stassi G (2008) Apoptosis resistance in epithelial tumors is mediated by tumor-cell-derived interleukin-4. Cell Death Differ 15:762–772

Tolcher AW, Mita M, Meropol NJ, von Mehren M, Patnaik A, Padavic K, Hill M, Mays T, McCoy T, Fox NL, Halpern W, Corey A, Cohen RB (2007) Phase I pharmacokinetic and biologic correlative study of mapatumumab, a fully human monoclonal antibody with agonist activity to tumor necrosis factor-related apoptosis-inducing ligand receptor-1. J Clin Oncol 25:1390–1395

Trauzold A, Siegmund D, Schniewind B, Sipos B, Egberts J, Zorenkov D, Emme D, Roder C, Kalthoff H, Wajant H (2006) TRAIL promotes metastasis of human pancreatic ductal adenocarcinoma. Oncogene 25:7434–7439

Truneh A, Sharma S, Silverman C, Khandekar S, Reddy MP, Deen KC, McLaughlin MM, Srinivasula SM, Livi GP, Marshall LA, Alnemri ES, Williams WV, Doyle ML (2000) Temperature-sensitive differential affinity of TRAIL for its receptors. DR5 is the highest affinity receptor. J Biol Chem 275:23319–23325

Uckun FM, Jaszcz W, Ambrus JL, Fauci AS, Gajl-Peczalska K, Song CW, Wick MR, Myers DE, Waddick K, Ledbetter JA (1988) Detailed studies on express ion and function of CD19 surface determinant by using B43 monoclonal antibody and the clinical potential of anti-CD19 immunotoxins. Blood 71:13–29

Uno T, Takeda K, Kojima Y, Yoshizawa H, Akiba H, Mittler RS, Gejyo F, Okumura K, Yagita H, Smyth MJ (2006) Eradication of established tumors in mice by a combination antibody-based therapy. Nat Med 12:693–698

VanOosten RL, Earel JK Jr, Griffith TS (2007) Histone deacetylase inhibitors enhance Ad5-TRAIL killing of TRAIL-resistant prostate tumor cells through increased caspase-2 activity. Apoptosis 12:561–571

Van Valen F, Fulda S, Schafer KL, Truckenbrod B, Hotfilder M, Poremba C, Debatin KM, Winkelmann W (2003) Selective and nonselective toxicity of TRAIL/Apo2L combined with chemotherapy in human bone tumour cells vs. normal human cells. Int J Cancer 107:929–940

Vaux DL, Silke J (2005) IAPs – the ubiquitin connection. Cell Death Differ 12:1205–1207

Wagner KW, Punnoose EA, Januario T, Lawrence DA, Pitti RM, Lancaster K, Lee D, von Goetz M, Yee SF, Totpal K, Huw L, Katta V, Cavet G, Hymowitz SG, Amler L, Ashkenazi A (2007) Death-receptor O-glycosylation controls tumor-cell sensitivity to the proapoptotic ligand Apo2L/TRAIL. Nat Med 13:1070–1077

Wajant H, Pfizenmaier K, Scheurich P (2002) TNF-related apoptosis inducing ligand (TRAIL) and its receptors in tumor surveillance and cancer therapy. Apoptosis 7:449–459

Walczak H, Degli-Esposti MA, Johnson RS, Smolak PJ, Waugh JY, Boiani N, Timour MS, Gerhart MJ, Schooley KA, Smith CA, Goodwin RG, Rauch CT (1997) TRAIL-R2: a novel apoptosis-mediating receptor for TRAIL. EMBO J 16:5386–5397

Walczak H, Miller RE, Ariail K, Gliniak B, Griffith TS, Kubin M, Chin W, Jones J, Woodward A, Le T, Smith C, Smolak P, Goodwin RG, Rauch CT, Schuh JC, Lynch DH (1999) Tumoricidal activity of tumor necrosis factor-related apoptosis-inducing ligand in vivo. Nat Med 5:157–163

Wiley SR, Schooley K, Smolak PJ, Din WS, Huang CP, Nicholl JK, Sutherland GR, Smith TD, Rauch C, Smith CA, et al (1995) Identification and characterization of a new member of the TNF family that induces apoptosis. Immunity 3:673–682

Wu GS, Burns TF, McDonald ER, 3rd, Jiang W, Meng R, Krantz ID, Kao G, Gan DD, Zhou JY, Muschel R, Hamilton SR, Spinner NB, Markowitz S, Wu G, el-Deiry WS (1997) KILLER/DR5 is a DNA damage-inducible p53-regulated death receptor gene. Nat Genet 17:141–143

Wu XX, Kakehi Y, Mizutani Y, Kamoto T, Kinoshita H, Isogawa Y, Terachi T, Ogawa O (2002) Doxorubicin enhances TRAIL-induced apoptosis in prostate cancer. Int J Oncol 20:949–954

Youle RJ, Strasser A (2008) The BCL-2 protein family: opposing activities that mediate cell death. Nat Rev Mol Cell Biol 9:47–59

Yue HH, Diehl GE, Winoto A (2005) Loss of TRAIL-R does not affect thymic or intestinal tumor development in p53 and adenomatous polyposis coli mutant mice. Cell Death Differ 12:94–97

Zerafa N, Westwood JA, Cretney E, Mitchell S, Waring P, Iezzi M, Smyth MJ (2005) Cutting edge: TRAIL deficiency accelerates hematological malignancies. J Immunol 175:5586–5590

Therapeutic Targeting of TWEAK/Fn14 in Cancer: Exploiting the Intrinsic Tumor Cell Killing Capacity of the Pathway

Jennifer S. Michaelson and Linda C. Burkly

Abstract TNF-like weak inducer of apoptosis (TWEAK) and FGF-inducible molecule 14 (Fn14) are a TNF superfamily ligand–receptor pair. Initially identified as an inducer of tumor cell killing, TWEAK has pleiotropic effects, mediating pro-inflammatory and pro-angiogenic activity as well as stimulation of invasion, migration, and survival through its widely recognized receptor, Fn14. Fn14 is expressed at relatively low levels in normal tissues, but is dramatically elevated locally in injured and diseased tissues, where it plays a role in tissue remodeling. Herein we review the link between the TWEAK/Fn14 pathway and cancer as well as discuss potential therapeutic strategies targeting this pathway for cancer treatment. Many of the activities associated with the TWEAK/Fn14 pathway are linked with tumorigenesis and could thereby provide a growth advantage to tumors, suggesting that inhibition of the pathway may be beneficial in the treatment of cancer. At the same time, the elevated expression of Fn14 by tumor cells as well as the intrinsic tumor cell killing capacity of this receptor represents a promising alternative of harnessing the intrinsic tumor cell killing capacity of Fn14 to treat cancer.

1 Introduction

1.1 Tweak

TWEAK was cloned from a macrophage library in 1997 and recognized to be a TNF ligand superfamily member based on the presence of the TNF homology domain (Chicheportiche et al. 1997). First described as a weak inducer of apoptosis of tumor cells, TWEAK has since been shown to mediate multiple cellular processes, including pro-inflammatory activity, angiogenesis, and cell proliferation

J.S. Michaelson and L.C. Burkly(✉)
Biogen Idec, 12 Cambridge Center, Cambridge, MA 02142
e-mail: Jennifer.Michaelson@biogenidec.com
e-mail: Linda.Burkly@biogenidec.com

Results Probl Cell Differ, doi:10.1007/400_2008_18

(recently reviewed in Burkly et al. 2007). TWEAK is a Type II transmembrane protein that is considered to act primarily as a secreted soluble cytokine due to efficient furin protease cleavage within the trans Golgi network. While TWEAK mRNA has been identified in a wide range of cell types, TWEAK protein has thus far been shown to be expressed by inflammatory cells and tumor cells (reviewed in Burkly et al. 2007).

1.2 Fn14

The receptor for TWEAK was definitively identified as Fn14 by expression cloning in 2001 (Wiley et al. 2001). Although Fn14 had previously been discovered as an FGF-inducible gene (Meighan-Mantha et al. 1999), it was not recognized as a TNF receptor family member at that time as it had only one cysteine-rich domain. Like other TNF receptor family members, Fn14 is a Type I transmembrane receptor and contains a well-recognized TNF receptor-associated (TRAF) binding sequence in its cytoplasmic tail, which confers signaling potential (see Sect. 1.3)

The possibility of alternative ligands for Fn14 or receptors for TWEAK has been suggested in a handful of studies (De Ketelaere et al. 2004; Dogra et al. 2007; Polek et al. 2003; Tanabe et al. 2003); however, no other partners were identified. Recently, a report showed TWEAK interaction with CD163, a scavenger receptor on macrophages (Bover et al. 2007), however, this report has not been confirmed.

Fn14 expression is detected in a variety of cell types, including epithelial, endothelial and mesenchymal cells, progenitor cells, and cells from the nervous system (reviewed in Burkly et al. 2007). However, Fn14 levels are generally low in normal cells, and increases locally only in the contexts of tissue injury and inflammatory diseases, and in cancer (see Sect. 2). There is growing support for the notion that the TWEAK/Fn14 pathway plays a physiological role in facilitating tissue repair after acute injury and a pathological role in tissue remodeling in the end organ sites of chronic inflammatory disease (reviewed in Burkly et al. 2007; Zheng and Burkly 2008).

1.3 TWEAK/Fn14 Signaling Pathways

Fn14 has the shortest cytoplasmic domain, only 28 amino acids, among all TNF receptor superfamily members. While it lacks a death domain, Fn14 contains the PIEET canonical TRAF-binding sequence in its cytoplasmic domain. Direct association between the cytoplasmic domain of Fn14 and TRAF-1, 2, 3, and 5 has been demonstrated in vitro (Brown et al. 2003), and thus it is believed that recruitment of TRAF adapter proteins by Fn14 is responsible for the activation of downstream signaling pathways, including NF-κB (Brown et al. 2003; Saitoh et al. 2003) as well as the MAPK and AKT pathways (Ando et al. 2006). Although PIEET is the widely recognized TRAF-binding motif, it appears that an adjacent KAFF sequence is also important in NF-κB signaling, as a KAFF-deleted mutant resulted in total abrogation of NF-κB activation (T.S. Zheng, unpublished).

TWEAK has been shown to induce prolonged NF-κB activation through temporally regulated biphasic activation of both canonical and noncanonical NF-κB pathways (Brown et al. 2003; Saitoh et al., 2003), in contrast to TNF which mainly activates the canonical pathway. Activation of the NF-κB pathway by TWEAK has been demonstrated in many contexts in vitro (reviewed in Burkly et al. 2007), and there is now evidence for TWEAK-induced NF-κB activation in vivo (Campbell et al. 2006). The ability of TWEAK/Fn14 to trigger NF-κB activation and the resulting pro-inflammatory activities correlate well with, and indeed have been linked to, the documented role of this pathway in inflammatory disease (reviewed in Burkly et al. 2007). However, given that Fn14 lacks a death domain, the signaling pathways responsible for cell killing mediated through Fn14 are less clear, as discussed later (see Sect. 4.4).

1.4 TWEAK and Fn14: Link to Cancer

Significantly, elevated levels of Fn14 are observed in the context of tumors, as described in detail further (see Sect. 2). Many of the activities associated with TWEAK, such as the ability to induce proliferation, survival, invasion, and angiogenesis, are characteristic properties of a tumor. Thus, Fn14 upregulation may provide an advantage to tumors, and inhibition of the TWEAK pathway could be beneficial in the treatment of cancer. At the same time, the fact that Fn14 expression is elevated in human tumors as compared to normal tissue suggests that it may be a potential tumor antigen, and therefore, on the basis of expression alone, a valuable therapeutic target. Indeed, novel tumor-specific antigens continue to be sought in the oncology arena. Moreover, the differential expression of Fn14 in cancer presents an opportunity to exploit the intrinsic tumor cell killing capacity of this pathway as an attractive therapeutic approach to targeting cancer. Importantly, in conceptualizing the potential for the TWEAK/Fn14 pathway from the perspective of cancer therapeutics, there are several possibilities to consider – and these may not be mutually exclusive.

2 TWEAK/Fn14 Pathway Expression in Cancer

2.1 Fn14 Expression in Human Tumors

There is increasing evidence that Fn14 expression is elevated in a variety of human tumors (Table 1). In an examination of human primary breast tumors, we detected mRNA expression in 58% (35/60) by in situ hybridization, and protein expression

Table 1 Fn14 expression in human tumors and correlation with clinical disease

Tumor type	mRNA expression	Protein expression	Correlation of expression with clinical disease	References
Breast	↑	↑	Correlation with lymph node status, metastasis Correlation with HER2+/ ER- status Correlation with invasive tumor type	Michaelson et al. (2005); Willis et al. (2008)
Glioma	↑	↑	Correlation with increasing tumor grade Inverse correlation with patient survival	Tran et al. (2003, 2006)
Esophageal	↑	↑	Correlation with disease progression Correlation with invasive tumor cells	Wang et al. (2006); Watts et al. (2007)
Pancreatic	↑	↑	ND	Han et al. (2002, 2005); JSM and LCB,unpublished
Hepatocellular	↑	ND	ND	Feng et al. (2000)
Testicular	↑	ND	ND	Almstrup et al. (2004)

ND, Not Done

in 52% (10/19) by immunohistochemistry (Michaelson et al. 2005). In contrast, Fn14 mRNA was not detected in normal breast samples (0/10), and only 10% (1/10) of normal breast samples had detectable Fn14 protein expression (Michaelson et al. 2005). In glioblastoma, Tran et al. reported fivefold or more over-expression of Fn14 mRNA in 68% of specimens examined (Tran et al. 2003). More recently, in an analysis of 27 glioblastoma samples for Fn14 protein expression, 1 tumor was negative (4%), 3 were weakly positive (11%), 20 were moderately positive (74%), and 3 were strongly positive (11%) (Tran et al. 2006). In a genome wide survey of gene expression, Fn14 was identified as one of the most highly up-regulated genes in pancreatic tumor cell lines (Han et al. 2002). This finding was confirmed in an immunohistochemical study of human pancreatic tumors, where 38% (16/42) stained positively for Fn14 compared with adjacent normal stromal tissue (Han et al. 2005). In a recent immunohistochemical analysis, we observed Fn14 over-expression in 50% (10/20) of human pancreatic tumor samples relative to normal pancreatic tissue (J.S.M. and L.C.B. unpublished). Fn14 mRNA was also identified to be elevated in esophageal tumors (Wang et al. 2006), with subsequent confirmation of upregulation of protein expression (Watts et al. 2007). In addition, Fn14 over-expression has been detected in hepatocellular carcinoma, though only a small number of specimens was examined exclusively at the mRNA level (Feng et al. 2000). Finally, transcriptional profiling studies identified Fn14 as highly expressed in testicular carcinoma in situ (Almstrup et al. 2004). Interestingly, in certain tumor types, for example, lung and prostate cancer, expression of Fn14 may not be elevated relative to normal tissue (J.S.M. unpublished).

2.2 Correlation of Fn14 Expression with Tumor Progression

While increased expression of Fn14 in tumors is suggestive of a link to cancer, even more notable is that in some tumor types there appears to be a significant correlation between Fn14 expression and tumor grade and/or prognosis (Table 1). In glioma, Tran et al. first reported that while anaplastic astrocytomas showed modest over-expression of Fn14 relative to normal brain, highest levels of over-expression were observed in glioblastoma specimens (Tran et al. 2003). This observation was recently confirmed with a larger data set, and further validated by immunohistochemical analysis, with highest levels of Fn14 expression in glioblastoma, as compared to modestly elevated levels of expression in oligoastrocytomas and anaplastic astrocytoma, relative to control brain samples (Tran et al. 2006). Moreover, Fn14 mRNA levels in glioblastoma samples were shown to be inversely correlated with patient survival (Tran et al. 2006). Similarly, in a microarray analysis of varying pathological stages of esophageal neoplasia, Fn14 was one of the 12 genes that most specifically correlated with disease progression, with the highest levels of expression detected in esophageal adenocarcinoma (EAC) (Wang et al. 2006). This was recently confirmed in clinical samples by immunohistochemistry (Watts et al. 2007). While 12% of normal squamous cells were positive for Fn14 expression, positive staining was observed in 88% of EAC cells, and expression was strongest in the invasive tumor cells of advanced EAC. Finally, in breast cancer, 42% of invasive lobular carcinoma samples scored strongly positive for Fn14, while all ductal carcinoma in situ samples scored negatively or weakly (Willis et al. 2008). In addition, a positive correlation was identified between Fn14 expression and metastasis, positive lymph node status, and HER2-positive/ER-negative tumors (Willis et al. 2008). Taken together, there appears to be an increasing body of evidence that Fn14 expression positively correlates with tumor progression in multiple cancer types.

2.3 TWEAK Expression in Cancer

There are several reports suggesting that the ligand for Fn14, namely TWEAK, is expressed in the context of human tumors. Ho et al. examined TWEAK mRNA expression in a panel of tumor and normal paired specimens, and detected expression in all tumor and normal samples at variable levels, though particularly notable was the enhanced expression in kidney tumors relative to normal tissue (Ho et al. 2004). In breast cancer, TWEAK expression was evident in more than half of the tumor samples we examined, in contrast to normal tissue where only weak expression was observed in a small minority of samples (Michaelson et al. 2005). In a differential gene expression profiling study of breast tumors, TWEAK was identified as up-regulated in invasive lobular as compared to ductal carcinomas (Zhao et al. 2004). With respect to colon cancer, TWEAK mRNA and protein was detected in 4/4 human colonic adenocarcinoma cell lines (Kawakita et al. 2005). Interestingly, in glioma, TWEAK mRNA expression was reported as low in glioblastoma samples relative to normal brain tissue (Tran et al. 2003), and there was no correlation between TWEAK mRNA levels and tumor grade (Tran et al. 2006). It is important to point out that the source of TWEAK in tumors may not necessarily reflect expression in the tumor cells, but rather may derive from TWEAK expression in stromal cells such as tumor infiltrating macrophages and fibroblasts.

3 Pro-Tumorigenic Effects of TWEAK/Fn14 and Rationale for Therapeutic Inhibition of the Pathway

Up-regulation of Fn14 and/or TWEAK expression in tumors is perhaps not surprising, considering that multiple activities induced by this pathway are likely beneficial for a tumor cell and thereby may provide a differential growth advantage.

3.1 TWEAK Promotes Cell Survival

There is strong evidence in glioma that TWEAK has the capacity to promote tumor cell survival. When glioma cells are cultured with cytotoxic agents, TWEAK protects from apoptosis by inducing Bcl-2 family members (Tran et al. 2005). In fact, TWEAK was shown to directly induce expression of Bcl-X_L and BCL-W survival proteins, and the ability of TWEAK to promote survival being specifically dependent upon induction of Bcl-2 family members was confirmed by siRNA studies (Tran et al. 2005). Recent evidence suggests that this process requires stabilization of Bcl-2 Antagonist of cell Death (BAD), which in turn is dependent on AKT function (Fortin et al. 2007). These data imply that blocking the TWEAK pathway may sensitize tumor cells to chemotherapeutic agents. This may be an important paradigm for the utility of TWEAK inhibition in cancer treatment, in that combination with chemotherapeutic agents may maximize their therapeutic potential.

3.2 TWEAK Promotes Migration and Invasion

TWEAK can also promote migration and invasion. In glioma cell lines, TWEAK treatment enhances cell migration through Rac1 and NF-κB activation, and endogenous levels of migration are blocked by addition of soluble Fn14-Fc protein (Tran et al. 2003). Correlating with this, Fn14 expression is enhanced in glioma cells migrating at the rim as compared with core cells and, likewise, is up-regulated in patient-derived glioblastoma cells isolated from the tumor rim as compared with matched core (Tran et al. 2006). In a study of breast cancer cell lines, overexpression of Fn14 markedly induced invasion, whereas RNAi knockdown reduced the invasive capacity (Willis et al. 2008). We have similarly used a trans-well assay to show that TWEAK can promote invasion in pancreatic tumor cell lines (JSM and L.C.B., unpublished). The mechanism by which TWEAK promotes invasion may be through its ability to up-regulate matrix expression of MMP-9 (Michaelson et al. 2005; Winkles et al. 2006).

3.3 TWEAK Promotes Proliferation and Inhibits Differentiation

Another pro-tumorigenic activity of TWEAK is its potential to induce cell proliferation; however, there is limited data with regard to effects specifically in the context of tumor cell proliferation. In one report, TWEAK was shown to promote proliferation of multiple hepatocellular carcinoma cell lines in vitro (Kawakita et al. 2005). A related pro-tumorigenic feature of TWEAK is its capacity to inhibit differentiation (Ando et al. 2006, Girgenrath et al. 2006, Michaelson et al. 2005). Inhibition of functional differentiation by TWEAK was observed in several in vitro model systems, including Eph4 mammary epithelial cells grown in 3D culture (Michaelson et al. 2005), C2C12 muscle progenitor cells (Girgenrath et al. 2006), BMP-induced MC3T3-E1 osteoblast progenitors (Ando et al. 2006), primary osteoblasts and mesenchymal stem cells (Perper et al. 2006), and 3T3L1 pre-adipocytes (Burkly et al. 2007). Inhibition of progenitor cell differentiation is particularly intriguing in the context of cancer and tumor stem cells. It is now well documented that at least certain tumors contain a population of so-called *tumor stem cells* that are slow cycling and have the capacity to both self-renew and give rise to more differentiated tumor cells (Buzzeo et al. 2007, Pardal et al. 2003). The tumor stem cell population is therefore considered to be relatively resistant to existing therapeutic modalities. It is tempting to speculate that the ability of TWEAK to inhibit progenitor cell differentiation may also be at play in impeding differentiation of a progenitor tumor cell population, and therefore blocking the TWEAK pathway might promote a more differentiated and therefore less tumorigenic population and/or more therapeutically sensitive tumor cell population.

3.4 TWEAK Promotes Angiogenesis

In addition to its effects on tumor cells, TWEAK may be pro-tumorigenic through its pro-angiogenic effects on the tumor vasculature. The role of TWEAK in angiogenesis is well described, including the ability of TWEAK to promote endothelial cell proliferation, survival, and migration in vitro (Lynch et al. 1999; Donohue et al. 2003, Jakubowski et al. 2002). Moreover, in vivo studies showed that TWEAK can induce neovascularization in a rat cornea model comparable to VEGF (Lynch et al. 1999), while soluble Fn14-Fc could inhibit FGF-2 stimulated angiogenesis in the mouse cornea (Wiley et al. 2001), and anti-TWEAK mAbs can inhibit angiogenesis in a mouse model of RA (Kamata et al. 2006; Perper et al. 2006). More pertinent to cancer, TWEAK derived from colonic adenocarcinoma cells was specifically shown to induce endothelial cell tube formation in matrigel (Kawakita et al. 2005). Moreover, in an in vivo model relevant to tumor angiogenesis, HEK293 cells overexpressing TWEAK exhibited increased growth in nude mice, likely due to enhanced neovascularization (Ho et al. 2004). A further involvement of TWEAK and the vasculature is the ability of TWEAK to enhance permeability of the blood brain barrier (Polavarapu et al. 2005). The therapeutic implications of these vascular effects of TWEAK are that inhibition of TWEAK may normalize the leaky and highly permeable vasculature characteristic of tumors (Jain 2005), thereby normalizing interstitial pressure and improving the penetration and efficacy of co-administered chemotherapeutic agents.

3.5 TWEAK and Tumor Surveillance

An additional mechanism by which it is speculated that TWEAK may be protumorigenic is by repressing tumor surveillance. Recently it was reported that TWEAK deficient mice exhibit increased NK cell numbers and activity and are resistant to B16.F10 melanoma as a result of an enhanced innate and adaptive antitumor immunity (Maecker et al. 2005). However, it should be noted that independently derived TWEAK deficient mice do not exhibit increased NK cell numbers or enhanced adaptive immunity after neoantigen immunization (LCB., unpublished), nor has the effect of TWEAK on NK cells been reproduced by others (H. Yagita, Juntendo Univ. School of Medicine, pers. comm.) and therefore these results must be interpreted with caution.

3.6 Rationale for Therapeutic Inhibition of TWEAK in Cancer

Taken together, it is evident that there are multiple mechanisms by which TWEAK may be capable of promoting tumorigenesis, indicating that it may be therapeutically beneficial to block the TWEAK/Fn14 pathway as a modality to cancer. Inhibition of the TWEAK pathway could directly impact the tumor cells, alter the microenvironment, and potentially enhance the efficacy of chemotherapeutic agents. One recent abstract describes efficacy of an anti-TWEAK mAb in significantly inhibiting growth of CT26 colorectal tumor cells grown in vivo (Nakamura et al. 2008). Additional strategies for therapeutic inhibition of the pathway were recently reviewed (Winkles 2008). However, whether inhibition of the pathway will prove to be therapeutically beneficial in a broad range of tumors is yet to be determined.

4 The Other Side of the Coin: TWEAK/Fn14 and Tumor Cell Death

4.1 Fn14 as a Novel Tumor Antigen

Tumor antigens are sought as a means of specifically targeting cancer cells, and the finding that Fn14 expression is elevated in tumors relative to normal tissues suggests the identification of a novel tumor antigen target. Since Fn14 is expressed at relatively low levels in normal tissues, employment of a monoclonal antibody (mAb) targeting Fn14 could mediate ADCC, CDC, and/or enhance Fn14-mediated tumor cell killing via Fc-dependent receptor hyper-crosslinking, by means of interaction of the mAb with Fc receptors on immune effector cells to specifically target tumor cells (Fig. 1). This is analogous to the presumed mechanism(s) of action of several mAbs in the clinic, including Rituximab (anti-CD20) in lymphoma. The differential expression of Fn14 in tumor cells may be of even greater value in light of the opportunity to exploit the intrinsic cell killing activity of the pathway.

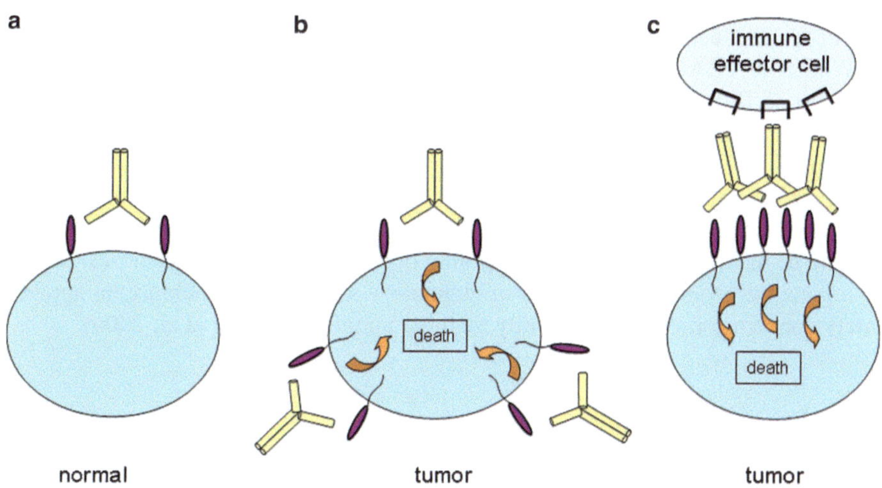

Fig. 1 Model for differential targeting of tumor cells as compared to normal cells to Fn14 agonist antibody mediated killing. As compared to normal cells (**a**), tumor cells express relatively elevated levels of Fn14 (**b**). In addition, tumor cells may be preferentially sensitive to Fn14-mediated cell killing, as suggested by Vince et al. (2008). Fc receptor engagement by immune effector cells (**c**) may further enhance the efficacy of an Fn14 agonist antibody via ADCC and/or receptor hypercrosslinking to augment the apoptotic signal

4.2 Precedence for Targeting Tumor Cell Killing Via TNF Receptors in Cancer

The TNF superfamily has historically represented an attractive approach for cancer therapy, given the associated tumor cell killing activity, dating back to early work with TNF and Fas/Apo1. However, while TNF and Fas/Apo1 ligands can effectively induce tumor cell apoptosis through activation of their respective death-domain receptors, toxicities related to systemic exposure have severely limited their development for cancer therapy. Nevertheless, alternative therapeutic strategies, for example, employing TNF fusion proteins or local delivery, are still being pursued (Daniel and Wilson 2008). More recently, TNF-Related Apoptosis Inducing Ligand (TRAIL/Apo2L) has emerged as a promising therapeutic approach (reviewed in Buchsbaum et al. 2006; Daniel and Wilson 2008). Treatment of mice bearing human tumors with recombinant soluble TRAIL or agonist antibodies to the TRAIL receptors, TRAIL-R1 (DR4) and TRAIL-R2 (DR5), have shown efficacy in tumor xenograft models, with enhanced activity when used in combination with chemotherapeutic agents or radiation therapy. Ongoing clinical trials with soluble TRAIL and antibodies to TRAIL receptors thus far demonstrate acceptable safety and tolerability and promising early signs of antitumor activity. Mechanistically speaking, TRAIL receptor stimulation results in recruitment of the adapter molecule Fas-associated death domain (FADD), thereby promoting recruitment of procaspases 8 and 10, leading to an assembly of the death inducing signaling complex (DISC) (reviewed in Daniel and Wilson 2008). Importantly, apoptosis does not appear to be activated in normal tissues, perhaps due to sensitization of cancer cells by oncogenes such as RAS and MYC (reviewed in Ashkenazi et al. 2008).

While the TRAIL receptors are characterized by an intracellular death domain, targeting of TNF family member receptors that lack a death domain, such as CD40, CD30, 41BB and LTβR, has also shown preclinical efficacy (Gladue et al. 2006; Lukashev et al. 2006; Wahl et al. 2002; Lynch 2008). In Phase I trials, CD40 agonist antibodies are showing early promise in that they appear to be well tolerated and biologically active (Vonderheide et al., 2001). Of note, modifying the administration scheme to include an initial, lower loading dose followed by higher doses looks to be a potential strategy to mitigate inflammatory symptoms, which are an apparent safety concern for some TNF family member agonists (Advani et al. 2006).

4.3 Fn14 and Tumor Cell Killing

Fn14 is likewise a non-death domain containing TNF family member receptor, whereby stimulation of the receptor nevertheless results in cell death. TWEAK's ability to weakly induce apoptosis of HT29 adenocarcinoma cells in vitro was initially described (Chicheportiche et al. 1997). Subsequently, it was shown that in some cell lines, such as Kym-1, the cell death is indirectly mediated by TNF, whereas in other cell lines, including HSC3, HT-29, and KATO-III, the activity is TNF-independent (Nakayama et al., 2002; Schneider et al. 1999). Moreover, the cell death induced by TWEAK may occur through multiple pathways, including caspase-dependent and caspase-independent, with features of both apoptosis and cathepsin-B dependent necrosis (Wilson and Browning 2002; Nakayama et al. 2003), and is wholly mediated by Fn14 (Nakayama et al. 2003). Cell death mediated by TWEAK requires extended incubation periods and generally requires co-incubation with sensitizing agents such as IFNγ; however, a reasonably broad range of tumor cell lines are sensitive to TWEAK-induced killing (Vince et al. 2008; JSM and LCB, unpublished). Notably, there appears to be differential sensitivity of transformed cells as compared to normal cells in their response to TWEAK. MEFs immortalized with SV40 large T antigen showed a 14-fold increase in cell death in response to TWEAK as compared to primary MEFs despite comparable Fn14 expression (Vince et al. 2008). These data support the intriguing possibility that targeting Fn14 will selectively kill tumor cells and spare normal cells, not only because of the upregulated Fn14 expression by tumor cells but also because of their differential sensitivity to Fn14-mediated cell death. Further studies are needed to further test this hypothesis. We have shown that primary, passaged human umbilical cord endothelial cells (HUVEC) are not killed in response to TWEAK (JSM and LCB, unpublished). There have been only a limited number of reports where TWEAK was shown to induce cell death in normal, primary cell types in vitro, usually requiring IFNγ, including human peripheral blood monocytes and NK cells and murine cortical neurons (reviewed in Burkly et al. 2007). There are only two examples of TWEAK induced cell death in vivo, namely in the context of stroke (Potrovita et al. 2004), although neuronal cell death after cerebral ischemia may be secondary to the TWEAK-dependent inflammatory response, and in the process of mammary gland involution (J.S.M. and L.C.B., unpublished).

4.4 Mechanisms of Fn14-Induced Cell Death

An obvious conundrum yet to be unraveled pertains to the mechanism by which Fn14, a non-death domain containing receptor, induces tumor cell death. While TWEAK has been shown to activate both canonical and noncanonical NF-κB pathways (Brown et al. 2003; Saitoh et al. 2003), as well as the MAPK and AKT pathways (Ando et al. 2006), these signaling pathways are generally not associated with induction of apoptosis. Stimulation of the NF-κB signaling pathway largely results in transcription of pro-inflammatory target genes (Karin 2006), and in the case of TNF-R1, NF-κB signaling is understood to oppose the death signal mediated through FADD (Daniel and Wilson 2008). Similarly, activation of the MAPK and AKT pathways is primarily associated with proliferative signals (Leicht et al. 2007). Thus, the question remains as to how the death signal is transduced. It is tempting to speculate that perhaps the capacity of TWEAK to activate the noncanonical NF-κB pathway (Hacker and Karin 2002; Wang et al. 2002), that is, induction of NF-κB2 p100, accounts for its apoptotic effects. Interestingly, the ability of TWEAK/Fn14 to trigger sustained NF-κB activation through the alternative pathway is consistent with the extended incubation periods required for induction of apoptosis. A recent paper sheds some new light on the mechanistic aspects of TWEAK and apoptosis (Vince et al. 2008). Vince and colleagues show that TWEAK induces recruitment of a complex consisting of TRAF2 and cellular Inhibitor of Apoptosis (cIAP), as well as NF-κB dependent up-regulation of TNF. The TRAF2–cIAP complex is subsequently degraded in a lysosomal compartment, resulting in sensitization of the cells to killing by TNF. However, one must be cautious in interpreting the TNF dependence of TWEAK-induced cell death. Although TWEAK-induced killing of Kym-1 cells, one of the same cell lines used in Vince et al. (2008), was TNF dependent (Schneider et al. 1999), Nakayama et al. (2002) examined several additional tumor cell lines and concluded that TWEAK induced apoptosis is generally not mediated by TNF, and we have similarly observed TNF-independent cell death in several cell lines (J.S.M. unpublished). Nevertheless, Vince et al. (2008) provide new insight into a putative TRAF2–cIAP complex being involved in the mechanism of TWEAK mediated cell killing, and further experiments are warranted to establish the relevance of TRAF2–cIAP2 complexes to TWEAK-induced killing across a spectrum of tumor cell lines.

4.5 *Agonist Approach to Therapeutic Targeting of Fn14 in Cancer*

Intriguingly, emerging data suggests that activation of the TWEAK pathway to promote tumor cell killing may be effective in vivo. A monoclonal antibody to Fn14, capable of inducing tumor cell apoptosis in vitro, was efficacious in a range of tumor xenograft models, including colorectal, breast, renal, and head/neck as recently reported (Culp et al. 2008). We have independently shown that administration of adenovirally delivered TWEAK, an Fc-TWEAK fusion protein, or an agonist antibody to Fn14 resulted in dramatic reduction in tumor growth in xenograft models (J.S.M. and L.C.B. unpublished). These exciting preliminary findings suggest that therapeutic activation of the TWEAK/Fn14 pathway may represent a novel modality to inhibit tumor growth (Fig. 1). In considering the mechanistic potential of targeting Fn14 with an agonist antibody to trigger intrinsic tumor cell killing activity, there may be added efficacy afforded by the Fc portion of the antibody in mediating antitumor activity (e.g., via ADCC, CDC, and/or receptor hypercrosslinking). The contribution of ADCC activity to the observed in vivo tumor inhibition was suggested (Culp et al. 2008). Finally, the exciting possibility that transformed cells are preferentially sensitive to TWEAK killing as compared to nontransformed cells (Vince et al. 2008) is encouraging in considering the tolerability of systemic administration of an Fn14 agonist therapeutic. Further studies to support this exciting new therapeutic approach, that is, targeting cancer with agonists to the TWEAK/Fn14 pathway, are in progress.

5 Conclusions

There is increasing evidence that Fn14 expression is elevated in a wide variety of tumors, including breast, pancreas, esophageal, and glioma. Likely, overexpression in tumors arises precisely since it may be beneficial for tumors to up-regulate this pathway to promote growth, survival, mobility and angiogenesis. As such, there is rationale for inhibition of the TWEAK/Fn14 pathway as a therapeutic modality for treating cancer. However, perhaps more exciting is the notion of exploiting the differential expression of Fn14 in human cancers to specifically target killing of tumor cells. This powerful therapeutic approach involves harnessing the intrinsic apoptotic capacity of this pathway to induce tumor killing through Fn14. Triggering with an agonist antibody may have increased benefit through Fc-dependent antitumor activity. Notably, the fact that Fn14 expression correlates with disease progression and metastatic status gives the prospect for a new therapeutic in advanced disease, where new therapies are most needed. Moreover, the suggestion of differential specificity of killing toward tumor rather than normal cells may be promising for the possibility of having a therapeutic window for treatment with a pathway agonist. Indeed, preliminary in vivo proof-of-concept preclinical studies show exciting promise in this area. Further investigation will undoubtedly be necessary to tease out whether therapeutic targeting of the TWEAK/Fn14 pathway in cancer will ultimately prove to be clinically beneficial.

Acknowledgments We acknowledge our colleagues at Biogen Idec, who have contributed to our conceptual understanding of the TWEAK/Fn14 pathway in cancer and provided unpublished data to support our conclusions: Aldo Amatucci, Jeff Browning, Ellen Garber, Yen-Ming Hsu, Timothy Zheng, and the Oncopharmacology group.

References

Advani R, Forero-Rorres A, Furman RR, Rosenblatt JD, Younes A, Shankles B, Harrop K, Drachman JG (2006) SGN-40 (Anti-huCD40 mAb) monotherapy induces durable objective responses in patients with relapsed aggressive non-hodgkin's lymphoma: evidence of antitumor activity from a Phase I study (2006). Blood (ASH Annual Meeting Abstracts) 108:Abstract 695

Almstrup K, Hoei-Hansen CE, Wirkner U, Blake J, Schwager C, Ansorge W, Nielsen JE, Skakkebaek NE, Rajpert-De Meyts E, Leffers H (2004) Embryonic stem cell-like features of testicular carcinoma in situ revealed by genome-wide gene expression profiling. Cancer Res 64:4736–4743

Ando T, Ichikawa J, Wako M, Hatsushika K, Watanabe Y, Sakuma M, Tasaka K, Ogawa H, Hamada Y, Yagita H, Nakao A (2006) TWEAK/Fn14 interaction regulates RANTES production, BMP-2-induced differentiation, and RANKL expression in mouse osteoblastic MC3T3-E1 cells. Arthrit Res Ther 8:R146

Ashkenazi A, Holland P, Eckhardt SG (2008) Ligand-based targeting of apoptosis in cancer: the potential of recombinant human apoptosis ligand2/Tumor Necrosis Factor-Related Apoptosis inducing Ligand. J Clin Oncol 26:3621–3630

Bover LC, Cardo-Vila M, Kuniyasu A, Sun J, Rangel R, Takeya M, Aggarwal BB, Arap W, Pasqualini R (2007) A previously unrecognized protein-protein interaction between TWEAK and CD163: potential biological implications. J Immunol 178:8183–8194

Brown SA, Richards CM, Hanscom HN, Feng SL, Winkles JA (2003) The Fn14 cytoplasmic tail binds tumour-necrosis-factor-receptor-associated factors 1, 2, 3 and 5 and mediates nuclear factor-kappaB activation. Biochem J 371:395–403

Buchsbaum DJ, Zhou T, Lobuglio AF (2006) TRAIL receptor-targeted therapy. Fut Oncol (London, England) 2:493–508

Burkly LC, Michaelson JS, Hahm K, Jakubowski A, Zheng TS (2007) TWEAKing tissue remodeling by a multifunctional cytokine: role of TWEAK/Fn14 pathway in health and disease. Cytokine 40:1–16

Buzzeo MP, Scott EW, Cogle CR (2007) The hunt for cancer-initiating cells: a history stemming from leukemia. Leukemia 21:1619–1627

Campbell S, Burkly LC, Gao HX, Berman JW, Su L, Browning B, Zheng T, Schiffer L, Michaelson JS, Putterman C (2006) Proinflammatory effects of TWEAK/Fn14 interactions in glomerular mesangial cells. J Immunol 176:1889–1898

Chicheportiche Y, Bourdon PR, Xu H, Hsu YM, Scott M, Hession C, Garcia I, Browning JL (1997) TWEAK, a new secreted ligand in the tumor necrosis factor family that weakly induces apoptosis. J Biol Chem 272:32401–32410

Daniel D, Wilson NS (2008) Tumor necrosis factor: renaissance as a cancer therapeutic? Curr Cancer Drug Targets 8:124–131

De Ketelaere A, Vermeulen L, Vialard J, Van De Weyer I, Van Wauwe J, Haegeman G, Moelans I (2004) Involvement of GSK-3beta in TWEAK-mediated NF-kappaB activation. FEBS Lett 566:60–64

Dogra C, Hall SL, Wedhas N, Linkhart TA, Kumar A (2007) Fibroblast growth factor inducible 14 (Fn14) is required for the expression of myogenic regulatory factors and differentiation of myoblasts into myotubes. Evidence for TWEAK-independent functions of Fn14 during myogenesis. J Biol Chem 282:15000–15010

Donohue PJ, Richards CM, Brown SA, Hanscom HN, Buschman J, Thangada S, Hla T, Williams MS, Winkles JA (2003) TWEAK is an endothelial cell growth and chemotactic factor that also potentiates FGF-2 and VEGF-A mitogenic activity. Arterioscler Thromb Vasc Biol 23:594–600

Feng SL, Guo Y, Factor VM, Thorgeirsson SS, Bell DW, Testa JR, Peifley KA, Winkles JA (2000) The Fn14 immediate-early response gene is induced during liver regeneration and highly expressed in both human and murine hepatocellular carcinomas. Am J Path 156:1253–1261

Fortin S, Tran N, Drake K, Savitch B, Ennis M, Winkles J, Loftus J, Berens M (2007) Fn14 induced glioma cell survival is dependent upon Akt2 function. AACR Meeting Abstracts 281.

Girgenrath M, Weng S, Kostek CA, Browning B, Wang M, Brown SA, Winkles JA, Michaelson JS, Allaire N, Schneider P, Scott ML, Hsu YM, Yagita H, Flavell RA, Miller JB, Burkly LC, Zheng TS (2006) TWEAK, via its receptor Fn14, is a novel regulator of mesenchymal progenitor cells and skeletal muscle regeneration. EMBO J 25:5826–5839

Hacker H, Karin M (2002) Is NF-kappaB2/p100 a direct activator of programmed cell death? Cancer Cell 2:431–433

Han H, Bearss DJ, Browne LW, Calaluce R, Nagle RB, Von Hoff DD (2002) Identification of differentially expressed genes in pancreatic cancer cells using cDNA microarray.[erratum appears in Cancer Res 2002 Aug;62(15):4532]. Cancer Res 62:2890–2896

Ho DH, Vu H, Brown SA, Donohue PJ, Hanscom HN, Winkles JA (2004) Soluble tumor necrosis factor-like weak inducer of apoptosis overexpression in HEK293 cells promotes tumor growth and angiogenesis in athymic nude mice. Cancer Res 64:8968–8972

Jain RK (2005) Normalization of tumor vasculature: an emerging concept in antiangiogenic therapy. Science 307:58–62

Jakubowski A, Browning B, Lukashev M, Sizing I, Thompson JS, Benjamin CD, Hsu YM, Ambrose C, Zheng TS, Burkly LC (2002) Dual role for TWEAK in angiogenic regulation. J Cell Sci 115:267–274

Kamata K, Kamijo S, Nakajima A, Koyanagi A, Kurosawa H, Yagita H, Okumura K (2006) Involvement of TNF-like weak inducer of apoptosis in the pathogenesis of collagen-induced arthritis. J Immunol 177:6433–6439

Karin M (2006) Nuclear factor-kappaB in cancer development and progression. Nature 441:431–436
Kawakita T, Shiraki K, Yamanaka Y, Yamaguchi Y, Saitou Y, Enokimura N, Yamamoto N, Okano H, Sugimoto K, Murata K, Nakano T (2005) Functional expression of TWEAK in human colonic adenocarcinoma cells. Int J Oncol 26:87–93
Leicht DT, Balan V, Kaplun A, Singh-Gupta V, Kaplun L, Dobson M, Tzivion G (2007) Raf kinases: function, regulation and role in human cancer. Biochim Biophys Acta 1773:1196–1212
Lukashev M, LePage D, Wilson C, Bailly V, Garber E, Lukashin A, Ngam-ek A, Zeng W, Allaire N, Perrin S, Xu X, Szeliga K, Wortham K, Kelly R, Bottiglio C, Ding J, Griffith L, Heaney G, Silverio E, Yang W, Jarpe M, Fawell S, Reff M, Carmillo A, Miatkowski K, Amatucci J, Crowell T, Prentice H, Meier W, Violette SM, Mackay F, Yang D, Hoffman R, Browning JL (2006) Targeting the lymphotoxin-beta receptor with agonist antibodies as a potential cancer therapy. Cancer Res 66:9617–9624
Lynch CN, Wang YC, Lund JK, Chen YW, Leal JA, Wiley SR (1999) TWEAK induces angiogenesis and proliferation of endothelial cells. J Biol Chem 274:8455–8459
Lynch DH (2008) The promise of 4–1BB (CD137)-mediated immunomodulation and the immunotherapy of cancer. Immunol Rev 222:277–286
Maecker H, Varfolomeev E, Kischkel F, Lawrence D, LeBlanc H, Lee W, Hurst S, Danilenko D, Li J, Filvaroff E, Yang B, Daniel D, Ashkenazi A (2005) TWEAK attenuates the transition from innate to adaptive immunity. Cell 123:931–944
Meighan-Mantha RL, Hsu DK, Guo Y, Brown SA, Feng SL, Peifley KA, Alberts GF, Copeland NG, Gilbert DJ, Jenkins NA, Richards CM, Winkles JA (1999) The mitogen-inducible Fn14 gene encodes a type I transmembrane protein that modulates fibroblast adhesion and migration. J Biol Chem 274:33166–33176
Michaelson JS, Cho S, Browning B, Zheng TS, Lincecum JM, Wang MZ, Hsu YM, Burkly LC (2005) Tweak induces mammary epithelial branching morphogenesis. Oncogene 24:2613–2624
Nakamura S, Sho M, Koyama F, Nomi T, Akahori T, Enomoto K, Yamato I, Nakagawa T, Uchimoto K, Otsuki W, Wakatsuki K, Yagita H, Nakajima Y(2008) Clinical importance and therapeutic potential of TWEAK/Fn14 pathway in human colorectal cancer. AACR Meeting Abstracts 984
Nakayama M, Ishidoh K, Kayagaki N, Kojima Y, Yamaguchi N, Nakano H, Kominami E, Okumura K, Yagita H (2002) Multiple pathways of TWEAK-induced cell death. J Immunol 168:734–743
Nakayama M, Harada N, Okumura K, Yagita H (2003) Characterization of murine TWEAK and its receptor (Fn14) by monoclonal antibodies. Biochem Biophys Res Comm 306:819–825
Pardal R, Clarke MF, Morrison SJ (2003) Applying the principles of stem-cell biology to cancer. Nat Rev Cancer 3:895–902
Perper SJ, Browning B, Burkly LC, Weng S, Gao C, Giza K, Su L, Tarilonte L, Crowell T, Rajman L, Runkel L, Scott M, Atkins GJ, Findlay DM, Zheng TS, Hess H (2006) TWEAK is a novel arthritogenic mediator. J Immunol 177:2610–2620
Polavarapu R, Gongora MC, Winkles JA, Yepes M (2005) Tumor necrosis factor-like weak inducer of apoptosis increases the permeability of the neurovascular unit through nuclear factor-kappa B pathway activation. J Neurosci 25:10094–10100
Polek TC, Talpaz M, Darnay BG, Spivak-Kroizman T (2003) TWEAK mediates signal transduction and differentiation of RAW264.7 cells in the absence of Fn14/TweakR. Evidence for a second TWEAK receptor. J Biol Chem 278:32317–32323
Potrovita I, Zhang W, Burkly L, Hahm K, Lincecum J, Wang MZ, Maurer MH, Rossner M, Schneider A, Schwaninger M (2004) Tumor necrosis factor-like weak inducer of apoptosis-induced neurodegeneration. J Neurosci 24:8237–8244
Saitoh T, Nakayama M, Nakano H, Yagita H, Yamamoto N, Yamaoka S (2003) TWEAK induces NF-κB2 p100 processing and long lasting NF-κB activation. J Biol Chem 278:36005–36012
Schneider P, Schwenzer R, Haas E, Muhlenbeck F, Schubert G, Scheurich P, Tschopp J, Wajant H (1999) TWEAK can induce cell death via endogenous TNF and TNF receptor 1. Eur J Immunol 29:1785–1792

Tanabe K, Bonilla I, Winkles JA, Strittmatter SM (2003) Fibroblast growth factor-inducible-14 is induced in axotomized neurons and promotes neurite outgrowth. J Neurosci 23:9675–9686
Tran NL, McDonough WS, Donohue PJ, Winkles JA, Berens TJ, Ross KR, Hoelzinger DB, Beaudry C, Coons SW, Berens ME (2003) The human Fn14 receptor gene is up-regulated in migrating glioma cells in vitro and overexpressed in advanced glial tumors. Am J Pathol 162:1313–1321
Tran NL, McDonough WS, Savitch BA, Sawyer TF, Winkles JA, Berens ME (2005) The tumor necrosis factor-like weak inducer of apoptosis (TWEAK)-fibroblast growth factor-inducible 14 (Fn14) signaling system regulates glioma cell survival via NFkappaB pathway activation and BCL-XL/BCL-W expression. J Biol Chem 280:3483–3492
Tran NL, McDonough WS, Savitch BA, Fortin SP, Winkles JA, Symons M, Nakada M, Cunliffe HE, Hostetter G, Hoelzinger DB, Rennert JL, Michaelson JS, Burkly LC, Lipinski CA, Loftus JC, Mariani L, Berens ME (2006) Increased fibroblast growth factor-inducible 14 expression levels promote glioma cell invasion via Rac1 and nuclear factor-{kappa}B and correlate with poor patient outcome. Cancer Res 66:9535–9542
Vince JE, Chau D, Callus B, Wong WW, Hawkins CJ, Schneider P, McKinlay M, Benetatos CA, Condon SM, Chunduru SK, Yeoh G, Brink R, Vaux DL, Silke J (2008) TWEAK-FN14 signaling induces lysosomal degradation of a cIAP1-TRAF2 complex to sensitize tumor cells to TNFalpha. J Cell Biol 182:171–184
Vonderheide RH, Dutcher JP, Anderson JE, Eckhardt SG, Stephans KF, Razvillas B, Garl S, Butine MD, Perry VP, Armitage RJ, Ghalie R, Caron DA, Gribben JG (2001) Phase I study of recombinant human CD40 ligand in cancer patients. J Clin Oncol 19:3280–3287
Wahl AF, Klussman K, Thompson JD, Chen JH, Francisco LV, Risdon G, Chace DF, Siegall CB, Francisco JA (2002) The anti-CD30 monoclonal antibody SGN-30 promotes growth arrest and DNA fragmentation in vitro and affects antitumor activity in models of Hodgkin's disease. Cancer Res 62:3736–3742
Wang S, Zhan M, Yin J, Abraham JM, Mori Y, Sato F, Xu Y, Olaru A, Berki AT, Li H, Schulmann K, Kan T, Hamilton JP, Paun B, Yu MM, Jin Z, Cheng Y, Ito T, Mantzur C, Greenwald BD, Meltzer SJ (2006) Transcriptional profiling suggests that Barrett's metaplasia is an early intermediate stage in esophageal adenocarcinogenesis. Oncogene 25:3346–3356
Wang Y, Cui H, Schroering A, Ding JL, Lane WS, McGill G, Fisher DE, Ding HF (2002) NF-kappa B2 p100 is a pro-apoptotic protein with anti-oncogenic function. Nature Cell Biol 4:888–893
Watts GS, Tran NL, Berens ME, Bhattacharyya AK, Nelson MA, Montgomery EA, Sampliner RE (2007) Identification of Fn14/TWEAK receptor as a potential therapeutic target in esophageal adenocarcinoma. Int J Cancer 121:2132–2139
Wiley SR, Cassiano L, Lofton T, Davis-Smith T, Winkles JA, Lindner V, Liu H, Daniel TO, Smith CA, Fanslow WC (2001) A novel TNF receptor family member binds TWEAK and is implicated in angiogenesis. Immunity 15:837–846
Willis AL, Tran NL, Chatigny JM, Charlton N, Vu H, Brown SA, Black MA, McDonough WS, Fortin SP, Niska JR, Winkles JA, Cunliffe HE (2008) The fibroblast growth factor-inducible 14 receptor is highly expressed in HER2-positive breast tumors and regulates breast cancer cell invasive capacity. Mol Cancer Res 6:725–734
Wilson CA, Browning JL (2002) Death of HT29 adenocarcinoma cells induced by TNF family receptor activation is caspase-independent and displays features of both apoptosis and necrosis. Cell Death Diff 9:1321–1333
Winkles JA (2008) The TWEAK-Fn14 cytokine-receptor axis: discovery, biology and therapeutic targeting. Nat Rev Drug Discov 7:411–425
Winkles JA, Tran NL, Berens ME (2006) TWEAK and Fn14: new molecular targets for cancer therapy? Cancer Lett 235:11–17
Zhao H, Langerod A, Ji Y, Nowels KW, Nesland JM, Tibshirani R, Bukholm IK, Karesen R, Botstein D, Borresen-Dale AL, Jeffrey SS (2004) Different gene expression patterns in invasive lobular and ductal carcinomas of the breast. Mol Biol Cell 15:2523–2536
Zheng T, Burkly LC (2008) No end in site: TWEAK/Fn14 activation and autoimmunity associated- end-organ pathologies. J Leukoc Biol 84:1–10

APRIL in B-cell Malignancies and Autoimmunity

Fiona C. Kimberley, Jan Paul Medema, and Michael Hahne

Abstract A Proliferation Inducing Ligand (APRIL) was first identified as a cytokine expressed predominantly by tumour tissues and was not found in most normal tissues. The activity of this new cytokine, in terms of its ability to stimulate tumour cell proliferation in vivo, determined the catchy acronym of yet another TNF family cytokine: APRIL. Reports showing an association between APRIL and cancer have since been prolific, in particular, those showing a link with B cell malignancies. Evidence is accumulating that APRIL is also a player in several autoimmune diseases, including systemic lupus erythematosus, rheumatoid arthritis, and Sjoegren's syndrome. However, we now know that APRIL also plays an important role in the immune system and in lymphocyte biology. In this chapter we outline the physiological role of APRIL in immunity and describe what is known regarding the role of APRIL in B cell malignancies and autoimmune disease.

1 The APRIL Protein

APRIL is expressed as a type-II transmembrane protein, but unlike most other TNF family members, it is never expressed at the cell membrane. Instead, it is cleaved in the Golgi apparatus by a furin convertase to release a soluble active form, which is subsequently secreted (Lopez-Fraga et al. 2001). Like other TNF family members APRIL assembles as a non-covalently linked homo-trimer with similar structural homology in protein fold to a number of other TNF family ligands

M. Hahne(✉)
Institut de Génétique Moléculaire de Montpellier (IGMM), CNRS UMR5535,
1919, route de Mende, 34293 Montpellier cedex 5, France
e-mail: hahne@igmm.cnrs.fr

F.C. Kimberley and J.P. Medema
LEXOR (Lab of Exp. Oncology and Radiobiology), Center for Experimental Molecular Medicine,
Academic Medical Center (AMC), University of Amsterdam, Room G1-114, 1105 AZ,
Amsterdam, The Netherlands
e-mail: F.C.Kimberley@amc.uva.nl

Results Probl Cell Differ, doi:10.1007/400_2008_19

(Wallweber et al. 2004). The human APRIL gene contains six exons, which can be transcribed as three alternatively spliced mRNAs, encoding β, γ and δ forms of the protein, which result in a shorter form, a slightly longer form (seven amino acids) or a membrane-bound non-cleavable form of APRIL, respectively (Kelly et al. 2000; Bossen and Schneider 2006). However, the relevance of these different forms is not known and similar sites are not found within the mouse gene. In addition, a unique intergenic splicing event creates a fusion between the transmembrane domain of TWEAK (a weak homologue of TNF) and the extracellular domain of APRIL, resulting in a membrane bound form of APRIL that is not cleaved inside the cell (Pradet-Balade et al. 2002; Kolfschoten et al. 2003). Unfortunately, this so-called TWE-PRIL is rather understudied and consequently there is little known about its expression and localisation, or when and where it is functionally relevant. It is possible that this form of APRIL is cleaved at the cell membrane by proteases and that APRIL generated in this fashion can also contribute to the soluble APRIL pool.

APRIL shows high homology (30%) to another member of the TNF superfamily, B cell activating factor belonging to the TNF family (BAFF or B Lymphocyte stimulator, BLyS). APRIL and BAFF share binding to two receptors of the TNF receptor superfamily: B cell maturation antigen (BCMA) and transmembrane activator and cyclophilin ligand interactor (TACI) (Fig. 1) (Marsters et al. 2000; Rennert et al. 2000; Wu et al. 2000; Yu et al. 2000). Since APRIL and BAFF bind the same receptors there is consequently large overlap in their function and it has been suggested that they can even form mixed trimers (Roschke et al. 2002). Such mixed trimers were found to occur at a higher prevalence in the serum of rheumatoid arthritis patients, though the significance of this is not yet known. BAFF also binds with high affinity to a unique receptor, BAFF-receptor (BAFF-R) (Thompson et al. 2001; Yan et al. 2001a) and APRIL has recently been shown to bind heparan sulfate proteoglycans (HSPGs) (Hendriks et al. 2005; Ingold et al. 2005). The HSPG binding domain of APRIL is located at the N-terminus and consists of two regions: (1) the consensus HSPG binding motif (QKQKKQ) and (2) three basic amino acids: R146, H220 and R189 (Hendriks et al. 2005; Ingold et al. 2005).

APRIL is predominantly expressed by immune cell subsets such as monocytes, macrophages, dendritic cells, neutrophils and T-cells, many of which also express BAFF. Significantly, both ligands are expressed on activated B cells both in vitro and in vivo and by resting B1 cells and developing B cells in the bone marrow. The expression of APRIL by many of these cells occurs in response to cytokines, via the engagement of CD40 or via the activation of pattern recognition receptors (PRRs) (Litinskiy et al. 2002; Craxton et al. 2003). Monocyte derived dendritic cells (MoDCs) have been shown to secrete large amounts of APRIL in response to CpG (Hardenberg et al. 2007). APRIL can also be expressed by some non-immune cells such as osteoclasts, epithelial cells and a variety of tumour tissues (Moreaux et al. 2005; He et al. 2007). Since APRIL has been shown to bind and stimulate several cell lines that do not express either BCMA or TACI, it is possible that the HSPG interaction may represent a third receptor, or function as an important co-factor

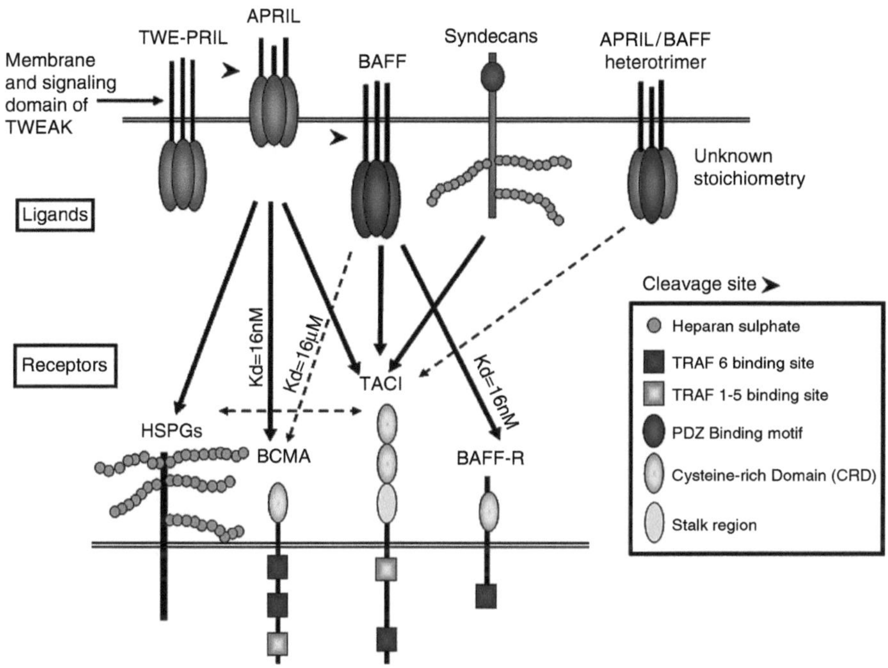

Fig. 1 APRIL and BAFF Interactions. APRIL and BAFF are homologous members of the TNF ligand superfamily. Both bind to the TNF receptors BCMA (B cell maturation antigen) and TACI (Transmembrane activator and calcium modulator and cyclophilin ligand interactor), though with differing affinities. BAFF also binds its unique receptor, BAFF-receptor, and APRIL has been shown to interact with HSPGs. In addition, it has been shown that certain syndecans (prototypic HSPGs) can both bind and activate signalling via TACI. Little is known about the role of these interactions in the context of APRIL or BAFF signalling. *Affinities are taken from Day et al. Biochemistry. 2005; 44(6):1919–1931

for an as yet unidentified interaction (Rennert et al. 2000). However, recent studies of our own have shown that the HSPG binding domain of APRIL serves as a platform for natural ligand cross-linking and multimerisation, which means that HSPGs in the extracellular matrix or other proteoglycans serve to cross-link APRIL for effective signalling (Kimberley et al., manuscript submitted). Without this binding capacity APRIL was non-functional in a number of different assays. So, HSPGs are necessary for natural cross-linking to present APRIL to the receptors, but it could be imagined that in areas of dense HSPG-deposition, such as certain extracellular matrix, APRIL can be super-clustered and thus presented to the receptor as multimers.

Despite homology between BAFF and APRIL, receptor specificity may be partly achieved by differential expression patterns of the ligands and different triggers for production, by differences in the affinity of the ligands for each of the two receptors or by the requirements of the receptors themselves for the way in which the ligand is presented. It has been suggested that TACI has a specific requirement for oligomerised ligands (Bossen et al. 2008). Since BAFF was crystallised as both a trimer and also a higher multimeric form, which reportedly occurs physiologically, it may be that these higher forms of BAFF preferentially signal via TACI. This would also suggest that TACI responds preferentially to membrane-bound ligand, which would suggest a more prominent role for TWE-PRIL in TACI-dependent responses or the requirement for 'super-clustering' of soluble APRIL by HSPGs, though further work is necessary to make this distinction. A reliance on HSPGs for effective positioning is illustrated by FGF, where binding to heparin induces the formation of dimers, which are the biologically active form (Moy et al. 1997).

2 The Role of APRIL in the Immune System

2.1 APRIL Receptors and Signalling

BCMA and TACI are both type II transmembrane proteins, which lack a signal peptide. Structurally they are similar to other TNFRs with characteristic cysteine-rich domains (CRDs) in the extracellular region. TACI consists of two CRDs: CRD2 is ligand binding while CRD1 has been shown to form a pre-ligand assembly domain (PLAD), first described for Fas and TNFR1 (Chan et al. 2000; Siegel et al. 2000). The PLAD holds the receptor in pre-assembled complexes to await the incoming ligand. BCMA is a small receptor consisting of only one CRD, which is also ligand binding, and so it is unlikely that this single CRD also constitutes a PLAD, though this has not yet been studied. APRIL has been crystallised as a soluble ligand as well as in complex with BCMA, and the CRD2 (ligand binding portion) of TACI (Hymowitz et al. 2005). These structures revealed that the receptor binding interface occurs at the C-terminal end of each APRIL in the trimer, leaving the HSPG domain free from receptor interactions at the other end. Unlike other TNFR–ligand complexes, such as DR5 and TRAIL, TACI and BCMA receptor monomers bind directly to one APRIL strand within the trimer, rather than binding at the interface between monomers in each trimer.

Ligation of both BCMA and TACI leads to downstream activation of the classical NF-κB pathway, and it is activation of this pathway that is thought to be at the crux of the proliferative signals delivered to malignant B-cells (Fig. 2). Signalling via both BCMA and TACI occurs via the recruitment of intracellular adaptors, termed TNF-receptor associated factors (TRAFS). TACI was originally shown to be a CAML interacting receptor, which can signal via the NFAT/AP-1 pathway, but it was later shown that ligation of TACI can also signal activation of NF-κB and c-Jun NH2-terminal Kinase (JNK) (Xia et al. 2000). Using yeast two-hybrid assays, the intracellular portion of TACI was shown to bind TRAF -2, -5 and -6; it was also found that the TRAF- and CAML-binding sites are distinct (von Bulow et al. 2000;

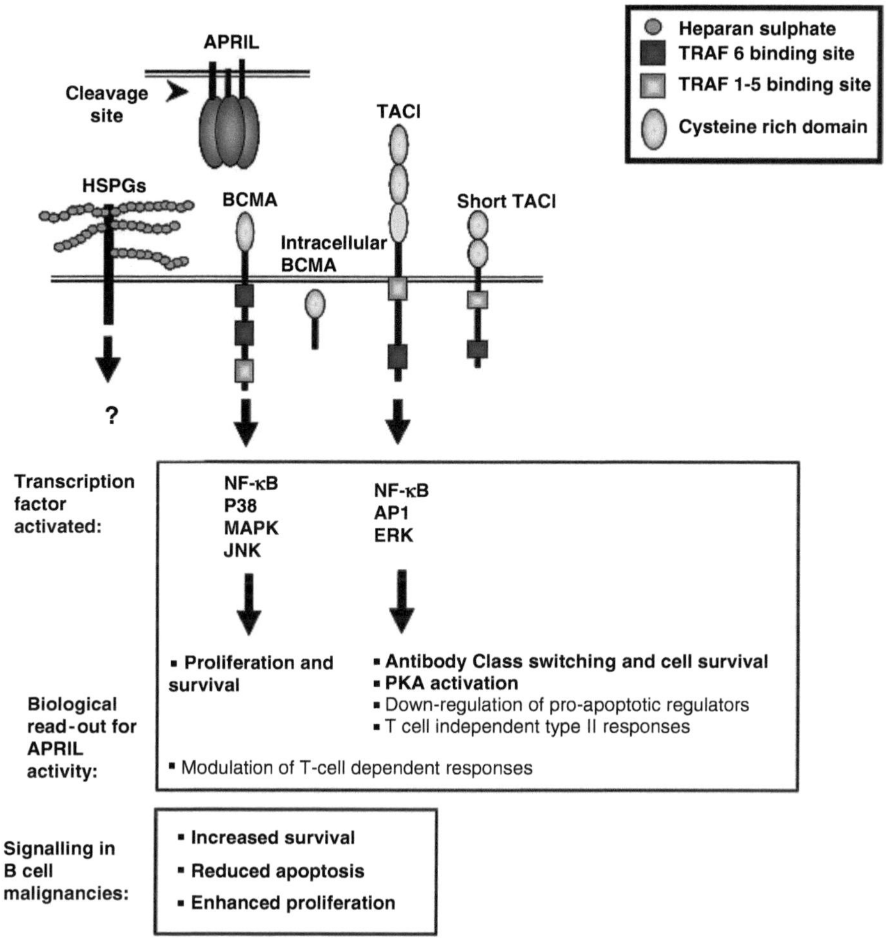

Fig. 2 APRIL Signalling. APRIL activates the canonical NF-KB pathway via both BCMA and TACI to stimulate proliferation and survival. Specific signalling via BCMA has been shown to drive activation of p38, MAP kinase and JNK pathways, again to stimulate either survival and/or proliferation. APRIL signalling via TACI was shown to be responsible for class switch recombination of IgG and IgA, as well as cell survival, PKA activation and the down-regulation of several apoptotic regulators

Xia et al. 2000). BCMA has also been shown to activate p38 mitogen activating protein kinase (MAPK) and c-Jun NH2-terminal Kinase (JNK) (Hatzoglou et al. 2000). However, the detailed complex formation following receptor ligation has not been well-studied for either receptor, and indeed the specific signalling components required to deliver distinct signals via the same receptor are also ill-defined.

The expression profile for the APRIL receptors is still not fully clear, but it appears that they are expressed on B cells at various levels depending on their maturation and activation state. TACI is expressed on distinct B cell populations and is up-regulated following B cell stimulation and has been found on some T cell subsets (von Bulow et al. 2000). BCMA expression is more restricted to the differentiated B cells such as plasma cells, plasmablasts and tonsillar GC B cells and seems to be crucial for the survival of long-lived plasma B cells (O'Connor et al. 2004; Darce et al. 2007; Benson et al. 2008). BAFF-R is expressed on the peripheral B cells and is therefore involved in the survival of these subsets and explains why BAFF is so essential for the maintenance of the B cell population. Indeed, mice lacking BAFF or BAFF receptor fail to produce or maintain a mature B cell pool (Batten et al. 2000; Thompson et al. 2000; Schiemann et al. 2001; Thompson et al. 2001). Recently, TACI was shown to be expressed by human memory B cells, plasma cells and a sub-population of CD27 negative B cells (Darce et al. 2007). TACI expression was found to be inducible early upon B cell activation and this expression was mediated via ERK1/2 signalling pathways. BCMA expression was found to be acquired in the memory B cells following loss of BAFF-R expression.

2.2 *APRIL in B Cell Biology*

Knockout and transgenic mice for APRIL, BAFF and their receptors, have been instrumental in revealing the pivotal role of these cytokines in B cell biology. BAFF knockout mice did not develop mature B cells, and thus had a decreased antibody-mediated response (Gross et al. 2001; Schiemann et al. 2001). In contrast, BAFF transgenic mice developed B cell hyperplasia, hyperglobulinemia, splenomegaly and autoimmunity (Mackay et al. 1999; Stohl et al. 2005). Likewise, mice expressing a non-functional BAFF-R (A/WySnJ) were found to have a severely decreased peripheral B cell population, indicating a critical role for BAFF in the maintenance of B cell survival (Yan et al. 2001a).

APRIL transgenic mice were generated using a T-cell driven promoter, since APRIL was found to be up-regulated in activated T cells (Stein et al. 2002). These mice were found to develop normally, though the T cell population showed enhanced survival, and IgM antibodies were found at elevated concentrations. In addition, T cell independent type II responses were enhanced in these mice. Of significance was the finding that aged APRIL transgenic mice displayed extreme enlargement and re-organisation of the lymph system and enlarged spleen due to infiltration of CD5 positive B cells (Planelles et al. 2004). This development of a B1 neoplasm closely resembled the human B-CLL phenotype (see later discussion).

APRIL deficient mice were generated by two independent research groups (Castigli et al. 2004; Varfolomeev et al. 2004). In one case the mice were found to develop normally and did not show any abnormalities of the immune system in aspects including T-cell dependent and independent responses to antigens (Varfolomeev et al. 2004). However, the second strain was found to have decreased levels of circulating IgA (Castigli et al. 2004). These mice showed decreased serum IgA responses when challenged with a T-dependent antigen. This data pointed towards a role for APRIL in humoral immunity and indeed later work showed that

APRIL, along with BAFF, is responsible for class-switch recombination (CSR) of antibodies to both IgG and IgA (Litinskiy et al. 2002). Class-switch is primarily driven by CD40–CD40L interactions in response to antigenic stimuli; CD40 ligand is up-regulated on T cells and then engages CD40 on B cells. However, it was shown that APRIL and BAFF could drive class-switch independently of the CD40 T-dependent signal (Litinskiy et al. 2002). This effect was shown to occur via either TACI in the case of APRIL-driven effects, or BAFF-R (Castigli et al. 2005a). Both cytokines were found to be up-regulated in response to a number of different pathogenic stimuli, suggesting a role in early innate immunity.

Recently, it was shown that commensal bacteria in the gut stimulate dendritic and intestinal epithelial cells to secrete APRIL, which in turn drives the production of IgA2 immunoglobulin (Tezuka et al. 2007). APRIL was found to be directly responsible for the switch from IgA1 producing cells to IgA2 secreting cells. The benefit is that IgA2 is less sensitive to the various proteases present in the gut and so this switch is an important event to maintain antibody defence in the intestine.

While APRIL is less critical than BAFF at the level of B cell maintenance, it does have a role in B cell signalling and has been shown to drive both proliferation and survival of human and murine B cells in vitro (Yu et al. 2000; Craxton et al. 2003). APRIL has also been shown to have a co-stimulatory effect on B cells treated with IL-4 and IL-6, by stimulating the up-regulation of certain co-stimulatory molecules (CD40, MHC-II, B7.1 and B7.2), which act to enhance antigen presentation (Yang et al. 2005). This event is thought to be mediated primarily via BCMA, which is up-regulated in response to these interleukins (Yang et al. 2005). However, a more recent study used TACI$^{-/-}$ murine B cells and demonstrated that the up-regulation of MHC class II in response to APRIL is a TACI-dependent event (Bossen et al. 2007). However, this study used very low doses of IL-4 and IL-6 and required antibody-mediated cross-linking of soluble APRIL to see a response. Therefore, it may be that at higher doses of interleukins the expression of BCMA dominates. In addition, different forms of APRIL used in vitro appear to have differences in activity and different requirements for cross-linking.

In humans, mutations in TACI are associated with about 5–10% of patients who develop common variable immunodeficiency (CVID), a disease characterised by recurrent bacterial infections, autoimmune complications, lymphoproliferation and splenomegaly (Castigli et al. 2005b; Salzer et al. 2005). Patients with CVID have impaired production of certain antibodies following challenge by vaccination or exposure. In similarity with other diseases caused by mutations in TNF receptors, patients are heterozygous for the specific point mutation in TACI. However, it has been shown that the C104R mutation leads to loss of ligand binding (both BAFF and APRIL) (Castigli et al. 2005a, b; Salzer et al. 2005). Since the receptors pre-associate on the cell surface via the PLAD, the result is a dominant negative effect on TACI signalling due to the formation of mixed complexes between wild-type and mutant receptors (Garibyan et al. 2007). Thus, the lack of antibodies in CVID patients appears to be due to impaired APRIL signalling.

Recently, higher than average levels of BAFF, APRIL and TACI were reported in the serum of CVID patients, but this did not correlate with the disease severity or any particular mutation (Knight et al. 2007). Intriguingly, TACI knockout mouse manifest with the opposite phenotype to CVID patients; the mice have elevated levels of B-2/Marginal Zone (MZ) B cells, fatal lymphoproliferation and autoimmunity, suggesting that TACI plays a negative regulatory role on murine B cells (Yan et al. 2001b).

Examination of BCMA-deficient mice initially revealed an apparently normal phenotype, but upon further investigation the mice were found to have decreased levels of long-lived plasma B cells, which fits with the expression pattern of the receptor in the later stages of B cell development (Xu and Lam 2001; O'Connor et al. 2004). Recently, two studies confirmed that long-lived bone marrow plasma B cells depend on APRIL signalling and that this is mediated via BCMA (Belnoue et al. 2008; Benson et al. 2008). It was also shown that memory B cells survive independently of both BAFF and APRIL (Benson et al. 2008).

2.3 *APRIL in T Cell Biology*

The role of APRIL in T cell biology has been the subject of much controversy. This has been somewhat confused by conflicting reports with regards to the expression of TACI on T cells of mice and humans. BCMA is exclusively expressed on B cells, but TACI expression has been shown on CD3-positive synovioctes from RA patients and on T cells from the peritoneal cavity of mice (Yu et al. 2000) (Stein et al. 2002). Several studies have suggested that both APRIL and BAFF can co-stimulate CD3-activated T-cells in vitro (Huard et al. 2001; Ng et al. 2004). Studies of APRIL transgenic mouse also indicated a co-stimulatory role, but this result was not found using mouse APRIL to treat isolated murine and human T-cells.

The APRIL deficient and the APRIL transgenic mice showed no gross abnormalities in T cell activity and repertoire (Castigli et al. 2004; Planelles et al. 2004; Varfolomeev et al. 2004). Likewise, the TACI transgenic and TACI knockout mice did not show any adverse effects on the T cell population, other than a subtle increase in the $CD4^+$ T cell population in the deficient mice (Gross et al. 2001; von Bulow et al. 2001; Yan et al. 2001b). More recently, functional data using viral challenge and various immune models suggest that a role for APRIL and BAFF in T cell maintenance is limited (Hardenberg et al., 2008).

3 The Role of APRIL in Cancer: B Cell Malignancies

The first association of APRIL with cancer was the initial description of this protein in 1998 (Hahne et al. 1998). High expression levels of APRIL mRNA were found in a panel of tumour cell lines as well as human primary tumours, such as colon and a lymphoid carcinoma. In addition, APRIL transfected murine fibroblast NIH-3T3 cells were shown to grow out more rapidly than parental cells in immunodeficient mice. A subsequent study confirmed this finding and also described that tumour growth of lung and colon carcinomas endogenously expressing APRIL can be blocked by a soluble APRIL receptor (Rennert et al. 2000). Since this, there have been multiple reports associating APRIL with cancer and in particular B cell malignancies.

3.1 Chronic Lymphocytic Leukaemia

Chronic Lymphocytic Leukaemia (CLL) is the most prevalent leukaemia in Western countries, and is characterised by a gradual accumulation of CD5 B cells due to a survival advantage. The first evidence for an involvement of APRIL in CLL came from reports that detected APRIL and APRIL-receptor transcripts in tumour cells (Novak et al. 2002b; Kern et al. 2004). APRIL was also detected at the protein level on the surface of CLL tumour cells (Kern et al. 2004). It was shown that recombinant APRIL protected CLL cells against spontaneous and drug-induced apoptosis and stimulated NF-κB activation. A later study used flow cytometry to confirm that the majority of CLL tumour cells express detectable levels of BCMA and TACI on the cell surface, and in addition found that all the cells tested expressed BAFF-R (Endo et al. 2007). It was also shown that signalling via both BCMA and TACI in these cell lines led to activation of the canonical NF-κB pathway. In contrast, BAFF-R triggering induced the non-canonical NF-κB pathway.

Treatment of CLL cells with soluble APRIL or BAFF induced I-κB phosphorylation and degradation, as well as translocation of the NF-κB p65 subunit to the nucleus (Endo et al. 2007). Blocking of the canonical NF-κB pathway abrogated the capacity of BAFF and APRIL to promote CLL cell survival in vitro. In view of these results, therapeutic application of IKKβ inhibitors, to block canonical NF-κB signalling, has been promoted (Endo et al. 2007). A further study of CLL cells found that APRIL or BAFF did not activate the mitogen-activated protein-kinases (MAPK) ERK1/2 or the serine/threonine kinase PKB/AKT in CLL cells (Kipps 1997; Nishio et al. 2005).

A retrospective study was performed in our laboratory on 95 CLL patients (Planelles et al. 2007). We found increased levels of APRIL in serum of the majority of samples tested. Importantly, we found that APRIL serum levels correlated with disease progression and overall patient survival, with a poorer prognosis for patients with high APRIL serum levels. This contrasts with another report in which elevated APRIL serum levels were found only in some of the samples tested, with no significant difference to control samples (Kolb et al. 2003; Haiat et al. 2006). However, it is noteworthy that various sources of APRIL have been identified, including non-hematopoietic cells. Therefore, it is likely that not only the CLL tumour cells contribute to the measured APRIL serum levels. Indeed, it has been described that nurse-like cells, which differentiate from CD14-positive cells originating from CLL patients when cultured with CLL B cells, are important producers of APRIL and BAFF (Nishio et al. 2005). Survival of CLL cells was rapidly reduced when nurse-like cells were removed from cell cultures, suggesting that APRIL and/or BAFF are essential for their survival. The viability of CLL cells was more affected by the addition of BCMA-Fc than BAFFR-Fc, pointing to a pivotal role for APRIL in CLL survival. However, addition of BCMA-Fc had no effect on the viability of CLL cells when cultured in the absence of nurse-like cells. This suggests that APRIL is solely provided by the supporting cells and stimulates CLL tumour growth in a paracrine fashion. However, this finding contrasts with other reports describing that addition of BCMA-Fc or anti-APRIL antibodies actually enhance B-CLL apoptosis (Novak et al. 2002a; Kern et al. 2004). The expression of various pro- and anti-apoptotic proteins was also analysed and it was found that the anti-apoptotic BCL-2 family member Mcl-1 is up-regulated in CLL cells when cultured in the presence of APRIL or BAFF (Nishio et al. 2005).

Further support for a role of APRIL in B cell malignancies came from our own studies of aged transgenic mice, where we observed the development of lymphoid tumours, originating from expansion of the peritoneal B-1 cell population (Planelles et al. 2004). In particular, CD5-positive B-1 cells infiltrated lymphoid and non-lymphoid organs, initiating hyperplasia. The significance of this finding was that this phenotype closely resembles human CLL and can therefore be used as a potential model to further study the disease. Nevertheless, an important difference is that the CD5-positive B cells accumulating in human patients are considered to be antigen-experienced and are therefore different to those found in the transgenic mice.

3.2 Hodgkin's Lymphoma

Hodgkin's lymphoma (HL) is characterised by the clonal expansion of Reed–Sternberg (RS) cells derived from B cells. Malignant B cells constitute only about 10% of the tumour mass. It was found that both the RS cells and the inflammatory cells express APRIL and BAFF, and therefore stimulation occurs in both an autocrine and paracrine fashion (Chiu et al. 2007). Stimulation of BCMA and TACI on RS cells was found to trigger NF-κB activation, Bcl-2, Bcl-xL and c-Myc up-regulation, as well as Bax down-regulation. RS cells were negative for BAFF-R. APRIL was found to co-localise in RS cell lines with BCMA, TACI and the adaptor molecule TRAF2 in the cytoplasm, in contrast to BAFF that co-localised with BCMA, TACI and TRAF2 only on the cell surface. Chiu et al. also analysed whether HPSGs contribute to APRIL signalling. Treatment of RS cells with either heparinase or heparinitase, enzymes that cleave side chains of HPSGs, abrogated APRIL binding and attenuated spontaneous and APRIL-stimulated proliferation. Administration of the HPSG analogue heparin also abrogated APRIL binding, but increased spontaneous and APRIL-stimulated proliferation. This data suggests that both membrane-anchored and extracellular matrix HPSGs enhance APRIL signalling on RS cells. A recent report confirmed that HSPGs are involved in the binding of

APRIL to RS cells (Mhawech-Fauceglia et al. 2006). However, in this study, the tumour-infiltrating neutrophils were found to be the main source of soluble APRIL, rather than the RS cells, which were found to be negative for APRIL expression at both the RNA and protein level.

3.3 Non-Hodgkin's Lymphoma

Non-Hodgkin's lymphomas (NHL) comprise a diverse group of diseases including T and B cell lymphoma, such as CLL, Burkitt's lymphoma and diffuse large B cell lymphoma (DLBCL). Based on their aggressiveness, NHL can be categorised into indolent (low-grade), aggressive (intermediate-grade) and highly aggressive (high-grade). Expression of BAFFR, BCMA and TACI was reported on NHL cells and it was shown that treatment with APRIL or BAFF increased survival via NF-κB activation, up-regulation of Bcl-2 and Bcl-xL and down-regulation of Bax (He et al. 2004). In addition, APRIL transcripts were detected in several NHL cell types, including CLL, DLBCL and Burkitt lymphoma.

An extensive study of APRIL protein expression on tumour tissues from NHL patients revealed high numbers of APRIL-producing cells. Increased levels of secreted APRIL was found in about 50% of the DLBCL and 20% of the Burkitt lymphoma samples tested, but not in lower-grade B-cell lymphomas (Schwaller et al. 2007, #216). A retrospective clinical analysis of 39 DLBCL patients showed that high APRIL expression in cancer lesions correlated with a poor survival rate. The main cellular source of APRIL in DLBCL was found to be the tumour-infiltrating neutrophils, and secreted APRIL was found to localise to proteoglycans on DLBCL cells that also expressed BCMA and TACI. It was suggested that on both DLBCL and RS cells, HSPGs function as both a co-receptor for APRIL as well as a means by which to establish a concentrated supply of ligand ready for signalling.

3.4 Multiple Myeloma

Multiple myeloma (MM) is a progressive and incurable disease characterised by the build-up of plasma cells in the bone marrow. Expression of the APRIL receptors in MM cell lines was analysed, and transcripts of BCMA, TACI and BAFFR were found in the majority of MM cell lines and purified primary myeloma cells tested (Moreaux et al. 2004). In MM cell lines, APRIL and BAFF were found to induce activation of NF-κB, PI-3 kinase/Akt pathway and MAP kinases, as well as induce up-regulation of Mcl-1 and Bcl-2. Moreover, APRIL and BAFF were able to rescue IL-6 dependent MM cell lines from IL-6 deprivation, as well as rescue primary MM cells from dexamethasone-induced apoptosis. A remarkable finding was that both APRIL and BAFF were increased fivefold in the serum of MM patients compared to normal controls. It was also found that MM cells undergo apoptosis when cultured in the presence of TACI-Fc, suggesting an autocrine role for APRIL. However, a separate study of MM patients reported APRIL production by neutrophils and monocytes in the bone marrow environment; the most important producers in the bone marrow environment were found to be osteoclasts. While this finding points more towards a paracrine effect of APRIL and BAFF in MM, it is possible that in reality both play a role since APRIL that is secreted directly from the tumour cells themselves may stick to HSPGs so rapidly that it will never be measured in the serum. However, the significance of osteoclasts as APRIL producers was also highlighted in a study in which TACI-Fc significantly enhanced apoptosis of MM cells in osteoclast-co-cultures (Abe et al. 2006). This finding points towards the use of TACI-Fc or APRIL and BAFF blocking antibodies as therapeutic agents in the treatment of MM.

3.5 Solid Tumours

APRIL was initially discovered as a ligand over-expressed in a number of different cell lines, many of which were derived from solid tumours (Hahne et al. 1998). Indeed, APRIL was able to stimulate in vitro proliferation of a number of these cell lines, which were later found to lack expression of both known APRIL receptors, BCMA and TACI (Hahne et al. 1998; Rennert et al. 2000). Despite its discovery 10 years ago, we are still no further in understanding how APRIL delivers a proliferative signal in many of these cell lines, though we now know that interactions with HSPGs are possible and may be of greater relevance to APRIL signalling in certain cells.

Because of the expression pattern of APRIL and its receptors, primarily on immune cells, a link with APRIL and most solid tumours is not intuitive. Nevertheless, there have been several studies reporting stimulatory capacity and over-expression of APRIL at the mRNA and protein level for a number of solid tumour cell lines. APRIL was shown to stimulate proliferation of several malignant glioblastoma cell lines, and expression of BCMA and TACI in these cells was also found (Roth et al. 2001; Deshayes et al. 2004). Another study reported APRIL expression in colon, cervix, hepatocellular carcinomas and melanomas, using an antibody that binds to the portion of APRIL left after furin cleavage and thus specifically detects cells endogenously expressing APRIL (Mhawech-Fauceglia et al. 2006).

In addition, as already mentioned, it is important to remember that APRIL can also be delivered to the tumour by infiltrating cells, such as neutrophils and dendritic cells. In certain cases it has been shown that APRIL found in histological samples is derived from the infiltrating cells, rather than the tumours themselves (Schwaller et al. 2007). Most recently, the expression of APRIL was studied in human breast cancer biopsies using immunohistochemical techniques (Pelekanou et al. 2008). Intriguingly, APRIL expression was higher in normal breast tissue compared to the tumour biopsies, and its expression declined with an increase in tumour grade. Thus, this is the first evidence indicating a potential role for APRIL in normal breast tissue development.

4 APRIL and Autoimmune Disease

4.1 Systemic Lupus Erythematosus

Systemic lupus erythematosus (SLE) is a chronic autoimmune disease with a high incidence in the Caucasian population: 3.9 cases per 100,000 Caucasian females and 0.4 cases per 100,000 Caucasian males (Planelles et al., submitted). SLE carries a significant risk of mortality and long-term morbidity, despite improvements in prognosis and treatments such as corticosteroids and immunosuppressive drugs. The pathogenesis of SLE has been linked to a unique form of B-cell hyper-proliferation. B-cells from SLE patients generate excessive quantities of IgG auto-antibodies, which include anti-nuclear and anti-dsDNA antibodies and patients show characteristic features of an antigen-driven T cell dependent immune response (Maddison 1999). These auto-antibodies form immune complexes that ultimately lead to lupus nephritis and renal failure. Auto-reactive B cells are therefore an attractive therapeutic target in this disease.

Constitutive over-expression of BAFF in mice generates a SLE-like disease (reviewed in Dillon et al. (2006)) and mice that naturally develop SLE harbour elevated circulating levels of BAFF. Treatment of these mice with BAFF antagonists was shown to halt disease progression and enhance survival (Gross et al. 2000). The level of BAFF was also found to be elevated in the serum of SLE patients (Cheema et al. 2001; Zhang et al. 2001; Stohl et al. 2003) and several companies are currently testing the therapeutic potential of BAFF antagonistic antibodies. However, while there is a consensus in the literature that increased BAFF in the serum of SLE patients reflects a contribution to disease, reports regarding the level of APRIL in the serum are contradictory. An initial report analysed APRIL in the serum of 68 SLE patients, which were followed-up for a median of 369 days (Stohl et al. 2004). However, they observed no correlation between the serum levels of APRIL and BAFF. Notably, short courses of corticosteroids decreased serum levels of BAFF while APRIL levels remained unchanged, suggesting different regulatory mechanisms for the two cytokines. The mRNA levels of APRIL and BAFF in the blood cells of SLE patients were also measured and found to parallel the observed serum levels. Nevertheless, any conclusions drawn from serum levels must be made with caution as evidence is increasing that non-hematopoietic cells are also important producers of APRIL (reviewed in Dillon et al. (2006)).

Patients with positive titres of anti-dsDNA antibodies were shown to display a modest, but significant, inverse correlation between the serum levels of APRIL and anti-dsDNA antibodies (Stohl et al. 2004). This same report described a modest, yet significant inverse correlation of APRIL serum level and disease activity, as assessed by the SLE Disease Activity Index (SLEDAI) that contains 24 descriptors (clinical, biochemical and serological parameters including anti-nuclear (ANA) and anti-dsDNA antibodies) with pre-assigned severity weighting (Bombardier et al. 1992). Both inverse correlations, however, were only obtained when patients were analysed in aggregates, that is when serum samples from all patients at the different time points were included. It was proposed that APRIL can act as a down modulator of autoimmunity in SLE, although an exact mechanism was not proposed.

This hypothesis was challenged in a subsequent report in which APRIL serum levels were compared between healthy subjects, patients with SLE and patients with rheumatoid arthritis (Koyama et al. 2005). The level of APRIL in the serum was found to be significantly higher in SLE patients than in patients with RA and healthy donors. A retrospective analysis revealed that APRIL serum levels tended to correlate with anti-dsDNA antibody titres. Disease activity was scored following two different indices: The disease activity index of the British Isles Lupus Assessment Group (BILAG) and the SLEDAI. The major difference between both indices is that the BILAG scores disease activity in different organs separately, in contrast to the SLEDAI in which a global score for disease activity is calculated (Hay et al. 1993). In the study carried out by Koyama et al., serum APRIL levels did not correlate with the SLEDAI, but instead were found to correlate with the BILAG index of musculoskeletal disease, mostly in arthritis. Based on these observations, opposing therapeutic strategies have been proposed (Stohl et al. 2004; Koyama et al. 2005).

In our laboratory, a study was also performed on 43 SLE patients that were positive for anti-dsDNA antibodies at least once in their medical record (Morel et al. manuscript submitted). In agreement with the study by Stohl et al., we found that APRIL serum levels showed a modest inverse correlation with anti-dsDNA titres and that patients with high levels of serum APRIL tended to have a lower incidence of renal involvement. In addition, we also found that APRIL levels in SLE patients inversely correlated with levels of BAFF, suggesting opposing roles for APRIL and BAFF in SLE. This would support the use of specific BAFF-antagonizing agents for the treatment of SLE, rather than the use of soluble TACI or BCMA receptors, which antagonize both BAFF and APRIL.

One study followed APRIL and BAFF serum levels over a period of 6 months in a small number of SLE patients (n = 10) treated with the B cell depleting anti-CD20 antibody rituximab (Vallerskog et al. 2006). CD20 is exclusively expressed on B cells and rituximab efficiently depletes temporary circulating B cells. No significant differences were observed between circulating APRIL in untreated SLE patients and controls, although BAFF was found to be elevated in SLE patients. During B cell depletion serum levels of APRIL significantly decreased in contrast to those of BAFF, which actually increased. When B cell numbers had recovered, the APRIL and BAFF levels returned to the normal levels before treatment. The different evolution of serum levels during B cell depletion is rather opposing for the two ligands in SLE. At present, there is no clear explanation for the different findings regarding APRIL levels in SLE patients, though it is possible that the different ELISA systems employed have differing sensitivities and are responsible for these discrepancies.

Several reports have investigated whether polymorphisms in APRIL and its receptor TACI occur in SLE. The sequence of TACI was analysed in 119 unrelated SLE patients and four variants were detected (Salzer et al. 2007). The frequency of these variants was not significantly different to those of healthy subjects and none of the mutations found in the TACI coding regions appeared to be associated with disease. Two studies analysed the coding regions of the APRIL gene in Japanese SLE patients (n = 148, and 266, respectively) and described two single-nucleotide polymorphisms (SNPs) at codons 67 (glycine/arginine) and 96 (asparagine/serine), with the codon for glycine at position 67 found to be significantly elevated in patients with SLE susceptibility (Kawasaki et al. 2007). This finding was confirmed in a separate study, which also identified that the haplotype 67 glycine + 96 asparagine confers risk, whereas the haplotype 67 arginine + 96 serine is protective (Kawasaki et al. 2007). The SNP at APRIL codon 67A was also found to be associated with risk of SLE in other ethnic groups (204 European-American, 103 African-American and 41 Hispanic) (Lee et al. 2007).

Several reports have described the introduction of adenovirus-encoded or purified soluble TACI-Fc fusion protein into lupus prone mice, in order to block BAFF and APRIL activity (Gross et al. 2000; Liu et al. 2004; Ramanujam et al. 2004). Treatment of lupus prone mice with soluble TACI-Fc fusion protein inhibited disease development and prolonged survival. One report compared the effect of TACI-Fc fusion protein with that of BAFF receptor -Fc (blocking only BAFF) in lupus prone NZB/WF1 mice (Ramanujam et al. 2006). Both reagents showed a comparable efficacy in NZB/WF1 mice. However, TACI-Fc treatment resulted in reduced IgM serum levels, decreased frequency of splenic plasma cells and inhibited the IgM response to a T cell-dependent antigen. This was likely due to APRIL effects on B cell production of neutralising anti-IgM antibodies. Thus a specific BAFF blocking agent would be more effective. It was concluded that a reagent specifically blocking BAFF might be favourable, since agents blocking both BAFF and APRIL may interfere with the development of neutralizing IgM antibodies, leaving patients more susceptible to infection.

4.2 Rheumatoid Arthritis

Arthritis is an inflammation of one or several joints and symptoms include swelling, pain, and restriction of motion. Many different forms of inflammatory arthritis exist though rheumatoid arthritis (RA) is the most common. RA affects the tissue of the joint lining called synoviu, which is normally a relatively acellular structure with a delicate intimal lining one or two cell layers deep. Rheumatoid synovial tissue is characterised by hyperproliferation of fibroblast-like synoviocytes (FLS) in the intimal lining layer and infiltration of the sub-lining by macrophages and T and B cells. This infiltration promotes inflammation as well as destruction of the bone and cartilage. RA is also characterized by the production of auto-antibodies.

The first clue that there may be a link between APRIL and RA came from mouse studies (Gross et al. 2001; Wang et al. 2001). Immunization of susceptible mouse strains with type II collagen (CII) from heterologous species leads to a pathology similar to human RA. This is now a common model used to study RA and is called collagen-induced arthritis (CIA) (Courtenay et al. 1980). Both CII antibodies and CD4+ T cells are required for the development of CIA. Inhibition of APRIL and BAFF with TACI-Fc in the CIA mouse model was found to prevent disease progression and lower disease scores, compared with controls (Gross et al. 2001; Wang et al. 2001). It was later reported that the synovial fluid of patients with inflammatory arthritis had significantly increased APRIL levels compared with those patients suffering from non-inflammatory arthritis such as osteoarthritis (Tan et al. 2003; Stohl 2006). BAFF levels were also found to be higher in patients with inflammatory arthritis, but did not correlate with APRIL levels. It is likely that both cytokines are involved in regulating pathogenic B or T cells in inflamed joints. Interestingly, one study analysed serum samples from patients with a range of different systemic immune-based rheumatic diseases (SLE, RA, Reiter's syndrome, psoriatic arthritis, polymyositis and ankylosing spondylitis) and found significantly increased APRIL levels in patients, in addition to the presence of APRIL/BAFF hetero-trimers (Roschke et al. 2002). However, taken the small number of samples analyzed, it will be interesting to see whether the concentration of APRIL/BAFF hetero-trimers are more pronounced in any particular rheumatic disease and if they have functional relevance.

A further study analysed serum from 16 RA patients and detected increased APRIL levels compared to controls (Nagatani et al. 2007). Significantly, they found that synovial fibroblasts in RA patients express BCMA, but could not find expression in the same cells from patients with osteoarthritis. APRIL treatment was found to enhance proliferation and induce the production of pro-inflammatory cytokines, such as IL-6 and TNF-alpha, suggesting that APRIL is one of the main regulators in the pathogenesis of RA.

Further evidence that APRIL plays a role in RA comes from two other studies. APRIL and BAFF serum levels were followed over a period of 6 months in nine RA patients treated with rituximab (Vallerskog et al. 2006). APRIL, but not BAFF serum levels, were elevated in the RA patients compared with controls. During B cell depletion, serum levels of APRIL remained unchanged whereas, as observed in SLE patients, concentrations of BAFF were significantly increased during B cell depletion.

APRIL was also found to be important following an examination of synovial biopsies from 72 RA patients for B cell function and expression of APRIL and BAFF (Seyler et al. 2005). Synovitis artnntis, that is inflamed synovial tissues, were classified according to their lymphoid micro-architecture: ectopic germinal centres (GCs), T cell–B cell aggregates lacking germinal centre reactions and unorganized diffuse infiltrates. It has been shown that about half of the RA patients have sinusitis with GCs or GC-negative T cell–B cell aggregates, whereas the other half of the patients have scattered synovial infiltrates (Weyand et al. 2003). APRIL was found to be exclusively expressed in $CD83^+$ dendritic cells, with the highest expression levels in GC synovitis; BAFF expression was similar in the different types of synovitis and restricted to $CD68^+$ macrophages. $CD138^+$ plasma cells were found to express TACI. In addition, a subset of T cells expressed TACI in aggregate and diffuse synovitis, but not in GC synovitis. To test for a possible functional difference between the different compartments, synovium-SCID mouse chimeras were treated with TACI-Fc. The result was a disappearance of GCs in synovial tissue, decreased Ig production and decreased production of IFN-gamma. The response to TACI-Fc treatment in aggregate and diffuse synovitis was rather contrary, with unaltered Ig levels and increased IFN-gamma production. It therefore appears that, depending on the type of synovitis, TACI-ligands trigger either stimulatory or suppressive signals.

4.3 *Sjögren's Syndrome*

Sjögren's syndrome (SS) is an autoimmune disease that affects exocrine glands and often occurs in association with RA and SLE. Typical anti-nuclear antibodies found in patients with SS are SSA/Ro and SSB/La, the latter being more specific for SS. Symptoms resembling SS were described in BAFF transgenic mice (Groom et al. 2002), but not in APRIL transgenic mice. Nevertheless, increased serum levels of both APRIL and BAFF have been found in patients with SS, in particular in those positive for SSA/Ro (Jonsson et al. 1986). Moreover, there was a positive correlation between serum levels of both cytokines, though the precise role in this disease remains to be determined.

5 Conclusion

APRIL is an intriguing ligand that seems to play a crucial role in the survival and proliferative capacity of several B-cell malignancies, as well as some solid tumours. APRIL is also emerging as a key player in autoimmunity, especially rheumatoid arthritis where it appears to play an active role in disease pathology. Thus, strategies to antagonise APRIL are an obvious therapeutic goal for a number of diseases. At the in-vitro level, soluble receptor Fc-fusion proteins function as efficient blockers of APRIL activity. However, for in-vivo blocking of APRIL, the challenge is to design molecules that target APRIL specifically, and leave the activity of BAFF unaffected. Although BAFF is a contributing factor to some of the same conditions as APRIL, inhibition of BAFF systemically may be detrimental to normal B-cell maintenance. Therefore, the development of APRIL-specific blocking antibodies or APRIL-specific receptor Fc-fusion proteins will be an important advancement. Such molecules would be a realistic therapy for a number of conditions in which circulating soluble APRIL correlates with disease activity and progression. Further understanding of the HSPG interactions of APRIL will also prove important in this field, both in terms of its effect on canonical TNF receptor signalling, and also in solid tumours, where BCMA and TACI appear not to play a role. Indeed, HSPG interactions may allow APRIL to be readily captured onto HSPG-rich extracellular matrix within the tumour bundle, potentially before systemic therapeutic agents can take effect.

Acknowledgments Work cited in this chapter from our own laboratory is supported by the Fondation de France and Association pour la Recherche sur le Cancer and the Dutch Cancer Society grants nr 2003-2812 and 2007-3750 and Stiching Vanderes.

References

Abe M, Kido S et al (2006) BAFF and APRIL as osteoclast-derived survival factors for myeloma cells: a rationale for TACI-Fc treatment in patients with multiple myeloma. Leukemia 20(7):1313–1315

Batten M, Groom J et al (2000) BAFF mediates survival of peripheral immature B lymphocytes. J Exp Med 192(10):1453–1466

Belnoue E, Pihlgren M et al (2008) APRIL is critical for plasmablast survival in the bone marrow and poorly expressed by early life bone marrow stromal cells. Blood 111(5):2755–2764

Benson MJ, Dillon SR et al (2008) Cutting edge: the dependence of plasma cells and independence of memory B cells on BAFF and APRIL. J Immunol 180(6):3655–3659

Bombardier C, Gladman DD et al (1992) Derivation of the SLEDAI. A disease activity index for lupus patients. The Committee on Prognosis Studies in SLE. Arthritis Rheum 35(6):630–640

Bossen C, Schneider P (2006) BAFF, APRIL and their receptors: structure, function and signaling. Semin Immunol 18(5):263–275

Bossen C, Cachero TG et al (2008) TACI, unlike BAFF-R, is solely activated by oligomeric BAFF and APRIL to support survival of activated B cells and plasmablasts. Blood 111:1004–1012

Castigli E, Scott S et al (2004) Impaired IgA class switching in APRIL-deficient mice. Proc Natl Acad Sci USA 101(11):3903–3908

Castigli E, Wilson SA et al (2005a) TACI and BAFF-R mediate isotype switching in B cells. J Exp Med 201(1):35–39

Castigli E, Wilson SA et al (2005b) TACI is mutant in common variable immunodeficiency and IgA deficiency. Nat Genet 37(8):829–834

Chan FK, Chun HJ et al (2000) A domain in TNF receptors that mediates ligand-independent receptor assembly and signaling. Science 288(5475):2351–2354

Cheema GS, Roschke V et al (2001) Elevated serum B lymphocyte stimulator levels in patients with systemic immune-based rheumatic diseases. Arthritis Rheum 44(6):1313–1319

Chiu A, Xu W et al (2007) Hodgkin lymphoma cells express TACI and BCMA receptors and generate survival and proliferation signals in response to BAFF and APRIL. Blood 109(2):729–739

Courtenay JS, Dallman MJ et al (1980) Immunisation against heterologous type II collagen induces arthritis in mice. Nature 283(5748):666–668

Craxton A, Magaletti D et al (2003) Macrophage- and dendritic cell – dependent regulation of human B-cell proliferation requires the TNF family ligand BAFF. Blood 101(11):4464–4471

Darce JR, Arendt BK et al (2007) Regulated expression of BAFF-binding receptors during human B cell differentiation. J Immunol 179(11):7276–7286

Deshayes F, Lapree G et al (2004) Abnormal production of the TNF-homologue APRIL increases the proliferation of human malignant glioblastoma cell lines via a specific receptor. Oncogene 23(17):3005–3012

Dillon SR, Gross JA et al (2006) An APRIL to remember: novel TNF ligands as therapeutic targets. Nat Rev Drug Discov 5(3):235–246

Endo T, Nishio M et al (2007) BAFF and APRIL support chronic lymphocytic leukemia B-cell survival through activation of the canonical NF-kappaB pathway. Blood 109(2):703–710

Garibyan L, Lobito AA et al (2007) Dominant-negative effect of the heterozygous C104R TACI mutation in common variable immunodeficiency (CVID). J Clin Invest 117(6):1550–1557

Groom J, Kalled SL et al (2002) Association of BAFF/BLyS overexpression and altered B cell differentiation with Sjogren's syndrome. J Clin Invest 109(1):59–68

Gross JA, Johnston J et al (2000) TACI and BCMA are receptors for a TNF homologue implicated in B-cell autoimmune disease. Nature 404(6781):995–999
Gross JA, Dillon SR et al (2001) TACI-Ig neutralizes molecules critical for B cell development and autoimmune disease impaired B cell maturation in mice lacking BLyS. Immunity 15(2):289–302
Hahne M, Kataoka T et al (1998) APRIL, a new ligand of the tumor necrosis factor family, stimulates tumor cell growth. J Exp Med 188(6):1185–1190
Haiat S, Billard C et al (2006) Role of BAFF and APRIL in human B-cell chronic lymphocytic leukaemia. Immunology 118(3):281–292
Hardenberg G, Fernandez L et al (2008) APRIL facilitates viral-induced erythroleukemia but is dispensable for T cell immunity and lymphomagenesis. J Leukoc Biol 84(2):380–388. Epub 2008 May15
Hardenberg G, Planelles L et al (2007) Specific TLR ligands regulate APRIL secretion by dendritic cells in a PKR-dependent manner. Eur J Immunol 37(10):2900–2911
Hatzoglou A, Roussel J et al (2000) TNF receptor family member BCMA (B cell maturation) associates with TNF receptor-associated factor (TRAF) 1, TRAF2, and TRAF3 and activates NF-kappa B, elk-1, c-Jun N-terminal kinase, and p38 mitogen-activated protein kinase. J Immunol 165(3):1322–1330
Hay EM, Bacon PA et al (1993) The BILAG index: a reliable and valid instrument for measuring clinical disease activity in systemic lupus erythematosus. Q J Med 86(7):447–458
He B, Chadburn A et al (2004) Lymphoma B cells evade apoptosis through the TNF family members BAFF/BLyS and APRIL. J Immunol 172(5):3268–3279
He B, Xu W et al (2007) Intestinal bacteria trigger T cell-independent immunoglobulin A(2) class switching by inducing epithelial-cell secretion of the cytokine APRIL. Immunity 26(6):812–826
Hendriks J, Planelles L et al (2005) Heparan sulfate proteoglycan binding promotes APRIL-induced tumor cell proliferation. Cell Death Differ 12(6):637–648
Huard B, Schneider P et al (2001) T cell costimulation by the TNF ligand BAFF. J Immunol 167(11):6225–6231
Hymowitz SG, Patel DR et al (2005) Structures of APRIL–receptor complexes: like BCMA, TACI employs only a single cysteine-rich domain for high affinity ligand binding. J Biol Chem 280(8):7218–7227
Ingold K, Zumsteg A et al (2005) Identification of proteoglycans as the APRIL-specific binding partners. J Exp Med 201(9):1375–1383
Jonsson H, Nived O et al (1986) Symptomatic secondary Sjogren's syndrome in patients with systemic lupus erythematosus (SLE). Relation to anti-SS-A and anti-SS-B autoantibodies. Scand J Rheumatol Suppl 61:166–169
Kawasaki A, Tsuchiya N et al (2007) Role of APRIL (TNFSF13) polymorphisms in the susceptibility to systemic lupus erythematosus in Japanese. Rheumatology (Oxford) 46(5):776–782
Kelly K, Manos E et al (2000) APRIL/TRDL-1, a tumor necrosis factor-like ligand, stimulates cell death. Cancer Res 60(4):1021–1027
Kern C, Cornuel JF et al (2004) Involvement of BAFF and APRIL in the resistance to apoptosis of B-CLL through an autocrine pathway. Blood 103(2):679–688
Kipps TJ (1997) Signal transduction pathways and mechanisms of apoptosis in CLL B-lymphocytes: their role in CLL pathogenesis. Hematol Cell Ther 39(Suppl 1):S17–S27
Knight AK, Radigan L et al (2007) High serum levels of BAFF, APRIL, and TACI in common variable immunodeficiency. Clin Immunol 124(2):182–189
Kolb JP, Kern C et al (2003) Re-establishment of a normal apoptotic process as a therapeutic approach in B-CLL. Curr Drug Targets Cardiovasc Haematol Disord 3(4):261–286
Kolfschoten GM, Pradet-Balade B et al (2003) TWE-PRIL; a fusion protein of TWEAK and APRIL. Biochem Pharmacol 66(8):1427–1432
Koyama T, Tsukamoto H et al (2005) Raised serum APRIL levels in patients with systemic lupus erythematosus. Ann Rheum Dis 64(7):1065–1067

Lee YH, Ota F et al (2007) APRIL polymorphism and systemic lupus erythematosus (SLE) susceptibility. Rheumatology (Oxford) 46(8):1274–1276
Litinskiy MB, Nardelli B et al (2002) DCs induce CD40-independent immunoglobulin class switching through BLyS and APRIL. Nat Immunol 3(9):822–829
Liu W, Szalai A et al (2004) Control of spontaneous B lymphocyte autoimmunity with adenovirus-encoded soluble TACI. Arthritis Rheum 50(6):1884–1896
Lopez-Fraga M, Fernandez R et al (2001) Biologically active APRIL is secreted following intracellular processing in the Golgi apparatus by furin convertase. EMBO Rep 2(10):945–951
Mackay F, Woodcock SA et al (1999) Mice transgenic for BAFF develop lymphocytic disorders along with autoimmune manifestations. J Exp Med 190(11):1697–1710
Maddison PJ (1999) Autoantibodies in SLE. Disease associations. Adv Exp Med Biol 455:141–145
Marsters SA, Yan M et al (2000) Interaction of the TNF homologues BLyS and APRIL with the TNF receptor homologues BCMA and TACI. Curr Biol 10(13):785–788
Mhawech-Fauceglia P, Kaya G et al (2006) The source of APRIL up-regulation in human solid tumor lesions. J Leukoc Biol 80(4):697–704
Moreaux J, Legouffe E et al (2004) BAFF and APRIL protect myeloma cells from apoptosis induced by interleukin 6 deprivation and dexamethasone. Blood 103(8):3148–3157
Moreaux J, Cremer FW et al (2005) The level of TACI gene expression in myeloma cells is associated with a signature of microenvironment dependence versus a plasmablastic signature. Blood 106(3):1021–1030
Moy FJ, Safran M et al (1997) Properly oriented heparin-decasaccharide-induced dimers are the biologically active form of basic fibroblast growth factor. Biochemistry 36(16):4782–4791
Nagatani K, Itoh K et al (2007) Rheumatoid arthritis fibroblast-like synoviocytes express BCMA and are stimulated by APRIL. Arthritis Rheum 56(11):3554–3563
Ng LG, Sutherland AP et al (2004) B cell-activating factor belonging to the TNF family (BAFF)-R is the principal BAFF receptor facilitating BAFF costimulation of circulating T and B cells. J Immunol 173(2):807–817
Nishio M, Endo T et al (2005) Nurselike cells express BAFF and APRIL, which can promote survival of chronic lymphocytic leukemia cells via a paracrine pathway distinct from that of SDF-1alpha. Blood 106(3):1012–1020
Novak AJ, Bram RJ et al (2002a) Aberrant expression of B-lymphocyte stimulator by B chronic lymphocytic leukemia cells: a mechanism for survival. Blood 100(8):2973–2979
Novak U, Oppliger Leibundgut E et al (2002b) A high-resolution allelotype of B-cell chronic lymphocytic leukemia (B-CLL). Blood 100(5):1787–1794
O'Connor BP, Raman VS et al (2004) BCMA is essential for the survival of long-lived bone marrow plasma cells. J Exp Med 199(1):91–98
Pelekanou V, Kampa M et al (2008) Expression of TNF-superfamily members BAFF and APRIL in breast cancer: Immunohistochemical study in 52 invasive ductal breast carcinomas. BMC Cancer 8(1):76
Planelles L, Carvalho-Pinto CE et al (2004) APRIL promotes B-1 cell-associated neoplasm. Cancer Cell 6(4):399–408
Planelles L, Castillo-Gutierrez S et al (2007) APRIL but not BLyS serum levels are increased in chronic lymphocytic leukemia: prognostic relevance of APRIL for survival. Haematologica 92(9):1284–1285
Pradet-Balade B, Medema JP et al (2002) An endogenous hybrid mRNA encodes TWE-PRIL, a functional cell surface TWEAK-APRIL fusion protein. Embo J 21(21):5711–5720
Ramanujam M, Wang X et al (2004) Mechanism of action of transmembrane activator and calcium modulator ligand interactor-Ig in murine systemic lupus erythematosus. J Immunol 173(5):3524–3534
Ramanujam M, Wang X et al (2006) Similarities and differences between selective and nonselective BAFF blockade in murine SLE. J Clin Invest 116(3):724–734
Rennert P, Schneider P et al (2000) A soluble form of B cell maturation antigen, a receptor for the tumor necrosis factor family member APRIL, inhibits tumor cell growth. J Exp Med 192(11):1677–1684

Roschke V, Sosnovtseva S et al (2002) BLyS and APRIL form biologically active heterotrimers that are expressed in patients with systemic immune-based rheumatic diseases. J Immunol 169(8):4314–4321
Roth W, Wagenknecht B et al (2001) APRIL, a new member of the tumor necrosis factor family, modulates death ligand-induced apoptosis. Cell Death Differ 8(4):403–410
Salzer U, Birmelin J et al (2007) Sequence analysis of TNFRSF13b, encoding TACI, in patients with systemic lupus erythematosus. J Clin Immunol 27(4):372–377
Salzer U, Chapel HM et al (2005) Mutations in TNFRSF13B encoding TACI are associated with common variable immunodeficiency in humans. Nat Genet 37(8):820–828
Schiemann B, Gommerman JL et al (2001) An essential role for BAFF in the normal development of B cells through a BCMA-independent pathway. Science 293(5537):2111–2114
Schwaller J, Schneider P et al (2007) Neutrophil-derived APRIL concentrated in tumor lesions by proteoglycans correlates with human B-cell lymphoma aggressiveness. Blood 109(1):331–338
Seyler TM, Park YW et al (2005) BLyS and APRIL in rheumatoid arthritis. J Clin Invest 115(11):3083–3092
Siegel RM, Frederiksen JK et al (2000) Fas preassociation required for apoptosis signaling and dominant inhibition by pathogenic mutations. Science 288(5475):2354–2357
Stein JV, Lopez-Fraga M et al (2002) APRIL modulates B and T cell immunity. J Clin Invest 109(12):1587–1598
Stohl W (2006) Therapeutic targeting of B lymphocyte stimulator (BLyS) in the rheumatic diseases. Endocr Metab Immune Disord Drug Targets 6(4):351–358
Stohl W, Metyas S et al (2003) B lymphocyte stimulator overexpression in patients with systemic lupus erythematosus: longitudinal observations. Arthritis Rheum 48(12):3475–3486
Stohl W, Metyas S et al (2004) Inverse association between circulating APRIL levels and serological and clinical disease activity in patients with systemic lupus erythematosus. Ann Rheum Dis 63(9):1096–1103
Stohl W, Xu D et al (2005) BAFF overexpression and accelerated glomerular disease in mice with an incomplete genetic predisposition to systemic lupus erythematosus. Arthritis Rheum 52(7):2080–2091
Tan SM, Xu D et al (2003) Local production of B lymphocyte stimulator protein and APRIL in arthritic joints of patients with inflammatory arthritis. Arthritis Rheum 48(4):982–992
Tezuka H, Abe Y et al (2007) Regulation of IgA production by naturally occurring TNF/iNOS-producing dendritic cells. Nature 448(7156):929–933
Thompson JS, Bixler SA et al (2001) BAFF-R, a newly identified TNF receptor that specifically interacts with BAFF. Science 293(5537):2108–2111
Thompson JS, Schneider P et al (2000) BAFF binds to the tumor necrosis factor receptor-like molecule B cell maturation antigen and is important for maintaining the peripheral B cell population. J Exp Med 192(1):129–135
Vallerskog T, Heimburger M et al (2006) Differential effects on BAFF and APRIL levels in rituximab-treated patients with systemic lupus erythematosus and rheumatoid arthritis. Arthritis Res Ther 8(6):R167
Varfolomeev E, Kischkel F et al (2004) APRIL-deficient mice have normal immune system development. Mol Cell Biol 24(3):997–1006
von Bulow GU, Russell H et al (2000) Molecular cloning and functional characterization of murine transmembrane activator and CAML interactor (TACI) with chromosomal localization in human and mouse. Mamm Genome 11(8):628–632
von Bulow GU, van Deursen JM et al (2001) Regulation of the T-independent humoral response by TACI. Immunity 14(5):573–582
Wallweber HJ, Compaan DM et al (2004) The crystal structure of a proliferation-inducing ligand, APRIL. J Mol Biol 343(2):283–290
Wang H, Marsters SA et al (2001) TACI-ligand interactions are required for T cell activation and collagen-induced arthritis in mice. Nat Immunol 2(7):632–637
Weyand CM, Kang YM et al (2003) The power of the third dimension: tissue architecture and autoimmunity in rheumatoid arthritis. Curr Opin Rheumatol 15(3):259–266

Wu Y, Bressette D et al (2000) Tumor necrosis factor (TNF) receptor superfamily member TACI is a high affinity receptor for TNF family members APRIL and BLyS. J Biol Chem 275(45):35478–35485
Xia XZ, Treanor J et al (2000) TACI is a TRAF-interacting receptor for TALL-1, a tumor necrosis factor family member involved in B cell regulation. J Exp Med 192(1):137–143
Xu S, Lam KP (2001) B-cell maturation protein, which binds the tumor necrosis factor family members BAFF and APRIL, is dispensable for humoral immune responses. Mol Cell Biol 21(12):4067–4074
Yan M, Brady JR et al (2001a) Identification of a novel receptor for B lymphocyte stimulator that is mutated in a mouse strain with severe B cell deficiency. Curr Biol 11(19):1547–1552
Yan M, Wang H et al (2001b) Activation and accumulation of B cells in TACI-deficient mice. Nat Immunol 2(7):638–643
Yang M, Hase H et al (2005) B cell maturation antigen, the receptor for a proliferation-inducing ligand and B cell-activating factor of the TNF family, induces antigen presentation in B cells. J Immunol 175(5):2814–2824
Yu G, Boone T et al (2000) APRIL and TALL-I and receptors BCMA and TACI: system for regulating humoral immunity. Nat Immunol 1(3):252–256
Zhang J, Roschke V et al (2001) Cutting edge: a role for B lymphocyte stimulator in systemic lupus erythematosus. J Immunol 166(1):6–10

Autophagy in Cancer and Chemotherapy

Shida Yousefi and Hans-Uwe Simon

Abstract Cancer cells often exhibit mutations in critical molecules of the apoptotic machinery, resulting in resistance to common anticancer therapies. In the absence of apoptosis, autophagic cell death can be an alternative form of cell death by excessive self-digestion. Therefore, autophagic cell death can be considered as a backup cell death mechanism when apoptotic cell death mechanisms fail. However, many tumors also exhibit deficiencies in autophagy that may result in both genomic instability and further anticancer drug resistance. This chapter summarizes our current understanding regarding the regulation of autophagy in tumors and discusses potential new anticancer drug treatment strategies.

1 Autophagy Definition

Autophagy is an ubiquitous cellular process in eukaryotic cells that results in the breakdown of cytoplasm within the lysosome in response to stress conditions and that allows the cells to adapt to environmental changes. Although initially considered as being a physiologic process, which largely results in the degradation of proteins and cell organelles, recent studies have revealed an integral role for autophagy in human pathophysiology (Klionsky 2007). When cells encounter environmental stresses, such as nutrient starvation and pathogen infection, their autophagic activity is increased, resulting in either adaptation and survival or cell death. A key conserved pathway regulating autophagy is mediated by the mammalian target of rapamycin (mTOR). Inactivation of mTOR induces autophagy, while TOR activation inhibits the process (Shintani and Klionsky 2004) (Fig. 1).

Autophagy is regulated by autophagy genes (Atgs), which are mostly involved in the process of autophagosome formation and generate two ubiquitin-like conjugation

S. Yousefi and H.-U. Simon(✉)
Institute of Pharmacology, University of Bern, Friedbühlstrasse 49,
CH-3010 Bern, Switzerland
e-mail: hus@pki.unibe.ch

Results Probl Cell Differ, doi:10.1007/400_2008_25

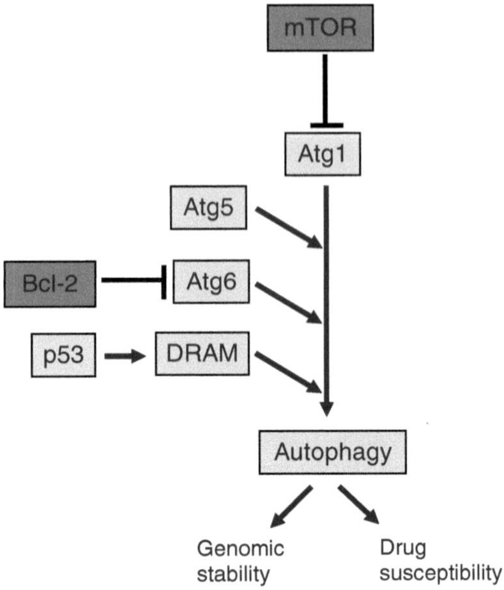

Fig. 1 Simplified view about the role of autophagy in cancer and its regulation. Key molecules mentioned in this article are displayed. mTOR and Bcl-2 inhibit autophagic activity by blocking different Atgs. Reduced autophagy is often associated with cancer. Novel therapeutic anticancer strategies aim to increase autophagic activity in cancer cells by either inhibiting mTOR (e.g., rapamycin) or Bcl-2 (e.g., ABT-737). Increased autophagic activity may increase genomic stability as well as the susceptibility towards anticancer drugs

systems: (1) the Atg12-Atg5 and (2) the Atg8 (LC3)-phosphatidylethanolamine (PE) systems. Atg12 covalently links to Atg5. The mode of conjugation of Atg12 to Atg5 is similar to that of ubiquitination, since Atg12 is first activated by Atg7 (=ubiquitin-activating enzyme E2) before it binds to Atg5. LC3 (microtubule-associated protein 1 light chain 3) is the mammalian ortholog of Atg8. In the process of autophagy, LC3 is cleaved by Atg4, resulting in the cytosolic form LC3-I, which is, similar to Atg12, first activated by Atg7, but then transferred to Atg3 and subsequently, interacts as LC3-II with PE. This results in the integration of LC3 into the membranes of pre-autophagosomes and phagosomes. Interestingly, this process is Atg5-dependent (Mizushima et al. 1998b; Ichimura et al. 2004).

Although autophagy plays an important role in cellular survival by providing energy during periods of starvation, but when stress conditions are excessive, increased autophagic activity above a certain level may mediate a cellular suicide pathway, operating by digestion of essential cellular proteins and structures (Gozuacik and Kimchi 2004). Such autophagic cell death is now considered as an alternative, non-apoptotic form of programmed cell death. Morphological change observed in this type of cell death involves double-membrane autophagic vacuoles, called autophagosomes, which degrade and recycle long-lived organelles and proteins by merging with lysosomes to form autolysosomes (Levine and Klionsky 2004).

2 The Role of Autophagy in Cancer

Increasing evidence indicates the importance of autophagy in cancer. Our understanding of the role of autophagy in cancer is at a very early stage, and even the most fundamental question, whether increased autophagic activity kills cancer cells or protects them from unfavorable conditions, has not been clearly defined (Kondo et al. 2005; Yousefi and Simon 2007). In some earlier studies, it was shown that amount of proteolysis or autophagic degradation was reduced in cancer cells compared with normal cells (Gunn et al. 1977; Kisen et al. 1993). One potential mechanism downregulating the autophagy machinery might be deletion and/or downregulation of key autophagy genes, such as Atg5 and Atg6 (beclin-1) (Fig. 1). The first hint that linked autophagy to cancer was suspected when researchers observed a deletion of one allele of Atg6 in a human breast tumor cell line (Liang et al. 1999). This discovery led to a whole new era of research aiming to understand the role of autophagy in cancer. Besides Atg6, levels of Atg5 may also control cancer development (Yousefi et al. 2006). In the following, we summarize the current knowledge on the expression and function of both Atg5 and Atg6 in cancer.

3 Atg5

Atg5 is an essential protein required for autophagy at the stage of autophagosome-precursor synthesis and its deletion in yeast or mammalian cells/mice effectively blocks autophagy (Kametaka et al. 1996; Mizushima et al. 1998a; Kuma et al. 2004). Besides regulating autophagosome formation, Atg5 may be important in apoptosis. Atg5 interacts with FADD (Fas-associated protein with death domain) and this interaction mediates interferon-γ (IFN-γ)-induced cell death. Downregulation of Atg5 in HeLa cells reduced cell death and vacuole formation induced by IFN-γ, and ectopic expression of Atg5-increased cell death. Atg5 did not modulate cell death caused by etoposide, staurosporine, or cisplatin. However, FADD-induced cell death was not affected by the reduced expression of Atg5 and by treatment with 3-MA, a class III PI-3 kinase inhibitor widely used to block autophagosome formation. Also, only cell death but not vacuole formation was blocked by caspase inhibition (Pyo et al. 2005). These data suggested that Atg5 may participate both in autophagy and certain forms of cell death, but that the two processes could be dissociated.

It was also shown that Atg5-overexpressing cells are sensitized towards various death stimuli and that silencing of Atg5 reduces anticancer drug-induced cell death. Death stimulation resulted in calpain activation, leading to Atg5 cleavage. Truncated Atg5 induced cytochrome *c* release and apoptosis, and both were blocked by high levels of Bcl-2. These data suggested that Atg5 represents a molecular switch between autophagy and apoptosis (Yousefi et al. 2006). These data indicated a direct link between Atg5 expression level and tumor cell susceptibility towards anticancer drugs. Interestingly, Atg5 may exhibit tumor suppressor activities, since cancer cells overexpressing Atg5 grow much slower in nude mice (Yousefi et al. 2006).

4 Atg6 (beclin1)

Atg6 was originally identified using yeast two-hybrid system with anti-apoptotic protein Bcl-2 as bait (Liang et al. 1998). Atg6 expression is low in mammary cancer cell line MCF-7 and, similar to Atg5, when overexpressed, it induced autophagy and decreased the tumorigenicity of cancer cells (Liang et al. 1999). In contrast, under-expression of Atg6 prevented autophagic cell death (Shimizu et al. 2004). A screen of 22 breast cell lines by Atg6-specific FISH analysis showed the presence of Atg6 allelic deletion in 41% (Aita et al. 1999). Additional studies showed that Atg6 protein expression was lost in eight out of 11 human breast carcinoma cell lines. The homozygotic deletion of Atg6 is lethal and heterozygotous deletion of Atg6 in mice leads to spontaneous tumors in liver, lungs, and B cells, suggesting that Atg6 is a haplo-insufficient tumor suppressor gene (Qu et al. 2003; Yue et al. 2003). The Atg6 gene is localized at chromosome 17q21, a locus that is deleted in 75% of ovarian, in 50% of breast, and in 40% of prostate cancers (Pattingre and Levine 2006). To date, Atg6 is the only member of the Atg family reported to be a tumor suppressor.

5 Atg and Bcl-2 Family Member Interactions

Bcl-2 and Bcl-x_L are anti-apoptotic proteins that have been reported to bind to Atg5 and Atg6, resulting in a block of both autophagy and autophagic cell death (Pattingre et al. 2005; Yousefi et al. 2006; Luo and Rubinsztein 2007). The interaction between Atg6 and Bcl-2 involves a BH3 domain within Atg6 (residues 114–123). Interestingly, the strength of molecular interaction is dependent on conditions known to modulate autophagic activity. Under nutrient-rich conditions, when autophagy is inhibited, Atg6 and Bcl-2 interact strongly. In contrast, during starvation, when autophagic rates are high, the interaction between Atg6 and Bcl-2 is weak. Recent data show that Bcl-2 phosphorylation by the kinase JNK1 is an important regulatory element. Under starvation conditions, JNK1 phosphorylates Bcl-2, and as a result, inhibits its interaction with Atg6, stimulating autophagy (Fig. 1). In contrast, under nutrient-rich conditions, Bcl-2 phosphorylation is inhibited, allowing Bcl-2 to interact with Atg6 and thus reducing autophagy (Wei et al. 2008). The ability of Bcl-2 to inhibit autophagy through a direct interaction with Atg6 is of a particular interest with respect to cancer. Pharmacological BH3 mimetic ABT-737 competitively inhibits the interaction between Atg6 and Bcl-2/Bcl-x_L that antagonizes autophagy inhibition by Bcl-2/Bcl-x_L and hence stimulates autophagy as well as apoptosis (Maiuri et al. 2007; Adams and Cory 2007). ABT-737 is currently used in several clinical trials to cure cancer.

6 mTOR Activity and Autophagy

mTOR is a conserved Ser/Thr protein kinase that regulates cell growth, cell cycle progression, nutrient import, protein synthesis, and autophagy (Pattingre et al. 2007). mTOR is sensitive to rapamycin, a macrolide originally used for immunosuppression. Rapamycin is now known as a specific inhibitor of mTOR activity (Loewith et al. 2002). The first evidence that mTOR has a role in regulating autophagy came from experiments involving rat hepatocytes. In this system, rapamycin partially reversed the inhibitory effects of amino acids on autophagic proteolysis (Blommaart et al. 1995). Atg1 and p70 S6 kinase (S6K) are two potential downstream targets for mTOR involved in controlling autophagy (Fig. 1). It has been suggested that S6K may contribute to the basal activity of autophagy via feed back inhibition of the class I PI3K-dependent insulin signaling pathway (Codogno and Meijer 2005). Moreover, Atg1 has recently been shown to inhibit cell growth in Drosophila and mammalian cells by down-regulating S6K (Klionsky et al. 2005; Lee et al. 2007). These studies indicate that a crosstalk exists between autophagy and cell growth regulation. Therefore, rapamycin and some analogs of rapamycin (CCI-779, RAD001, AP23573) are in clinical development for anticancer therapy (Faivre et al. 2006). It should be mentioned here that mTOR-independent stimulation of autophagy may also exist and has been reported to occur in response to LiCl treatment (Sarkar et al. 2007; Criollo et al. 2007).

7 p53 and Autophagy

The transcription factor p53 is an essential tumor suppressor and apoptosis inducer. Recent publications suggests that p53 has a positive role in cell survival in response to physiological (as opposed to genotoxic) stress by stimulating anti-oxidant pathways (Vousden and Lane 2007) and autophagy (Zeng et al. 2007). Furthermore, p53 has been shown to transactivate DRAM (damage regulated autophagy modulator), a lysosomal protein, that can stimulate the accumulation of autophagic vacuoles (Fig. 1). Knockdown of DRAM abolishes the accumulation of autophagic vacuoles induced by p53 and also reduced the induction of apoptosis (Crighton et al. 2006, 2007). Therefore, dysregulated autophagic activity in cancer cells might at least partially be due to cancer-associated dysregulation of p53 (Tasdemir et al. 2008).

8 Autophagy Protects Genome Stability

Recent publications conclude that the tumor suppressive function of Atg6 is linked to the ability of autophagy to protect cells from DNA damage. In this respect, reduced autophagic activity might trigger oncogenic events, such as chromosomal aberrations and mutations by deregulating the turnover of centomers (which results in multipolar division and chromosomal instability) or by compromising the quality control of mitochondria (which increase ROS production and ROS-mediated DNA damage) (Mathew et al. 2007a). Identification of this novel role of autophagy may be important for rational chemotherapy and therapeutic exploitation of autophagy inducers as potential chemopreventive agents (Mathew et al. 2007b).

9 Conclusion

Already in 1977, it was observed that the autophagic activity of cancer cell lines was often lower than their normal counterparts, especially under nutrient and/or serum deprivation (Gunn et al. 1977). This observation raised the possibility that the breakdown of the autophagy process may contribute to the development of cancer, but the direct evidence was missing. The identification of Atg6 as a novel tumor suppressor gene as well as the involvement of other well-established oncoproteins and tumor suppressor proteins in autophagy-related pathways indicate that autophagy plays an important role in cancer prevention.

One mechanism through which reduction of autophagic activity could contribute to malignant transformation emerges from the link of this fundamental process to apoptosis. Apoptosis is known to be the main type of programmed cell death that restrains abnormal cell proliferation (Green and Evan 2002). However, in case of blockage of apoptosis, which is common in cancer cells due to mutation or lack of essential apoptosis machinery, cell death still proceeds via activation of non-apoptotic pathways (Leist and Jaattela 2001).

Drugs that induce autophagic activities in tumors might be useful. Indeed, several preclinical and clinical studies indicate that pharmacological inhibition of mTOR, which leads to the induction of autophagy, is associated with anticancer effects (Hidalgo and Rowinsky 2000). Moreover, pharmacological inhibition of mTOR in combination with conventional anticancer therapies might be useful to induce apoptosis and/or autophagic cell death with high efficacy in cancer cells without mediating myelosuppression (Martinelli et al. 2006).

References

Adams JM, Cory S (2007) Bcl-2-regulated apoptosis: mechanism and therapeutic potential. Curr Opin Immunol 19(5):488–496

Aita VM, Liang XH et al (1999) Cloning and genomic organization of Beclin1, a candidate tumor suppressor gene on chromosome 17q21. Genomics 59(1):59–65

Blommaart EF, Luiken JJ et al (1995) Phosphorylation of ribosomal protein S6 is inhibitory for autophagy in isolated rat hepatocytes. J Biol Chem 270(5):2320–2326

Codogno P, Meijer AJ (2005) Autophagy and signaling: their role in cell survival and cell death. Cell Death Differ 12(2):1509–1518

Crighton D, Wilkinson S et al (2006) DRAM, a p53-induced modulator of autophagy, is critical for apoptosis. Cell 126(1):121–134

Crighton D, Wilkinson S et al (2007) DRAM links autophagy to p53 and programmed cell death. Autophagy 3(1):72–74
Criollo A, Maiuri MC et al (2007) Regulation of autophagy by the inositol triphosphate receptor. Cell Death Differ 14(5):1029–1039
Faivre S, Kroemer G et al (2006) Current development of mTOR inhibitors as anticancer agents. Nat Rev Drugs Discov 5(8):671–688
Green DR, Evan GI (2002) A matter of life and death. Cancer Cell 1(1):19–30
Gozuacik D, Kimchi A (2004) Autophagy as a cell death and tumor suppressor mechanism. Oncogene 12(23):2891–2906
Gunn JM, Clark MJ et al (1977) Reduced rates of proteolysis in transformed cells. Nature 266(5597):58–60
Hidalgo M, Rowinsky EK (2000) The rapamycin-sensitive signal transduction pathway as a target for cancer therapy. Oncogene 19(56):6680–6686
Ichimura Y, Imamura Y et al (2004) In vivo and in vitro reconstitution of Atg8 conjugation essential for autophagy. J Biol Chem 279(39):40584–40592
Kametaka S, Matsuura A et al (1996) Structural and functional analyses of APG5, a gene involved in autophagy in yeast. Gene 178(1–2):139–143
Kisen GO, Tessitore L et al (1993) Reduced autophagic activity in primary rat hepatocellular carcinoma and ascites hepatoma cells. Carcinogenesis 14(12):2501–2505
Klionsky DJ (2007) Autophagy: from phenomenology to molecular understanding in less than a decade. Nat Rev 8(11):931–937
Klionsky DJ, Meijer AJ et al (2005) Autophagy and p70S6 kinase. Autophagy 1(1):59–61
Kondo Y, Kanzawa T et al (2005) The role of autophagy in cancer development and response to therapy. Nat Rev 5(9):726–737
Kuma A, Hatano M et al (2004) The role of autophagy during the early neonatal starvation period. Nature 432(7020):1032–1036
Lee SB, Kim S et al (2007) Atg1, an autophagy regulator, inhibits cell growth by negatively regulating S6 kinase. EMBO Rep 8(4):360–365
Leist M, Jaattela M (2001) Four deaths and a funeral: from caspases to alternative mechanism. Nat Rev Mol Cell Biol 2(8):589–598
Levine B, Klionsky DJ (2004) Development by self-digestion: molecular mechanisms and biological functions of autophagy. Dev Cell 6(4):463–477
Liang XH, Kleeman LK et al (1998) Protection against fatal Sindbis virus encephalitis by beclin, a novel Bcl-2-interacting protein. J Virol 72(11):1029–1030
Liang XH, Jackson S et al (1999) Induction of autophagy and inhibition of tumorigenesis by beclin1. Nature 402(6762):672–676
Loewith R, Jacintho E et al (2002) Two TOR complexes, only one of which is rapamycin sensitive, have distinct roles in cell growth control. Mol Cell 10(3):457–468
Luo S, Rubinsztein DC (2007) Atg5 and Bcl-2 provide novel insights into the interplay between apoptosis and autophagy. Cell Death Diff 14(7):1247–1250
Martinelli S, Kostylina G et al (2006) Targeting surviving via PI3K but not c-akt/PKB by anticancer drugs in immature neutrophils. Oncogene 25(52):6915–6923
Maiuri MC, Toumelin GL et al (2007) Functional and physical interaction between Bcl-x_L and a BH3-like domain in Beclin-1. EMBO J 26(10):2527–2539
Mathew R, Kongara S et al (2007) Autophagy suppresses tumor progression by limiting chromosomal instability. Genes Dev 21(11):1367–1381
Mathew R, Karantza-Wadsworth V et al (2007). Role of autophagy in cancer. Nat Rev 7(12):961–967
Mizushima N, Sugita H et al (1998) A new protein conjugation system in human. J Biol Chem 273(51):33889–33892
Mizushima N, Noda T et al (1998) A protein conjugation system essential for autophagy. Nature 395(6700):395–398
Pattingre S, Tassa A et al (2005) Bcl-2 antiapoptotic proteins inhibit Beclin 1-dependent autophagy. Cell 122(6):927–939
Pattingre S, Levine B (2006) Bcl-2 inhibitionof autophagy: a route to cancer? Cancer Res 66(6):2885–2888

Pattingre S, Espert L et al (2007) Regulation of macroautophagy by mTOR and Beclin 1 complexes. Biochimie 90(2):313–323

Pyo JO, Jang MH et al (2005) Essential roles of Atg5 and FADD in autophagic cell death: dissection of autophagic cell death into vacuole formation and cell death. J Biol Chem 280(21): 20722–20729

Qu X, Yu J et al (2003) Promotion of tumorigenesis by heterozygous disruption of the *beclin1* autophagy gene. J Clin Invest 112(12):1809–1820

Sarkar S, Floto RA et al (2007) Lithium induces autophagy by inhibiting inositol monophosphatase. J Cell Biol 170(7):1101–1111

Shintani T, Klionsky DJ (2004) Autophagy in health and disease: a double-edged sword. Science 306(5698):990–995

Shimizu S, Kanaseki T et al (2004) Role of Bcl-2 family proteins in a non-apoptotic programmed cell death dependent on autophagy genes. Nat Cell Biol 6(12):1221–1228

Tasdemir E, Maiuri MC et al (2008) Regulation of autophagy by cytoplasmic p53. Nat Cell Biol 10(6):676–687

Vousden KH, Lane DP (2007) p53 in health and disease. Nat Rev Mol Cell Biol 8(4):275–283

Wei Y, Pattingre S et al (2008) JNK1-mediated phosphorylation of Bcl-2 regulates starvation-induced autophagy. Mol Cell 30(6):678–688

Yue Z, Jin S et al (2003) Beclin 1, an autophagy gene essential for early embryonic development, is a haploinsufficient tumor suppressor. PNAS 100(25):15077–15082

Yousefi S, Simon H-U (2007) Apoptosis regulation by autophagy gene 5. Crit Rev Oncol Hematol 63(3):241–244

Yousefi S, Perozzo R et al (2006) Calpain-mediated cleavage of Atg5 switches autophagy to apoptosis. Nat Cell Biol 8(10):1124–1132

Zeng X, Yan T et al (2007) DNA mismatch repair initiates 6-thioguanine-induced autophagy through p53 activation in human tumor cells. Clin Cancer Res 13(4):1315–1321

Glucocorticoid-Mediated Apoptosis Resistance of Solid Tumors

Ingrid Herr, Markus W. Büchler, and Jürgen Mattern

Abstract More than a quarter of a century ago, the phenomenon of glucocorticoid-induced apoptosis in the majority of hematological cells was first recognized. More recently, glucocorticoid-induced antiapoptotic signaling associated with apoptosis resistance towards cytotoxic therapy has been identified in cells of epithelial origin, most of malignant solid tumors and some other tissues. Despite these huge amounts of data demonstrating differential pro- and anti-apoptotic effects of glucocorticoids, the underlying mechanisms of cell type-specific glucocorticoid signaling are just beginning to be described. This review summarizes our present understanding of cell type-specific pro- and anti-apoptotic signaling induced by glucocorticoids. We shortly introduce mechanisms of glucocorticoid resistance of hematological cells. We highlight and discuss the emerging molecular evidence of a general induction of survival signaling in epithelial cells and carcinoma cells by glucocorticoids. We give a summary of our current knowledge of decreased proliferation rates in response to glucocorticoid pre- and combination treatment, which are suspicious to be involved not only in protection of normal tissues, but also in protection of solid tumors from cytotoxic effects of anticancer agents.

1 Introduction

For nearly 50 years, physicians have relied on glucocorticoids (GCs) – hormones normally secreted by the body in response to stress – to treat several types of cancer. With their ability to kill cancerous lympoid cells, GCs are important components of therapeutic approaches for treatment of lymphoid malignancies.[1] Because of several other beneficial effects, such as reduction of nausea and emesis, protection

I. Herr, M.W. Büchler and J. Mattern
Department of Surgery, University of Heidelberg, Germany

I. Herr(✉) and J. Mattern
Molecular OncoSurgery, University of Heidelberg and German Cancer Research Center, Heidelberg, Germany
e-mail: i.herr@dkfz.de

Results Probl Cell Differ, doi:10.1007/400_2008_20

of normal tissue from cytotoxic side effects as well as presumably a reduction of tissue reactions against the invasive malignant growth, these steroid hormones are also widely used as co-medication in chemotherapy of solid tumors.[2,3] However, and as reviewed here, GCs are highly suspicious to induce a resistant phenotype in cells of solid tumors while acting pro-apoptotic in lymphoid tumor cells. In addition, at the high doses usually required for co-medication in cancer chemotherapy, GCs suppress the immune system, weakening the patient's ability to fight disease. Thus, despite many benefits in cancer therapy, GCs may mediate a faster growth and metastases of solid tumors.

2 The Therapeutic Effect of GCs Depends on GR Levels

Mechanistically, GCs function by interacting with an intracellular receptor, the glucocorticoid (GR), a ligand-regulated transcription factor that positively or negatively alters the expression of specific target genes. Although some effects of GCs do not seem to be mediated by the GR, and are therefore designated nonspecific, there is growing evidence for rapid non-transcriptional actions of GR (Buttgereit and Scheffold 2002; Letuve et al. 2002). The non-transcriptional actions of GR involve the quick activation of protein kinases, such as the MAPK cascade, phosphatidylinositol-3 kinase (PI3-K), and Akt, leading to the activation of endothelial nitric oxide (NO) synthase (Limbourg and Liao 2003). However, a prerequisite in GC-induced apoptosis is binding of the GR to its ligand, followed by translocation to the nucleus, where it regulates a variety of target genes and molecular mechanisms (Cato et al. 2002). GR functions as a hormone-dependent protein in mediating the effects of GCs on growth regulation, both through stimulation and inhibition of expression of target genes which include other transcription regulatory systems (Norgaard and Poulsen 1991). The processes of receptor activation, receptor-DNA-binding, and the regulation of GR levels and transcription seem to be tissue-specific to a certain extent. Receptors for GCs have been found to be expressed in a large number of malignancies reflecting its wide distribution in human tissue (Table 1). Tumor receptor incidence is variable and may be dependent on either the histogenetic

Table 1 Glucocorticoid receptors in biopsy samples of human malignancies

Tumor type	Total No. of tumors	No. of tumors positive	(%)
Breast carcinoma [5–8]	376	169	45
Lung carcinoma [7, 9, 10]	112	52	46
Renal carcinoma [7]	7	4	57
Glioma [11–13]	28	14	50
Melanoma [14–16]	106	55	51
Sarcoma [17, 18]	33	13	39
Gastric carcinoma [19]	25	19	76

origin of the lesion or the method assay. Lu and colleagues (2005) investigated the effects of dexamethasone (DEX), a synthetic, clinically used GC, on growth and chemosensitivity in 14 carcinoma cell lines. The authors demonstrated that the GR content of cell lines in which DEX blocked cell cycle progression was significantly higher than in those cell lines unaffected by DEX. Hofmann and colleagues (1995) found that human lung cancer cell lines responsive to DEX had higher GR concentrations than the non-responders. The growth-inhibitory effect of DEX could be blocked by the GR agonist RU-486. Zibera and colleagues (1992) made obvious that DEX inhibited the growth of GR-positive and -negative cells only at high doses (5–125 μM). On the contrary, lower DEX concentrations (1.6–1,000 1987nM), resembling more plasma levels of patients receiving antiemetic GC therapy (Brady et al.), increased the proliferation of GR-positive but not of -negative cells (Freshney et al. 1980). Langeveld and colleagues (1992) investigated the effect of DEX on the proliferation of seven human glioma cell lines. In two of them, DEX consistently stimulated the proliferation, which may be associated with a relatively high expression level of GR. Braunschweiger and colleagues (1982, 1984) investigated the antiproliferative effects of DEX in solid tumors growing as xenografts in nude mice. The results from these kinetic and sequential treatment studies indicated that the duration of the DEX-induced cell cycle progression delay, and the timing of maximal S-phase cellularity after DEX was directly correlated with the level of GR in the treated tumors, that is, the higher the GR content the longer the progression delay. We ourselves could demonstrate that a human lung tumor line growing as xenografts in nude mice with high GR content responded to treatment with hydrocortisone, whereas another human tumor line with low GR content reacted only slightly to treatment with hormones (Mattern et al. 1985). In a clinical study with advanced non-small cell lung carcinomas, the authors found that outcome was significantly superior for patients whose tumors showed high GR expression compared to those with either low or nonexpression of GR. However, as almost all patients received GC treatment before chemotherapy as antiemetics or allergic preventive, they could not determine whether the better prognosis of NSCLC patients whose tumors showed high GR expression was related to the effect of GC or chemotherapy (Lu et al. 2006).

3 GC-Induced Anti-Apoptotic Signaling in Carcinoma Cells

GC-induced apoptosis critically depends on sufficient levels of GR and subsequent alterations in gene expression. The requirement for the receptor has been shown in thymocytes from genetically modified mice and human acute lymphoblastic leukemia (ALL) cell lines with mutated GR, and by conferring GC sensitivity to GC-resistant ALL cell lines by GR transgenesis (Schmidt et al. 2004). At least in leukemia cell lines, an autoregulatory upregulation of GR levels throughout the initiation and execution phase appears to be important, since removal of GCs (Brunet et al. 1998), or downregulation of tetracyclin-inducible GR expresssion (Ramdas et al. 1999),

Table 2 Molecules and pathways involved in GC-induced apoptosis in hematological cells

Name	GC-mediated effect
Mitochondria	Loss of mitochondrial membrane potential, caspase-9 activity, mitochondrial transloction of GR
Death receptor signaling	Secondary activation downstream of the mitochondrial death pathway
Bcl-2 family	Bim and Puma are upregulated and are the most promising candidates for pro-apoptotic BH3-only proteins promoting GC-induced apoptosis; the level of Bcl-2 and Bcl-xL may determine sensitivity to GCs
Caspases	Caspase-3, -8, and -9 are activated; antisense AP-4 is activated and downregulates caspase-9 expression
c-myc	Downregulated gene expression: conflicting signal
IkB	Induction of gene expression: conflicting signal by inhibition of the survival transcription factor NF-kB in lymphoid cells
Granzyme A	Is activated and may promote GC-induced apoptosis by inducing caspase-3 activity
TDAG8	Pro-apoptotic G protein-coupled receptor is activated and may promote GC-induced apoptosis
Lysosomes	Cathepsin B is released to activate caspases or to directly cleave nuclear substrates
Proteasomal degradation	c-IAP1 and XIAP are degraded at the protein level; Bcl-2 counteracts proteasome activation by GCs
Stress pathway	Repression of anti-apoptotic MKP-1 leads to JNK activity, involved in pro-apoptotic signaling
Other modulators	H2O2, PKC, IL-6, T-cell receptor, Neurotropin-1/B-cell stimulating factor-3, Ca+/Na+/K+-homeostasis and intracellular pH, synthesis of sphinomyelin, ceramide and sphingosine, metabolism

prevented GC-induced apoptosis in lymphoid cell lines. In contrast, impaired autoinduction was observed in GC-resistant but not in GC-sensitive subclones of the same ALL model (Kofler et al. 2003). These studies therefore suggest that basal level GR expression is inadequate to mediate GC-induced apoptosis in GC-sensitive T cells and that positive autoregulation is a necessary component of this process. Therefore, a prerequisite in GC-induced apoptosis in lymphoid cell lines is binding of the GR to its ligand, followed by translocation to the nucleus, where it regulates a variety of target genes and molecular mechanisms (Cato et al. 2002). A summary of mechanisms by which GC induce apoptosis in hematological cells is given in Table 2.

While lymphoid tumors are among the few cell types that can respond to GC exposure by initiating their apoptotic program, GCs have been reported to exhibit potent anti-apoptotic effects in epithelial cells (Amsterdam and Sasson 2002; Messmer et al. 2001), hepatocytes (Buchmann et al. 1999; Yamamoto et al. 1998), fibroblasts,(Gascoyne et al. 2003; Hammer et al. 2004), and activated T cells (Baumann et al. 2005; Cui et al. 1996; D'Adamio et al. 1997; Iseki et al. 1991; Wang et al. 2001; Webster et al. 2002; Yang et al. 1995, 2000; Yerramasetti et al. 2002; Zipp et al. 2000). The physiological function of GCs extends to the maintenance of lobular–alveolar structures in female mice during lactation (Topper and Freeman

1980). After cessation of lactation, apoptosis of mouse mammary epithelial cells and subsequent involution of the mammary gland can be inhibited by GC treatment (Feng et al. 1995; Lund et al. 1996). GC-induced protection from apoptosis has also been demonstrated in several cells of solid tumors when used with various anticancer drugs and with radiotherapy (Herr and Pfitzenmaier 2006). Such observations were made in established carcinoma cell lines cultured in vitro, in xenografts on nude mice, and in primary cells that had been isolated from fresh surgical samples of solid tumors. Tumors analyzed were derived from bladder (Zhang et al. 2006d), bone (Meyer et al. 2006; Zhang et al. 2006a), brain (Benedetti et al. 2003; Gorman et al. 2000; Rieger et al. 1999; Wolff et al. 1996; Wolff and Jurgens 1994; Zhang et al. 2006a), breast (Messmer et al. 2001; Mikosz et al. 2001; Pang et al. 2006; Sui et al. 2006; Wu et al. 2004; Zhang et al. 2006a), cervix (Herr et al. 2003; Kamradt et al. 2000, 2001; Rutz et al. 1998; Zhang et al. 2006a), colon or rectum (Zhang et al. 2005), liver (Evans-Storms and Cidlowski 2000; Zhang et al. 2005), lung (Bergman et al. 2001; Gassler et al. 2005; Wen et al. 1997), kidney (Zhang et al. 2006d), nerve tissue (Sengupta et al. 2000; Zhang et al. 2006a), ovary (Sui et al. 2006; Zhang et al. 2006c), pancreas (Zhang et al. 2006b), prostate (Kruit et al. 1999; Zhang et al. 2006d), skin (Zhang et al. 2006a), and testis (Zhang et al. 2006d). In the next parts of this review we investigate mechanisms of GC-induced anti-apoptotic signaling in cells derived from solid tumors, a possible influence of carcinomas' tissue formation on GC-induced survival signaling along with summarizing the network of GC-induced signaling pathways. This part is shortly presented in Table 3.

Table 3 Molecules and pathways involved in resistance to GC-induced apoptosis in cells derived from a tissue

Name	Factors that may be involved in GC-resistance
Downregulation of pro-apoptotic molecules	Death ligands such as CD95-L, TRAIL Adapter molecules such as FADD, Bid Caspases such as Caspase-8, -9, -3 Abnormal phosphorylation
Upregulation of anti-apoptotic molecules	cFLIP, Bcl-2, and IAP-family members – after prolonged exposure to GCs, these proteins are even depleted
Mitochondrial potential	Block of drug-induced depolarization
GR	Auto-upregulation and functionality of GR is required for GC-induced resistance
Tissue formation	GCs induce formation of cell–cell contacts including expression of E-cadherin, involved in adhesion of epithelial cells. This is followed by PI3-K/Akt kinase activity and Akt-mediated phospho-repression of GSK3 signaling
PI3-K/Akt	Is activated by GR and Cell-contacts, inactivates proapoptotic molecules e.g. caspases; blocks GSK3, inhibits FOXO transcription factors
SGK-1	Like Akt, SGK-1 is a downstream survival kinase of PI3-K and involved in survival signaling upon GCs, e.g. by inhibition of FOXO transcription factors

(continued)

Table 3 (continued)

Name	Factors that may be involved in GC-resistance
MKP-1	MKP-1 is a phoshpatase downstream of GR and involved in survival signaling by blocking JNK activity
JNK	The cellular stress pathway with its key molecules JNKs is inhibited by GR-induced MKP-1 and Akt-signaling thereby inhibiting apoptosis induction
Wnt/GSK3	Wnt signaling is activated by GCs, leads to phospho-repression of GSK3, and enhanced cell-contact
Transcription factors	AS-AP4 is induced by GCs and inhibits Caspase-9 and apoptosis in fibroblasts by downregulation of the transcription factor AP4
	NF-kB translocation to the nucleus and degradation of IkB may be involved in enhanced transcription of anti-apoptotic target genes such as Bcl-xL, caspase inhibitors and the pro-survival gene c-Myb
	FOXO3a is inhibited by GC-induced activation of SGK-1 and Akt-1 leading to inhibition of pro-apoptotic molecules and IGFBP-3

3.1 Inhibition of Core Apoptosis Signaling Pathways by GC-Treatment of Malignant Solid Tumor Cells

Apoptosis mechanisms that are induced by GCs in sensitive lymphoid cells have been found to be blocked in carcinoma cells, thereby mediating resistance towards chemo- and radiation therapy-induced apoptosis (Herr et al. 2003; Rutz and Herr 2004). A large number of dysregulated molecules within the apoptosis pathway were identified as underlying reason. A recent study used human cervical and lung carcinoma cells to experimentally address DEX-induced resistance disruption of critical cell death signaling following treatment of the cells with cisplatin, γ-irradiation, or ligation of death receptors (CD95, TRAIL) in vitro and in xenografts (Herr et al. 2003). By using microarray assays, Western blot analysis, and RT-PCR, several genes of the death receptor and mitochondrial apoptosis pathways were identified to be influenced by DEX. DEX antagonized basal and cisplatin-induced expression of key elements of the cell death receptor pathway such as CD95-L, TRAIL, FADD, and caspase-8 in carcinoma but not in human leukemic T cells (Herr et al. 2003). Correspondingly, GCs have been found to interfere with death receptor signaling in normal cells, since DEX increased cFLIP expression in hepatocytes (Oh et al. 2006). In this study, DEX administration into mice increased cFLIP expression in the liver and prevented CD95-induced hepatic injury by inhibiting caspase-8 and -3 activities (Oh et al. 2006). In functional assays, DEX-pretreatment prevented cisplatin-induced or serum-deprivation-induced depolarization of the mitochondrial membrane potential together with strong inhibition of expression and activity of caspase-9 in cervix carcinoma and hepatoma cells (Evans-Storms and Cidlowski 2000; Herr et al. 2003). These results indicate an action of GCs upstream of mitochondria. Also, the expression of BID, which links the death receptor pathway to mitochondria, was strongly downregulated in carcinomas (Herr et al. 2003). Proteins of the Bcl-2 and IAP family, which are preferentially located in mitochondrial membranes to stabilize them against permeabilization and to prevent cell death, were expressed in an anti-apoptotic manner after culture of carcinoma cells or primary cultured hepatocytes with GCs (Das et al. 2004; Messmer et al. 2001; Oh et al. 2006; Sasson and Amsterdam 2003; Sasson et al. 2001; Yamamoto et al. 1998). Involvement of IAP proteins is underlined by a recent patient study (Runnebaum and Bruning 2005) in which a chemo-naive patient with ovarian carcinoma received DEX for palliation to reduce ascites production. Ascites was collected from the ovarian cancer patient before and after oral administration of DEX. By this way, a significantly higher level of expression of cIAP2 mRNA was found after treatment with DEX in cancer cells isolated from ascites. Thus, downregulation of pro-apoptotic genes along with upregulation of anti-apoptotic molecules seems to be involved in GC-induced therapy resistance. On the other hand, upregulation of anti-apoptotic proteins has been shown to depend on the time of incubation of a GC, since, after prolonged exposure with DEX, for 48 h and longer, even depletion of the anti-apoptotic molecules Bcl-2, IAP, and cFLIP was observed (Herr et al. 2003), suggesting the involvement of several distinct mechanisms in GC-induced resistance. Because of repression of anti- as well as pro-apoptotic genes after prolonged exposure to GCs (Herr et al. 2003), one might speculate that the underlying reason for GC-induced resistance might be due to a broad spectrum inhibition of protein synthesis. However, since the well known global protein synthesis inhibitor cycloheximide did not mimic induction of therapy resistance in carcinoma cells (Herr et al. 2003), this assumption may be negligible. Nevertheless, GC-induced repression of the death receptor and mitochondrial pathway in carcinoma cells resulted in inhibition of caspase-8 and -9, leading to a decreased activity of the effector caspase-3 (Herr et al. 2003; Runnebaum and Bruning 2005). In line with these data, decrease of caspase-3 cleavage and apoptosis following exposure of primary human and rat hepatocytes to DEX has been demonstrated (Bailly-Maitre et al. 2001). Therefore, failure to activate caspases appears to be the main reason for DEX-induced apoptosis resistance in carcinoma cells. This suggestion is strongly supported by transfection experiments, showing that transfer of the genes as well as the active proteins of caspase-8 and -9 is able to restore the sensitivity of DEX-treated carcinoma cells to cisplatin-induced apoptosis in vitro and in vivo (Herr et al. 2003). Together, DEX-treatment of carcinoma cells results in inhibition of key elements of the death receptor and the mitochondrial apoptosis pathways, which may be sufficient to inhibit chemotherapy-induced apoptosis.

3.2 Anti-Apoptotic Signaling of GCs in Carcinoma Cells Depends on a Functional GR

DEX-mediated suppression of apoptosis in carcinoma cells is not due to a mutation of GR but rather involves transcriptional activity of the functional receptor. This conclusion was drawn by several investigators who found intact autoregulatory upregulation of GR expression and could partially rescue apoptosis sensitivity of

DEX-treated carcinoma cells by using the GR inhibitor RU486 (Herr et al. 2003; Messmer et al. 1999; Mikosz et al. 2001; Moran et al. 2000). Therefore, mutations of the GR may be excluded as a reason for DEX-mediated apoptosis inhibition in cancer cells from solid tumors. More likely, activated GR may transcriptionally affect gene expression in a cell type-specific manner, resulting in anti-apoptotic signaling in malignant cells of a solid tumor. The GR might negatively regulate promoter regions of pro-apoptotic genes in carcinomas but positively regulate them in lymphoid cells via direct or indirect effects. One example is the influence of GR on other transcription factors by cross-talk as described for AP-1, NF-AT, and NF-kB (Beato et al. 1995; Herrlich 2001). Recently, a competition between GR and NF-kB for control of the human CD95-L promoter has been described (Novac et al. 2006). A negative GR element was found at position −990 in the CD95-L promoter, which overlaps with a known NF-kB binding site. GR downregulates CD95-L promoter by competing with NF-kB for binding to the common response element. This type of repression may depend on the cellular levels of NF-kB and may well contribute to cell type-specific pro- and anti-apoptotic GR signaling. Alternatively, reduced expression or release of cytokines upon GC treatment may be interpreted as an apoptosis-signal in dependent cells such as leukemic T cells (Harmon and Thompson 1981). In contrast, more cytokine-independent cells such as peripheral T lymphocytes (Ray and Prefontaine 1994; Reichardt et al. 1998) or carcinoma cells survive.

3.3 Tissue Formation as a Mediator of GC-Induced Resistance in Carcinoma Cells

Resistance mechanisms have been functionally identified in malignant hematological and malignant solid tumor cells in unicellular models. In contrast to hematological cells, cells of a solid tumor are growing in a tissue formation with intercellular communication and attached to a matrix. Thus, implicit in unicellular models of resistant carcinoma cells is the lack of consideration of host–tumor cell interactions that may participate in the emergence of a resistant phenotype. Experimental data supporting this concept were provided by Teicher et al. (1990). These investigators showed that treatment of mice bearing breast cancer xenografts over a 6-month period with different chemotherapeutic agents resulted in the selection of a drug resistant phenotype in vivo. Conversely, the resistant phenotype was not detected in a unicellular tissue culture system, indicating the influence of the microenvironment in mediating the expression of a functional drug resistant phenotype. One important explanation for these results may be based on the phenomenon of anoikis: upon detachment from the extracellular matrix, epithelial cells enter into programmed cell death, ensuring that they are unable to survive in an inappropriate location (Rennebeck et al. 2005). In 1998, Murakami et al. suggested that GCs could be involved in the dynamic control of normal cell–cell contacts and cell migration of human gastric carcinoma cells and speculate that there might be a correlation to resistance and malignant behavior (Murakami et al. 1998). In these particular cells, DEX has been shown to inhibit cell migration by increasing cell adhesion to the matrix through stimulation of $\alpha 1\beta 1$ integrin expression. In parallel, GCs are reported to induce intercellular communications by the formation of tight junctions in a variety of epithelial cells (Stevenson et al. 1986; Woo et al. 1996). Although the influence of a tissue structure to GC-induced apoptosis resistance is only marginally examined until now, recent data from epithelial-like granulosa cells of ovarian follicles in which DEX induces apoptosis resistance point to this direction (Sasson and Amsterdam 2003). In these cells, GCs induced the development of a rigid network of actin cytoskeleton, increased incidence of adherence and gap junctions, together with a higher content of connexin 43 and total cadherins within 24 h of hormone stimulation. The same authors showed in another study that protein phosphorylation, cell–cell contact, and intracellular communication are important mediators in GC-induced protection against apoptosis (Sasson et al. 2003). Beneath direct cell contacts, attachment of epithelial cells to extracellular matrix is known to provide a survival signal, in which GC-induced activity of the survival kinase PI3-K/Akt is suggested to be involved (Khwaja et al. 1997). Binding of E-cadherin, a molecule involved in adhesion between epithelial cells, may play a role in preventing apoptosis, as examined in granulosa cells (Peluso et al. 2001) and in nasal epithelial cells, in which GCs induced expression of E-cadherin (Carayol et al. 2002). To determine whether disrupting E-cadherin binding leads to apoptosis, granulosa cells were cultured in serum in the presence of EGTA or an E-cadherin antibody. Disruption of all calcium-dependent contacts by EGTA, or preventing E-cadherin binding by the antibody, induced apoptosis. Exposure to EGTA reduced MEK and PI3-K/Akt kinase activity, while E-cadherin antibody only attenuated PI3-K/Akt kinase activity. Caspase-3 activity, controlled by PI3-K/Akt kinase, increased after serum depletion, or EGTA or E-cadherin antibody treatment (Peluso et al. 2001). These data are consistent with the concept that GC-induced and E-cadherin-mediated cell contact, either directly or indirectly, promotes PI3-K/Akt kinase activity, which results in the inhibition of apoptosis and survival signaling.

3.4 *GC-Induced PI3-K/Akt Survival Signaling and Resistance of Carcinoma Cells*

PI3-K/Akt kinase is known to be an important regulator of cell cycle progression and mediator of cellular survival (Song et al. 2005), and is assumed as a key molecule in signaling of GC-induced activities, especially in non-lymphoid cells. The detailed mechanism of PI3-K activation by GR is currently unknown, but seems to involve association of GR with the regulatory p85α subunit of PI3-K (Limbourg and Liao 2003). The importance of PI3-K/Akt signaling to survival has been shown repeatedly by both the ability of activated PI3-K to abrogate apoptosis and the ability of a dominant-negative PI3-K to enhance it. Further, a common method of inducing or enhancing apoptosis is by treatment with the PI3-K/Akt inhibitor LY 294002 (Cox and Der 2003). Finally, PI3-K/Akt activation upon GC-treatment was clearly demonstrated (Hafezi-Moghadam et al. 2002; Limbourg and Liao 2003), supporting the proposed signaling sequence of GC-induced PI3-K/ Akt activation using cadherins and cell–cell as well as cell–matrix contacts as intermediates. PI3-K/Akt in turn phosphorylates a number of substrates important for inhibition of apoptosis and for promotion of proliferation. Only limited information is available about how GR-induced PI3-K/Akt signaling may mediate resistance. However, there are a few data available demonstrating that GR-induced PI3-K/Akt signaling directly influences the activity of apoptosis molecules. PI3-K/Akt inactivates, for example, caspase-9 by phosphorylation on Ser196 (Cardone et al. 1998), resulting in inhibition of downstream caspases as shown for caspase-3 (Peluso et al. 2001). Another pro-apoptotic target inactivated by PI3-K/Akt is BAD, a member of the Bcl-2 family, which binds to Bcl-2 and Bcl-xL to inhibit their anti-apoptotic potential (Datta et al. 1997; del Peso et al. 1997). In addition, PI3-K/Akt has been found to upregulate expression of anti-apoptotic proteins via phosphorylation of cyclic AMP (cAMP)-response element binding protein (CREB) transcription factor, which in turn activates expression of the anti-apoptotic genes Bcl-2 and Mcl-1 (Wang et al. 1999). This szenario is described in detail in the following chapters and summarized in Figure 2 and Table 3.

3.5 GR-Mediated NF-kB Activity and Resistance in Carcinoma Cells

One prominent candidate gene for anti-apoptotic regulation by GCs is the transcription factor NF-kB, involved in the regulation of diverse apoptosis genes such as CD95-L, CD95, DR4, and DR5 that all carry NF-kB consensus sequences in their promoter regions (Herr and Debatin 2001). In carcinoma cells, DEX has been shown to enhance the activity of NF-kB by increasing the nuclear translocation of this transcription factor (Evans-Storms and Cidlowski 2000; Huang et al. 2000). NF-kB activation depends in most cases on the phosphorylation of the IkB kinase (IKK) complex and degradation of IkB, an inhibitor of NF-kB (Karin 1999). Machuca and colleagues (Machuca et al. 2006) found that DEX-mediated apoptosis protection is lost in MCF7 breast cancer cells expressing the dominant negative form of IkB. Through analyses of a number of apoptosis-associated genes or regulatory proteins, Fan and co-workers found that GCs antagonized paclitaxel-mediated NF-kB activation through induction of IkB synthesis (Fan et al. 2004). These results provide further support to the notion that GC-induced resistance may require NF-kB activation. GC-induced PI3-K/Akt signaling seems to contribute to NF-kB-induce resistance, since these kinases were shown to regulate IKK activity, both in a direct and an indirect manner (Kane et al. 1999). Subsequently, translocation and activation of NF-kB was found to be responsible for enhanced transcription of the anti-apoptotic target genes Bcl-xL, caspase inhibitors, and the pro-survival gene c-Myb (Barkett and Gilmore 1999; Lauder et al. 2001). However, NF-kB activation through PI3-K/

Akt is a controversial issue due to cell type-specific variations (Madge and Pober 2000; Pastorino et al. 1999). Although in some cells Akt acts upstream of NF-kB (Gustin et al. 2004; Ozes et al. 1999; Reddy et al. 2000), other authors describe DEX-induced therapy resistance in breast cancer cell lines to be completely independent from Akt function (Machuca et al. 2006). In consequence, and irrespective of Akt involvement, GK-induced translocation of NF-kB to the nucleus is one important candidate for anti-apoptotic regulation of apoptosis genes in carcinoma cells.

3.6 GC-Induced Expression and Enhanced Activity of SGK-1 may be Involved in Apoptosis Resistance of Carcinoma Cells

Recently, induction of serum and GC-regulated kinase-1 (SGK-1) was found to be required for anti-apoptotic signaling following GR activation in epithelial cells and breast cancer cell lines such as MDA-MB-231 (Mikosz et al. 2001; Wu et al. 2004). SGK-1 was originally identified as an immediate early response gene that is regulated primarily by transcriptional induction following various environmental stimuli including GCs, as well as by post-translational phosphorylation, ubiquitin-modification, and proteasomal degradation (Belova et al. 2006; Lang and Cohen 2001). Like Akt, the serine/threonine kinase SGK-1 is a downstream effector of PI3-K signaling (Kobayashi et al. 1999), and was shown previously to contribute to survival signaling in neurons (Brunet et al. 2001), mammary epithelial cells, and breast cancer cell lines (Leong et al. 2003; Mikosz et al. 2001). In contrast to Akt, which shares with SGK-1 about 50% amino acid homology in its kinase domain, SGK-1 is phosphorylated downstream of endogenous PI3-K activity (Leong et al. 2003). It has been well-documented that SGK-1 is transcriptionally upregulated following various stress stimuli including GCs, thereby providing a transient increase in SGK-1 protein levels, which provides a strong survival signal to epithelial cells (Belova et al. 2006; Leong et al. 2003; Wu et al. 2004). The link between Akt and SGK-1 signaling in GC-induced therapy resistance has been only marginally examined. However, a recent study in *Caenorhabditis elegans* demonstrated a function of both molecules in controlling the DAF-2 signaling pathway (Hertweck et al. 2004). The *daf-2* gene encodes an insulin-like receptor, and mutations in *daf-2* have been shown to double the lifespan of the worms. The gene is known to regulate reproductive development, ageing, resistance to oxidative stress, thermotolerance, resistance to hypoxia, and also resistance to bacterial pathogens. DAF-2 is the only insulin/IGF-1-like receptor in the worm. Insulin/IGF-1-like signaling is conserved from worms to humans. DAF-2 acts to negatively regulate the forkhead transcription factor DAF-16 through a phosphorylation cascade. Genetic analysis reveals that DAF-16 is required for daf-2-dependent lifespan extension. When not phosphorylated, DAF-16 is active and present in the nucleus. SGK-1 seems to act in parallel to Akt to mediate DAF-2 signaling, since loss of the *sgk-1* gene results in, for example, increased stress resistance and an extension of life span in the worm. These effects are suggested to be mediated by a protein complex formed by SGK-1 and Akt, which is activated by and strictly depends on pyruvate dehydrogenase kinase, isoenzyme 1 (PDK-1). All three kinases of this complex are able to directly phosphorylate and thereby inactivate DAF-16, resulting in anti-apoptotic signaling.

3.7 GC-Induced Expression and Activity of MKP-1 and Dependent Reduction of JNK Activity may be Involved in Apoptosis Resistance of Carcinoma Cells

In addition to activation of SGK-1, PI3-K/Akt, and NF-kB, GR-induced MKP-1 expression has recently been shown to be required for GC-mediated cell survival in breast cancer cell lines (Wu et al. 2005). MKP-1 is a MAPK phosphatase, shown to exhibit anti-apoptotic effects in prostate (Magi-Galluzzi et al. 1997) and breast cancer cells (Wu et al. 2005). Endogenous or GC-induced overexpression of MKP-1 has been demonstrated in primary human breast (Loda et al. 1996), prostate cancers (Loda et al. 1996; Magi-Galluzzi et al. 1997), prostate cancer cell lines (Srikanth et al. 1999), pancreatic cancer cell lines (Zhang et al. 2006b), and ovarian carcinoma cell lines (Steinmetz et al. 2004; Zhang et al. 2006c). Additional evidence for an important function of MKP-1 together with SGK-1 in anti-apoptotic signaling of GCs in carcinoma cells comes from studies with SGK-1 and MKP-1 siRNA. Each siRNA decreased protein expression and subsequently reversed DEX-induced survival in breast cancer cells (Wu et al. 2005). Thus, induction of both genes together may be required for cell survival signaling downstream of GR activation. Following GR activation, induction of MKP-1-activity is associated with the dephosphorylation and inactivation of extracellular regulated kinases 1, 2 and c-Jun NH_2-terminal kinases 1, 2 (JNK) (Engelbrecht et al. 2003; Wu et al. 2005). Since JNKs are induced by cytotoxic agents and are involved in induction of apoptosis (Herr and Debatin 2001), inhibition of these molecules by MKP-1-activity may contribute to GC-induced therapy resistance. Moreover, apoptosis triggered by activation of JNKs was found to be enhanced through repression of the JNK-phosphatase MKP-1. Vice versa, ectopic expression of MKP-1 inhibited DNA-damage-induced JNK activity and apoptosis (Hamdi et al. 2005). Independently of MKP-1, Akt activation may result in phosphorylation and thereby inactivation of three pro-apoptotic kinases upstream of JNK: Apoptosis signal-regulating kinase 1 (ASK1/MKKK5), mixed lineage kinase 3 (MLK3/MKKK11), and (SEK1/MKKK4) molecules (Barthwal et al. 2003; Kim et al. 2001; Park et al. 2002). Thus, members of the JNK pathway may be inhibited by GR-induced MKP-1 and Akt signaling and participate by this way in GR-induced pro-survival signaling. Similar to GCs, long term treatment of cells with insulin stimulated expression of MKP-1, followed by inhibition of JNK activity in which PI3-K was involved (Desbois-Mouthon et al. 2000). JNK activity recovered by the PI3-K inhibitors wortmannin and LY294002, while transient expression of an antisense MKP-1 RNA reduced the insulin inhibitory effect on JNKs. Vice versa, over-expression of a dominant negative JNK1 mutant increased insulin stimulation of DNA synthesis and mimicked the protective effect of insulin against serum withdrawal-induced apoptosis. The over-expression of wild-type JNK1 or antisense MKP-1 RNA reduced the proliferative and/or anti-apoptotic responses to insulin. Altogether, these results demonstrate that PI3-K/Akt and MKP-1 signaling together are involved in inhibition of the pro-apoptotic JNK/SAPK pathway, which may result in a switch to survival signaling following treatment of cells with GCs or insulin.

3.8 GR-Mediated Inhibition of FOXO Forkhead Transcription Factors and Resistance in Carcinoma Cells

While it is well known that ligand-bound GR interferes directly with transcription factors such as AP-1 or NF-kB, the indirect effects of the GR on other transcription factors regulating apoptosis molecules are not well understood. Recent data demonstrate the involvement of GR-mediated FOXO3a inactivation in mammary epithelial cell survival (Wu et al. 2006). This molecule is a member of the forkhead transcription factors (FOXO1, FOXO3a, FOXO4) exerting pro-apoptotic effects until phosphorylation by SGK-1 in coordination with Akt, which preferentially phosphorylates FOXO3a (Brunet et al. 2001). In addition, in *C. elegans*, SGK-1 activity is required for the inactivating phosphorylation of the FOXO3a homologue Daf 16 (Hertweck et al. 2004), suggesting a highly conserved pathway involving SGK-1 and FOXO3a. Phosphorylated FOXO proteins are thought to activate promoters of extracellular ligands, including CD95-L, TRAIL, and TRADD, as well as intracellular apoptosis molecules like Bim, a pro-apoptotic Bcl-2 family member, and Bcl-6 (Burgering and Medema 2003; Song et al. 2005). Therefore, inactivation of FOXO proteins upon GC-induced Akt and SGK-1 activity may result in transcriptional downregulation of the above described pro-apoptotic proteins and contribute to GR-induced apoptosis resistance. Correspondingly, a recent report demonstrated that activation of the GR mediates phosphorylation and thereby inactivation of FOXO3a in mammary epithelial cancer cells (Wu et al. 2006). To identify downstream target genes of FOXO3 inactivation, the authors of this study used a bioinformatic approach combining temporal gene expression data and FOXO3a binding motif analyses. This approach identified a group of known and novel FOXO3a target genes. Included among these genes are TRAIL, insulin growth factor binding protein-3 (IGFBP-3), and serine threonine kinase 11 (STK11), all of which encode pro-apoptotic proteins. While TRAIL is a known target of FOXO3a, IGFBP-3 has not been previously identified as a FOXO3a target gene. To validate this relationship, the authors used chromatin immunoprecipitation (ChIP) and showed that DEX completely inhibited TRAIL and IGFBP-3 promoter occupancy by FOXO3a, while the GR antagonist RU486 restored occupancy (Wu et al. 2006). These data have been confirmed by overexpression of FOXO3a that induces IGFBP-3 or by siRNA-mediated repression of SGK-1 that reversed

GR-induced IGFBP-3 downregulation (Wu et al. 2006). Collectively, these data suggest that GR-mediated FOXO3a inactivation is an important mechanism contributing to GC-mediated epithelial cell survival.

3.9 GR-Induced Activation of the Wnt/GSK3/β-Catenin Pathway

Several stimuli, including signaling by the GR, lead to Wnt activity. Wnt signaling is involved in various differentiation events during embryonic development and leads to tumor formation when aberrantly activated (Giles et al. 2003). Active Wnt signaling is also well understood for its ability to increase cell–cell contacts, adhesion, and proliferation (Mulholland et al. 2005), which may enforce by this way GC-induced apoptosis resistance. A main downstream target of Wnt is glycogen synthase kinase 3 (GSK3), a serine/threonine protein kinase, initially identified as an enzyme that inhibits glycogen synthesis through phosphorylation of glycogen synthase (Embi et al. 1980; Woodgett 1990). Recent studies revealed that GSK3 regulates a wide range of cellular functions, including development, gene expression, cytoskeletal organization, protein translation, cell cycle regulation, and apoptosis (Doble and Woodgett 2003; Harwood 2001; Pap and Cooper 1998), and the later involving inhibition of conformational activation of the pro-apoptotic Bax protein (Rathmell et al. 2003). GSK3 resides at the junction of the PI3-K/Akt and Wnt survival pathways and is phospho-inhibited by both (Doble and Woodgett 2003; Frame and Cohen 2001). Another function of GSK3 is regulation of metabolic pathways, for example, the import of nutrients into the cell, initially induced by enhanced levels of insulin. GCs mimic the effect of insulin and both lead to phosphorylation and activation of Akt, which has been implicated in a variety of insulin effects involving phospho-inhibition of forkhead transcription factors (Nakae et al. 1999), followed by anti-apoptotic signaling. Insulin-induced Akt activation allows cells to continue with the import of nutrients into the cell (Edinger and Thompson 2002; Vander Heiden et al. 2001), partially by phosphorylation and thereby inhibition of GSK3 (Cross et al. 1995). As a result of Wnt signaling, β-catenin, another promiscuous Wnt signaling member, is dephosphorylated and escapes the ubiquitylation-dependent destruction machinery (van Noort et al. 2002). Phosphorylation of β-catenin by GSK3 occurs within the destruction complex, which consists minimally of GSK3, β-catenin, axin/conductin, and APC (Dihlmann and von Knebel Doeberitz 2005; Hinoi et al. 2000). The GR is able to modulate the integrity of this β-catenin complex by involving Fascin, a negative regulator of epithelial adherens junctions in mammary epithelial tumor cells (Tao et al. 1996). Upon exposure to DEX, Fascin facilitates assembly of E-cadherin and β-catenin complexes (Guan et al. 2004; Wong et al. 1999), resulting in enhancement of tight and adherens junctions (Buse et al. 1995a, b; Zettl et al. 1992). With the ability to stabilize adherens junctions, GCs may be beneficial in restoring polarization, morphology, or adhesiveness of poorly differentiated cancer cells. In this context, mutations in β-catenin that prevent its phosphorylation by GSK3 have been found in cancers of the skin, colon, prostate, liver, endometrium, and ovary (Polakis 2000).

4 Cell Cycle Arrest by GCs may be Involved in Mediating Resistance

GCs are suggested to act cytostatic rather than cytotoxic. Although cytotoxicity of GCs has been demonstrated in some cell culture systems (Mealey et al. 1971; Wellington and Moon 1961), this effect may be due to the use of steroid concentrations exceeding the pharmacological levels found in patients. Where more physiological concentrations have been used, both stimulation of cell proliferation (Kawamura et al. 1998; Langeveld et al. 1992; Zibera et al. 1992) and inhibition have been observed in cultures of human tumor cells (Braunschweiger et al. 1982; Chen et al. 2006; Kudawara et al. 2001; Yamamoto et al. 2002). In most cases, DEX significantly inhibited growth at high concentrations but stimulated growth at low concentrations. These effects on cell proliferation seem also to be dependent on the cell line used and on the expression of GC receptors. Studies in a variety of cell lines in vitro have indicated that antiproliferative effects of GCs may be mediated by a reversible G1-block in cell cycle progression (Glick et al. 2000; Goya et al. 1993; Sanchez et al. 1993), since after withdrawal of corticosteroids cells synchronously progressed through the cell cycle (Braunschweiger et al. 1978; Goya et al. 1993). Investigations with glioblastoma cells (Kaup et al. 2001), osteosarcoma cells (Yamamoto et al. 2002), and ovarian cancer cells (Chen et al. 2006) have shown that DEX led to cell growth arrest in a time- and dose-dependent manner. Analyses of DNA content by flow cytometry, [^{3}H]-thymidine incorporation, and autoradiography of [^{3}H]-thymidine-labeled nuclei revealed that, after release from DEX, cells synchronously reinitiated cell cycle progression and entered S-phase with a peak between 12 and 36 h, depending on the dose of DEX, duration of treatment, and GC receptor content. These results were not only obtained in vitro (Goya et al. 1993; Zhang et al. 2007), but also in vivo using experimental solid tumors on mice (Braunschweiger et al. 1982, 1983, 1984; Kudawara et al. 2001). These animal studies indicated in addition that, upon cessation of DEX treatment, the subsequent cell cycle progression results in a relative enrichment of cells in the S-phase of the cell cycle after 24–48 h. Mechanistically, DEX-induced GR activity may block cell cycle progression by inhibition of p53-dependent functions, including upregulation of $p21^{WAF1/CIP1}$ as found in neuroblastoma cells (Sengupta et al. 2000). Similarly, GCs stimulated p21 gene expression by targeting multiple transcriptional elements within a steroid responsive region of the $p21^{WAF1/CIP1}$ promoter in rat hepatoma cells (Cha et al. 1998).

General agreement exists that cancer chemotherapy is most successful when applied to rapidly growing cells (Shackney et al. 1978; Valeriote and van Putten 1975). Experimental data obtained in a variety of systems ranging from mammalian cell cultures to transplanted rodent and human tumors show that proliferating cells are more sensitive to most anticancer drugs than resting cells (Drewinko et al. 1981; Mattern et al. 1984). Exposure of various tumor cell lines of different origin to DEX has shown that this GC interferes with the cytotoxic and antiproliferative actions of most chemotherapeutic drugs. These drugs include the major agents currently used for chemotherapy of solid tumors. The mechanisms of action vary widely and DEX-mediated cytoprotection is restricted neither to specific drugs nor to specific tumor types. This suggests that the observed therapy resistances are not the exception but a more common phenomenon. Most of the tumor cell lines were pretreated with DEX during 24–72 h of culture so that the proliferation in these cells may have been reduced or completely inhibited at the time point when the anticancer drugs are given. Since cytotoxicity is significantly less pronounced in nonproliferating compared with proliferating cells (Drewinko et al. 1981), it is predictable that the cells reacted less sensitive to the lethal activity of most anticancer agents after pretreatment with DEX.

Similar as has been found for cytotoxic drugs, it is well documented that radiation sensitivity varies with the position of the cells in the cell cycle (Sinclair 1968; Terasima and Tolmach 1961). When cells were irradiated at different phases of the cell cycle, the cells in M phase showed the greatest sensitivity. Cells at the transition between G1 and S are also quite radiosensitive, but less than M-phase cells. Accordingly, the effect of a single dose of radiation on a population of asynchronously dividing cells will depend on the distribution of cells according to their specific cell replication cycle. Thus, a radiation dose kills the more sensitive cells and results in partial synchrony such that most of surviving cells are in a relatively resistant phase. When cells are pretreated with GC, an early G1 block in cell cycle progression is induced, so that the cells are in a relatively resistant cell cycle phase for irradiation.

5 Investigation of GC-Induced Resistance in Clinical Trials

To our knowledge there are only few clinical trials that were designed to assess whether the addition of corticosteroids would enhance or reduce the therapeutic effectiveness of adjuvant chemotherapy. In a prospective trial with premenopausal breast cancer patients, Tormey and colleagues (1990) compared the therapeutic effectiveness of CMF, CMFP, and CMFPT (plus tamoxifen). The lack of a significant difference between the regimens in survival time and time to relapse and the associated toxicities of the CMFP overrides their recommendation for widespread routine use. The addition of prednisone to CMF was associated with less nausea/emesis and thrombocytopenia, but more gastrointestinal side effects, infections, neurologic side effects, edema, weight gain, thrombophlebitis, cardiac events, and hypertension. Thus, the results of this trial do not suggest an overall therapeutic benefit for adding prednisone to CMF adjuvant treatment. Postmus and colleagues (1995) evaluated retrospectively the results of a phase II study of teniposide in 80 patients with brain metastases of small-cell lung cancer. Of 43 patients who received corticosteroids, only 9 responded (21%), whereas 17 of 37 patients not treated with steroids had a response (46%). Treatment with corticosteroids during therapy with teniposide resulted in a subjective improvement in this parameter in 18 of 42 assessable patients (43%). Teniposide without corticosteroids resulted in subjective improvement in 20 of 34 assessable patients (59%). Patients who required corticosteroids for peritumoral edema had a significantly lower response rate than patients who did not receive corticosteroids. Marini and colleagues (1996) tested the hypothesis in patients with breast cancer that low-dose prednisone might influence the efficacy of the cytotoxic regimen used, the toxicity profiles, and the patterns of treatment failure (relapse, second malignancy, or death). They found that low-dose continuous prednisone added to adjuvant CMF chemotherapy enabled the use of higher doses of cytotoxic drugs. However, this increased dose had no beneficial effect on treatment outcome, but was associated with an increased risk of bone relapses and an increased incidence of second malignancies. Münstedt and colleagues (2004) evaluated retrospectively patients with ovarian carcinoma to elucidate the effects of corticosteroids on hematologic parameters during chemotherapy, on the response to chemotherapy, and on survival. Kaplan–Meier analyses showed no significant differences in survival between the groups. Patients who received GC treatment had significantly higher leukocyte values compared with patients who did not receive GCs. In conclusion, there was no evidence that GC treatment had a positive effect on outcomes in these patients; however, GCs exert protective effects on the bone marrow.

Many early investigations have been carried out on the effects of cortisone on experimental tumors. Most interest has centered around the action of cortisone on growth, heterotransplantability, and metastatic spread of certain experimental tumors. From these studies it seems that cortisone, depending on the dose, has a general inhibitory action that is slight and temporary on transplantable tumors of the lymphoid series and no action on transplantable tumors of an epithelial origin. The action of cortisone on the metastatic spread of tumors is well investigated with induced and transplanted tumors in mice. Agosin and colleagues (1952) and Molomut and colleagues (1952) have reported an increase in the incidence of metastases in cortisone-treated tumor-bearing mice. These observations could be confirmed by many authors (Baserga and Shubik 1954; Iversen 1957; Pomeroy 1954). However, cortisone treatment has also been found to favor an increased metastatic spread in patients with breast carcinoma (Iversen and Hjort 1958; Sherlock and Hartmann 1962), lung carcinoma (Wolf et al. 1960), and renal carcinoma (Rasmuson et al. 2001). Recently, Sorensen and colleagues (2004) compared in a large population-based follow-up study among 59.043 individuals who received prescriptions for GCs, the observed and expected number of cases of skin cancer and non-Hodgkin lymphoma. They found an elevated risk for squamous cell carcinomas and basal cell carcinomas of the skin and also for non-Hodgkin lymphoma in patients who received prescriptions for GCs. These results further indicate that immunosuppression by GCs may be a risk for these malignancies.

6 Conclusions and Future Perspective

The balance of GC-induced apoptosis vs. apoptosis resistance dramatically depends on the cell lineage. It seems that cells of hematopoietic origin such as monocytes, macrophages, lymphocytes, and lymphoma cells are very sensitive to GC stimulation of apoptosis. In contrast, cells of epithelial origin, such as mammary gland, ovarian follicular cells, and hepatocytes as well as the majority of transformed epithelial cells, are protected by GCs against various stimuli for apoptosis. A number of models have been postulated to explain the mechanism by which GCs induce apoptosis in most hematologic malignancies. There is convincing evidence indicating that resistance of hematological cells towards GC-induced apoptosis may concern defects in the GR itself; in GR binding partners, a dysregulation of GR target genes and transcription factors, it may be due to the activity of general resistance mechanisms, defects in the apoptosis pathway, or just reflect a shift of the balance of cellular signaling pathways to anti-apoptotic signaling. In contrast, GC-induced apoptosis resistance in most epithelial cells, including cells from solid malignant tumors, crucially depends on a functional GR. GC-mediated modification of cell–cell contacts, intracellular communication, and detachment to an extracellular matrix, unique for non-hematological and especially epithelial cells, seems to be crucially involved in survival signaling following GCs. Thus, GC-induced closer cell–cell contacts promote PI3-K/Akt kinase activity, which activates a variety of downstream pathways. These results may result in most cases in inhibition of apoptosis and survival signaling. In addition, the GR might negatively regulate promoter regions of pro-apoptotic genes in, for example, carcinoma cells but positively regulate them in, for example, lymphoid cells, dependent on tissue-specific factors. One example is the influence of GR to the transcription factor NF-kB by cross-talk. This type of repression may depend on the cellular levels of NF-kB and may well contribute to cell type specific pro- and anti-apoptotic GR signaling. Another important point that should be kept in mind is the documented capacity of GCs to exert antiproliferative effects in many normal and tumor cell systems. These antiproliferative effects are mediated by a G1-block in cell cycle progression. After withdrawal of the GCs, cells synchronously progress through the cell cycle. The duration and magnitude of the antiproliferative effects is dose- and receptor-dependent. Thus, cell cycle arrest induced by DEX pre- and combination treatment may be a mediator of chemo- and radiotherapy-resistance in many cases. Although GC-induced cell kinetic perturbations (e.g., synchronization of cells) have been observed in a variety of cell lines and experimental tumor systems, the therapeutic potential of these alterations have not yet been tested in the clinic. In the clinic, high doses of GCs are frequently given simultaneously with cell cycle-specific drugs in solid tumor chemotherapy. The cell kinetic and treatment resulting from many studies might indicate that, in solid tumors sensitive to GCs, the effectiveness of cycle active agents will be reduced when administered simultaneously with GC. Whether synchronization with steroid hormones and subsequent kinetically based cell cycle-specific chemotherapy would be more effective for tumor control and survival will require further studies. However, the in vitro and in vivo data reported here provide plenty of evidence to reevaluate the indication of GCs in the management of cancer patients.

References

Agosin M, Christen R, Badinez O, Gasic G, Neghme A, Pizarro O, Jarpa A (1952) Cortisone-induced metastases of adenocarcinoma in mice. Proc Soc Exp Biol Med 80:128–131

Amsterdam A, Sasson R (2002) The anti-inflammatory action of glucocorticoids is mediated by cell type specific regulation of apoptosis. Mol Cell Endocrinol 189:1–9

Bailly-Maitre B, de Sousa G, Boulukos K, Gugenheim J, Rahmani R (2001) Dexamethasone inhibits spontaneous apoptosis in primary cultures of human and rat hepatocytes via Bcl-2 and Bcl-xL induction. Cell Death Differ 8:279–288

Barkett M, Gilmore TD (1999) Control of apoptosis by Rel/NF-kB transcription factors. Oncogene 18:6910–6924

Barthwal MK, Sathyanarayana P, Kundu CN, Rana B, Pradeep A, Sharma C, Woodgett JR, Rana A (2003) Negative regulation of mixed lineage kinase 3 by protein kinase B/AKT leads to cell survival. J Biol Chem 278:3897–3902

Baserga R, Shubik P (1954) The action of cortisone on transplanted and induced tumors in mice. Cancer Res 14:12–16

Baumann S, Dostert A, Novac N, Bauer A, Schmid W, Fas SC, Krueger A, Heinzel T, Kirchhoff S, Schutz G, Krammer PH (2005) Glucocorticoids inhibit activation-induced cell death (AICD) via direct DNA-dependent repression of the CD95 ligand gene by a glucocorticoid receptor dimer. Blood 106:617–625

Beato M, Herrlich P, Schutz G (1995 Dec 15) Steroid hormone receptors: many actors in search of a plot. Cell 83:851–857

Belova L, Sharma S, Brickley DR, Nicolarsen JR, Patterson C, Conzen SD (2006) Ubiquitin/proteasome degradation of serum and glucocorticoid-regulated kinase-1 (SGK-1) is mediated by the chaperone-dependent E3 ligase CHIP. Biochem J 400(2):235–244

Benedetti S, Pirola B, Poliani PL, Cajola L, Pollo B, Bagnati R, Magrassi L, Tunici P, Finocchiaro G (2003) Dexamethasone inhibits the anti-tumor effect of interleukin 4 on rat experimental gliomas. Gene Ther 10:188–192

Bergman AM, Pinedo HM, Peters GJ (2001) Steroids affect collateral sensitivity to gemcitabine of multidrug-resistant human lung cancer cells. Eur J Pharmacol 416:19–24

Brady ME, Sartiano GP, Rosenblum SL, Zaglama NE, Bauguess CT (1987) The pharmacokinetics of single high doses of dexamethasone in cancer patients. Eur J Clin Pharmacol 32:593–596

Braunschweiger PG, Stragand JJ, Schiffer LM (1978) Effect of methylprednisolone on cell proliferation in C3H/HeJ spontaneous mammary tumors. Cancer Res 38:4510–4514

Braunschweiger PG, Ting HL, Schiffer LM (1982) Receptor-dependent antiproliferative effects of corticosteroids in radiation-induced fibrosarcomas and implications for sequential therapy. Cancer Res 42:1686–1691

Braunschweiger PG, Ting HL, Schiffer LM (1983) Correlation between glucocorticoid receptor content and the antiproliferative effect of dexamethasone in experimental solid tumors. Cancer Res 43:4757–4761

Braunschweiger PG, Ting HL, Schiffer LM (1984) Receptor-mediated antiproliferative effects of corticosteroids in Lewis lung tumors. Eur J Cancer Clin Oncol 20:427–433

Brunet A, Park J, Tran H, Hu LS, Hemmings BA, Greenberg ME (2001) Protein kinase SGK mediates survival signals by phosphorylating the forkhead transcription factor FKHRL1 (FOXO3a). Mol Cell Biol 21:952–965

Brunet CL, Gunby RH, Benson RS, Hickman JA, Watson AJ, Brady G (1998) Commitment to cell death measured by loss of clonogenicity is separable from the appearance of apoptotic markers. Cell Death Differ 5:107–115

Buchmann A, Willy C, Buenemann CL, Stroh C, Schmiechen A, Schwarz M (1999) Inhibition of transforming growth factor beta1-induced hepatoma cell apoptosis by liver tumor promoters: characterization of primary signaling events and effects on CPP32-like caspase activity. Cell Death Differ 6:190–200

Burgering BM, Medema RH (2003) Decisions on life and death: FOXO Forkhead transcription factors are in command when PKB/Akt is off duty. J Leukoc Biol 73:689–701

Buse P, Woo PL, Alexander DB, Cha HH, Reza A, Sirota ND, Firestone GL (1995a) Transforming growth factor-alpha abrogates glucocorticoid-stimulated tight junction formation and growth suppression in rat mammary epithelial tumor cells. J Biol Chem 270:6505–6514

Buse P, Woo PL, Alexander DB, Reza A, Firestone GL (1995b) Glucocorticoid-induced functional polarity of growth factor responsiveness regulates tight junction dynamics in transformed mammary epithelial tumor cells. J Biol Chem 270:28223–28227

Buttgereit F, Scheffold A (2002) Rapid glucocorticoid effects on immune cells. Steroids 67:529–534

Carayol N, Vachier I, Campbell A, Crampette L, Bousquet J, Godard P, Chanez P (2002) Regulation of E-cadherin expression by dexamethasone and tumour necrosis factor-alpha in nasal epithelium. Eur Respir J 20:1430–1436

Cardone MH, Roy N, Stennicke HR, Salvesen GS, Franke TF, Stanbridge E, Frisch S, Reed JC (1998) Regulation of cell death protease caspase-9 by phosphorylation. Science 282:1318–1321

Cato AC, Nestl A, Mink S (2002) Rapid actions of steroid receptors in cellular signaling pathways. Sci STKE 2002:RE9

Cha HH, Cram EJ, Wang EC, Huang AJ, Kasler HG, Firestone GL (1998) Glucocorticoids stimulate p21 gene expression by targeting multiple transcriptional elements within a steroid responsive region of the p21waf1/cip1 promoter in rat hepatoma cells. J Biol Chem 273:1998–2007

Chen YX, Li ZB, Diao F, Cao DM, Fu CC, Lu J (2006) Up-regulation of RhoB by glucocorticoids and its effects on the cell proliferation and NF-kappaB transcriptional activity. J Steroid Biochem Mol Biol 101:179–187

Cox AD, Der CJ (2003) The dark side of Ras: regulation of apoptosis. Oncogene 22:8999–9006

Cross DA, Alessi DR, Cohen P, Andjelkovich M, Hemmings BA (1995) Inhibition of glycogen synthase kinase-3 by insulin mediated by protein kinase B. Nature 378:785–789

Cui H, Sherr DH, el-Khatib M, Matsui K, Panka DJ, Marshak-Rothstein A, Ju ST (1996) Regulation of T-cell death genes: selective inhibition of FasL- but not Fas-mediated function. Cell Immunol 167:276–284

D'Adamio F, Zollo O, Moraca R, Ayroldi E, Bruscoli S, Bartoli A, Cannarile L, Migliorati G, Riccardi C (1997) A new dexamethasone-induced gene of the leucine zipper family protects T lymphocytes from TCR/CD3-activated cell death. Immunity 7:803–812

Das A, Banik NL, Patel SJ, Ray SK (2004) Dexamethasone protected human glioblastoma U87MG cells from temozolomide induced apoptosis by maintaining Bax:Bcl-2 ratio and preventing proteolytic activities. Mol Cancer 3:36

Datta SR, Dudek H, Tao X, Masters S, Fu H, Gotoh Y, Greenberg ME (1997) Akt phosphorylation of BAD couples survival signals to the cell-intrinsic death machinery. Cell 91:231–241

del Peso L, Gonzalez-Garcia M, Page C, Herrera R, Nunez G (1997) Interleukin-3-induced phosphorylation of BAD through the protein kinase Akt. Science 278:687–689

Desbois-Mouthon C, Cadoret A, Blivet-Van Eggelpoel MJ, Bertrand F, Caron M, Atfi A, Cherqui G, Capeau J (2000) Insulin-mediated cell proliferation and survival involve inhibition of c-Jun N-terminal kinases through a phosphatidylinositol 3-kinase- and mitogen-activated protein kinase phosphatase-1-dependent pathway. Endocrinology 141:922–931

Dihlmann S, von Knebel Doeberitz M (2005) Wnt/beta-catenin-pathway as a molecular target for future anti-cancer therapeutics. Int J Cancer 113:515–524

Doble BW, Woodgett JR (2003) GSK-3: tricks of the trade for a multi-tasking kinase. J Cell Sci 116:1175–1186

Drewinko B, Patchen M, Yang LY, Barlogie B (1981) Differential killing efficacy of twenty antitumor drugs on proliferating and nonproliferating human tumor cells. Cancer Res 41:2328–2333

Edinger AL, Thompson CB (2002) Akt maintains cell size and survival by increasing mTOR-dependent nutrient uptake. Mol Biol Cell 13:2276–2288
Embi N, Rylatt DB, Cohen P (1980) Glycogen synthase kinase-3 from rabbit skeletal muscle. Separation from cyclic-AMP-dependent protein kinase and phosphorylase kinase. Eur J Biochem 107:519–527
Engelbrecht Y, de Wet H, Horsch K, Langeveldt CR, Hough FS, Hulley PA (2003) Glucocorticoids induce rapid up-regulation of mitogen-activated protein kinase phosphatase-1 and dephosphorylation of extracellular signal-regulated kinase and impair proliferation in human and mouse osteoblast cell lines. Endocrinology 144:412–422
Evans-Storms RB, Cidlowski JA (2000) Delineation of an antiapoptotic action of glucocorticoids in hepatoma cells: the role of nuclear factor-kappaB. Endocrinology 141:1854–1862
Fan W, Sui M, Huang Y (2004) Glucocorticoids selectively inhibit paclitaxel-induced apoptosis: mechanisms and its clinical impact. Curr Med Chem 11:403–411
Feng Z, Marti A, Jehn B, Altermatt HJ, Chicaiza G, Jaggi R (1995) Glucocorticoid and progesterone inhibit involution and programmed cell death in the mouse mammary gland. J Cell Biol 131:1095–1103
Frame S, Cohen P (2001) GSK3 takes centre stage more than 20 years after its discovery. Biochem J 359:1–16
Freshney RI, Sherry A, Hassanzadah M, Freshney M, Crilly P, Morgan D (1980) Control of cell proliferation in human glioma by glucocorticoids. Br J Cancer 41:857–866
Gascoyne DM, Kypta RM, Vivanco Md (2003) Glucocorticoids inhibit apoptosis during fibrosarcoma developmentby transcriptionally activating Bcl-xL. J Biol Chem 278:18022–18029
Gassler N, Zhang C, Schnabel PA, Dienemann H, Debatin K-M, Mattern J, Herr I (2005) Dexamethasone-induced cisplatin and gemcitabine resistance in lung carcinoma samples treated ex vivo. Br J Cancer 92:1084–1088
Giles RH, van Es JH, Clevers H (2003) Caught up in a Wnt storm: Wnt signaling in cancer. Biochim Biophys Acta 1653:1–24
Glick RD, Medary I, Aronson DC, Scotto KW, Swendeman SL, La Quaglia MP (2000) The effects of serum depletion and dexamethasone on growth and differentiation of human neuroblastoma cell lines. J Pediatr Surg 35:465–472
Gorman AM, Hirt UA, Orrenius S, Ceccatelli S (2000) Dexamethasone pre-treatment interferes with apoptotic death in glioma cells. Neuroscience 96:417–425
Goya L, Maiyar AC, Ge Y, Firestone GL (1993) Glucocorticoids induce a G1/G0 cell cycle arrest of Con8 rat mammary tumor cells that is synchronously reversed by steroid withdrawal or addition of transforming growth factor-alpha. Mol Endocrinol 7:1121–1132
Guan Y, Rubenstein NM, Failor KL, Woo PL, Firestone GL (2004) Glucocorticoids control beta-catenin protein expression and localization through distinct pathways that can be uncoupled by disruption of signaling events required for tight junction formation in rat mammary epithelial tumor cells. Mol Endocrinol 18:214–227
Gustin JA, Ozes ON, Akca H, Pincheira R, Mayo LD, Li Q, Guzman JR, Korgaonkar CK, Donner DB (2004) Cell type-specific expression of the IkappaB kinases determines the significance of phosphatidylinositol 3-kinase/Akt signaling to NF-kappa B activation. J Biol Chem 279:1615–1620
Hafezi-Moghadam A, Simoncini T, Yang Z, Limbourg FP, Plumier JC, Rebsamen MC, Hsieh CM, Chui DS, Thomas KL, Prorock AJ, Laubach VE, Moskowitz MA, French BA, Ley K, Liao JK (2002) Acute cardiovascular protective effects of corticosteroids are mediated by non-transcriptional activation of endothelial nitric oxide synthase. Nat Med 8:473–479
Hamdi M, Kool J, Cornelissen-Steijger P, Carlotti F, Popeijus HE, van der Burgt C, Janssen JM, Yasui A, Hoeben RC, Terleth C, Mullenders LH, van Dam H (2005) DNA damage in transcribed genes induces apoptosis via the JNK pathway and the JNK-phosphatase MKP-1. Oncogene 24:7135–7134
Hammer S, Sauer B, Spika I, Schraut C, Kleuser B, Schafer-Korting M (2004) Glucocorticoids mediate differential anti-apoptotic effects in human fibroblasts and keratinocytes via sphingosine-1-phosphate formation. J Cell Biochem 91:840–851

Harmon JM, Thompson EB (1981) Isolation and characterization of dexamethasone-resistant mutants from human lymphoid cell line CEM-C7. Mol Cell Biol 1:512–521

Harwood AJ (2001) Regulation of GSK-3: a cellular multiprocessor. Cell 105:821–824

Herr I, Debatin K-M (2001) Cellular stress response and apoptosis in cancer therapy. Blood 98:2603–2614

Herr I, Pfitzenmaier J (2006) Glucocorticoid use in prostate cancer and other solid tumors: implications for effectiveness of cytotoxic treatment and metastases. Lancet Oncology 7:425–430

Herr I, Ucur E, Herzer K, Okouoyo S, Ridder R, Krammer PH, von Knebel Doeberitz M, Debatin KM (2003) Glucocorticoid co-treatment induces apoptosis resistance toward cancer therapy in carcinomas. Cancer Res 63:3112–3120

Herrlich P (2001) Cross-talk between glucocorticoid receptor and AP-1. Oncogene 20:2465–2475

Hertweck M, Gobel C, Baumeister R (2004) C. elegans SGK-1 is the critical component in the Akt/PKB kinase complex to control stress response and life span. Dev Cell 6:577–588

Hinoi T, Yamamoto H, Kishida M, Takada S, Kishida S, Kikuchi A (2000) Complex formation of adenomatous polyposis coli gene product and axin facilitates glycogen synthase kinase-3 beta-dependent phosphorylation of beta-catenin and down-regulates beta-catenin. J Biol Chem 275:34399–34406

Hofmann J, Kaiser U, Maasberg M, Havemann K (1995) Glucocorticoid receptors and growth inhibitory effects of dexamethasone in human lung cancer cell lines. Eur J Cancer 31A:2053–2058

Huang Y, Johnson KR, Norris JS, Fan W (2000) Nuclear factor-kappaB/IkappaB signaling pathway may contribute to the mediation of paclitaxel-induced apoptosis in solid tumor cells. Cancer Res 60:4426–4432

Iseki R, Mukai M, Iwata M (1991) Regulation of T lymphocyte apoptosis. Signals for the antagonism between activation- and glucocorticoid-induced death. J Immunol 147:4286–4292

Iversen HG (1957) The influence of cortisone on the frequency of tumour metastases. Acta Pathol Microbiol Scand 41:273–280

Iversen HG, Hjort GH (1958) The influence of corticoid steroids on the frequency of spleen metastases in patients with breast cancer. Acta Pathol Microbiol Scand 44:205–212

Kamradt MC, Mohideen N, Krueger E, Walter S, Vaughan AT (2000) Inhibition of radiation-induced apoptosis by dexamethasone in cervical carcinoma cell lines depends upon increased HPV E6/E7. Br J Cancer 82:1709–1716

Kamradt MC, Walter S, Koudelik J, Shafer L, Weijzen S, Velders M, Vaughan ATM (2001) Steroid-mediated inhibition of radiation-induced apoptosis in C4–1 cervical carcinoma cells is p53-dependent. Eur J Cancer 37:2240–2246

Kane LP, Shapiro VS, Stokoe D, Weiss A (1999) Induction of NF-kappaB by the Akt/PKB kinase. Curr Biol 9:601–604

Karin M (1999) How NF-kB is activated: The role of the IkB kinase (IKK) complex. Oncogene 18:6867–6874

Kaup B, Schindler I, Knupfer H, Schlenzka A, Preiss R, Knupfer MM (2001) Time-dependent inhibition of glioblastoma cell proliferation by dexamethasone. J Neurooncol 51:105–110

Kawamura A, Tamaki N, Kokunai T (1998) Effect of dexamethasone on cell proliferation of neuroepithelial tumor cell lines. Neurol Med Chir (Tokyo) 38:633–638; discussion 638–640

Khwaja A, Rodriguez-Viciana P, Wennstrom S, Warne PH, Downward J (1997) Matrix adhesion and Ras transformation both activate a phosphoinositide 3-OH kinase and protein kinase B/Akt cellular survival pathway. EMBO J 16:2783–2793

Kim AH, Khursigara G, Sun X, Franke TF, Chao MV (2001) Akt phosphorylates and negatively regulates apoptosis signal-regulating kinase 1. Mol Cell Biol 21:893–901

Kobayashi T, Deak M, Morrice N, Cohen P (1999) Characterization of the structure and regulation of two novel isoforms of serum- and glucocorticoid-induced protein kinase. Biochem J 344(Pt 1):189–197

Kofler R, Schmidt S, Kofler A, Ausserlechner MJ (2003) Resistance to glucocorticoid-induced apoptosis in lymphoblastic leukemia. J Endocrinol 178:19–27

Kruit A, Reyes-Moreno C, Newling DW, Geldof A, Koutsilieris M (1999) Response of PC-3 prostate cancer cells to combination therapy using irradiation with glucocorticoids or doxorubicin. Anticancer Res 19:3153–156

Kudawara I, Ueda T, Yoshikawa H, Miyama T, Yamamoto T, Nishizawa Y (2001) In vivo inhibition of tumour growth by dexamethasone in murine osteosarcomas. Eur J Cancer 37:1703–1708

Lang F, Cohen P (2001) Regulation and physiological roles of serum- and glucocorticoid-induced protein kinase isoforms. Sci STKE 2001:RE17

Langeveld CH, van Waas MP, Stoof JC, Sutanto W, de Kloet ER, Wolbers JG, Heimans JJ (1992) Implication of glucocorticoid receptors in the stimulation of human glioma cell proliferation by dexamethasone. J Neurosci Res 31:524–531

Lauder A, Castellanos A, Weston K (2001) c-Myb transcription is activated by protein kinase B (PKB) following interleukin 2 stimulation of Tcells and is required for PKB-mediated protection from apoptosis. Mol Cell Biol 21:5797–5805

Leong ML, Maiyar AC, Kim B, O'Keeffe BA, Firestone GL (2003) Expression of the serum- and glucocorticoid-inducible protein kinase, Sgk, is a cell survival response to multiple types of environmental stress stimuli in mammary epithelial cells. J Biol Chem 278:5871–5882

Letuve S, Druilhe A, Grandsaigne M, Aubier M, Pretolani M (2002) Critical role of mitochondria, but not caspases, during glucocorticosteroid-induced human eosinophil apoptosis. Am J Respir Cell Mol Biol 26:565–571

Limbourg FP, Liao JK (2003) Nontranscriptional actions of the glucocorticoid receptor. J Mol Med 81:168–174

Loda M, Capodieci P, Mishra R, Yao H, Corless C, Grigioni W, Wang Y, Magi-Galluzzi C, Stork PJ (1996) Expression of mitogen-activated protein kinase phosphatase-1 in the early phases of human epithelial carcinogenesis. Am J Pathol 149:1553–1564

Lu YS, Lien HC, Yeh PY, Yeh KH, Kuo ML, Kuo SH, Cheng AL (2005) Effects of glucocorticoids on the growth and chemosensitivity of carcinoma cells are heterogeneous and require high concentration of functional glucocorticoid receptors. World J Gastroenterol 11:6373–6380

Lu YS, Lien HC, Yeh PY, Kuo SH, Chang WC, Kuo ML, Cheng AL (2006) Glucocorticoid receptor expression in advanced non-small cell lung cancer: clinicopathological correlation and in vitro effect of glucocorticoid on cell growth and chemosensitivity. Lung Cancer 53:303–310

Lund LR, Romer J, Thomasset N, Solberg H, Pyke C, Bissell MJ, Dano K, Werb Z (1996) Two distinct phases of apoptosis in mammary gland involution: proteinase-independent and -dependent pathways. Development 122:181–193

Machuca C, Mendoza-Milla C, Cordova E, Mejia S, Covarrubias L, Ventura J, Zentella A (2006) Dexamethasone protection from TNF-alpha-induced cell death in MCF-7 cells requires NF-kappaB and is independent from AKT. BMC Cell Biol 7:9

Madge LA, Pober JS (2000) A phosphatidylinositol 3-kinase/Akt pathway, activated by tumor necrosis factor or interleukin-1, inhibits apoptosis but does not activate NFkappaB in human endothelial cells. J Biol Chem 275:15458–15465

Magi-Galluzzi C, Mishra R, Fiorentino M, Montironi R, Yao H, Capodieci P, Wishnow K, Kaplan I, Stork PJ, Loda M (1997) Mitogen-activated protein kinase phosphatase 1 is overexpressed in prostate cancers and is inversely related to apoptosis. Lab Invest 76:37–51

Marini G, Murray S, Goldhirsch A, Gelber RD, Castiglione-Gertsch M, Price KN, Tattersall MH, Rudenstam CM, Collins J, Lindtner J, Cavalli F, Cortes-Funes H, Gudgeon A, Forbes JF, Galligioni E, Coates AS, Senn HJ (1996) The effect of adjuvant prednisone combined with CMF on patterns of relapse and occurrence of second malignancies in patients with breast cancer. International (Ludwig) Breast Cancer Study Group. Ann Oncol 7:245–250

Mattern J, Wayss K, Volm M (1984) Effect of five antineoplastic agents on tumor xenografts with different growth rates. J Natl Cancer Inst 72:1335–1339

Mattern J, Klinga K, Runnebaum B, Volm M (1985) Influence of hormone therapy on human lung tumors transplanted into nude mice. Oncology 42:388–390

Mealey J, Jr, Chen TT, Schanz GP (1971) Effects of dexamethasone and methylprednisolone on cell cultures of human glioblastomas. J Neurosurg 34:324–334

Messmer UK, Winkel G, Briner VA, Pfeilschifter J (1999) Glucocorticoids potently block tumour necrosis factor-alpha- and lipopolysaccharide-induced apoptotic cell death in bovine glomerular endothelial cells upstream of caspase 3 activation. Br J Pharmacol 127:1633–1640

Messmer UK, Pereda-Fernandez C, Manderscheid M, Pfeilschifter J (2001) Dexamethasone inhibits TNF-alpha-induced apoptosis and IAP protein downregulation in MCF-7 cells. Br J Pharmacol 133:467–476

Meyer S, Eden T, Kalirai H (2006) Dexamethasone protects against cisplatin-induced activation of the mitochondrial apoptotic pathway in human osteosarcoma cells. Cancer Biol Ther5

Mikosz CA, Brickley DR, Sharkey MS, Moran TW, Conzen SD (2001) Glucocorticoid receptor-mediated protection from apoptosis is associated with induction of the serine/threonine survival kinase gene, sgk-1. J Biol Chem 276:16649–16654

Molomut N, Spain DM, Gault SD, Kreisler L (1952) Preliminary report on the experimental induction of metastases from a heterologous cancer graft in mice. Proc Natl Acad Sci USA 38:991–995

Moran TJ, Gray S, Mikosz CA, Conzen SD (2000) The glucocorticoid receptor mediates a survival signal in human mammary epithelial cells. Cancer Res 60:867–872

Mulholland DJ, Dedhar S, Coetzee GA, Nelson CC (2005) Interaction of nuclear receptors with Wnt/{beta}-catenin/Tcf signalling: Wnt you like to know? Endocr Rev 7:898–915

Munstedt K, Borces D, Bohlmann MK, Zygmunt M, von Georgi R (2004) Glucocorticoid administration in antiemetic therapy: is it safe? Cancer 101:1696–702

Murakami N, Fukuchi S, Takeuchi K, Hori T, Shibamoto S, Ito F (1998) Antagonistic regulation of cell migration by epidermal growth factor and glucocorticoid in human gastric carcinoma cells. J Cell Physiol 176:127–137

Nakae J, Park BC, Accili D (1999) Insulin stimulates phosphorylation of the forkhead transcription factor FKHR on serine 253 through a Wortmannin-sensitive pathway. J Biol Chem 274:15982–15985

Norgaard P, Poulsen HS (1991) Glucocorticoid receptors in human malignancies: a review. Ann Oncol 2:541–557

Novac N, Baus D, Dostert A, Heinzel T (2006) Competition between glucocorticoid receptor and NFkappaB for control of the human FasL promoter. Faseb J 20:1074–1081

Oh HY, Namkoong S, Lee SJ, Por E, Kim CK, Billiar TR, Han JA, Ha KS, Chung HT, Kwon YG, Lee H, Kim YM (2006) Dexamethasone protects primary cultured hepatocytes from death receptor-mediated apoptosis by upregulation of cFLIP. Cell Death Differ 13:512–523

Ozes ON, Mayo LD, Gustin JA, Pfeffer SR, Pfeffer LM, Donner DB (1999) NF-kappaB activation by tumour necrosis factor requires the Akt serine-threonine kinase. Nature 401:82–85

Pang D, Kocherginsky M, Krausz T, Kim SY, Conzen SD (2006) Dexamethasone decreases xenograft response to paclitaxel through inhibition of tumor cell apoptosis. Cancer Biol Ther 5(8):941–942

Pap M, Cooper GM (1998) Role of glycogen synthase kinase-3 in the phosphatidylinositol 3-Kinase/Akt cell survival pathway. J Biol Chem 273:19929–19932

Park HS, Kim MS, Huh SH, Park J, Chung J, Kang SS, Choi EJ (2002) Akt (protein kinase B) negatively regulates SEK1 by means of protein phosphorylation. J Biol Chem 277:2573–2578

Pastorino JG, Tafani M, Farber JL (1999) Tumor necrosis factor induces phosphorylation and translocation of BAD through a phosphatidylinositide-3-OH kinase-dependent pathway. J Biol Chem 274:19411–19416

Peluso JJ, Pappalardo A, Fernandez G (2001) E-cadherin-mediated cell contact prevents apoptosis of spontaneously immortalized granulosa cells by regulating Akt kinase activity. Biol Reprod 64:1183–1190

Polakis P (2000) Wnt signaling and cancer. Genes Dev 14:1837–1851

Pomeroy TC (1954) Studies on the mechanism of cortisone-induced metastases of transplantable mouse tumors. Cancer Res 14:201–204

Postmus PE, Smit EF, Haaxma-Reiche H, van Zandwijk N, Ardizzoni A, Quoix E, Kirkpatrick A, Sahmoud T, Giaccone G (1995) Teniposide for brain metastases of small-cell lung cancer: a phase II study. European Organization for Research and Treatment of Cancer Lung Cancer Cooperative Group. J Clin Oncol 13:660–665

Ramdas J, Liu W, Harmon JM (1999) Glucocorticoid-induced cell death requires autoinduction of glucocorticoid receptor expression in human leukemic T cells. Cancer Res 59:1378–1385

Rasmuson T, Ljungberg B, Grankvist K, Jacobsen J, Olsson T (2001) Increased serum cortisol levels are associated with high tumour grade in patients with renal cell carcinoma. Acta Oncol 40:83–87

Rathmell JC, Fox CJ, Plas DR, Hammerman PS, Cinalli RM, Thompson CB (2003) Akt-directed glucose metabolism can prevent Bax conformation change and promote growth factor-independent survival. Mol Cell Biol 23:7315–7328

Ray A, Prefontaine KE (1994) Physical association and functional antagonism between the p65 subunit of transcription factor NF-kB and the glucocorticoid receptor. Proc Natl Acad Sci USA 91:752–756

Reddy SA, Huang JH, Liao WS (2000) Phosphatidylinositol 3-kinase as a mediator of TNF-induced NF-kappa B activation. J Immunol 164:1355–1363

Reichardt HM, Kaestner KH, Tuckermann J, Kretz O, Wessely O, Bock R, Gass P, Schmid W, Herrlich P, Angel P, Schutz G (1998) DNA binding of the glucocorticoid receptor is not essential for survival. Cell 93:531–541

Rennebeck G, Martelli M, Kyprianou N (2005) Anoikis and survival connections in the tumor microenvironment: is there a role in prostate cancer metastasis? Cancer Res 65:11230–11235

Rieger J, Durka S, Streffer J, Dichgans J, Weller M (1999) Gemcitabine cytotoxicity of human malignant glioma cells: modulation by antioxidants, BCL-2 and dexamethasone. Eur J Pharmacol 365:301–308

Runnebaum IB, Bruning A (2005) Glucocorticoids inhibit cell death in ovarian cancer and up-regulate caspase inhibitor cIAP2. Clin Cancer Res 11:6325–6332

Rutz HP, Herr I (2004) Interference of glucocorticoids with apoptosis signaling and host-tumor interactions. Cancer Biol Ther 3:715–718

Rutz HP, Mariotta M, von Knebel Doeberitz M, Mirimanoff RO (1998) Dexamethasone-induced radioresistance occurring independent of human papilloma virus gene expression in cervical carcinoma cells. Strahlenther Onkol 174:71–74

Sanchez I, Goya L, Vallerga AK, Firestone GL (1993) Glucocorticoids reversibly arrest rat hepatoma cell growth by inducing an early G1 block in cell cycle progression. Cell Growth Differ 4:215–225

Sasson R, Amsterdam A (2003) Pleiotropic anti-apoptotic activity of glucocorticoids in ovarian follicular cells. Biochem Pharmacol 66:1393–1401

Sasson R, Tajima K, Amsterdam A (2001) Glucocorticoids protect against apoptosis induced by serum deprivation, cyclic adenosine 3¢,5¢-monophosphate and p53 activation in immortalized human granulosa cells: involvement of Bcl-2. Endocrinology 142:802–811

Sasson R, Shinder V, Dantes A, Land A, Amsterdam A (2003) Activation of multiple signal transduction pathways by glucocorticoids: protection of ovarian follicular cells against apoptosis. Biochem Biophys Res Commun 311:1047–1056

Schmidt S, Rainer J, Ploner C, Presul E, Riml S, Kofler R (2004) Glucocorticoid-induced apoptosis and glucocorticoid resistance: molecular mechanisms and clinical relevance. Cell Death Differ 11(Suppl 1):S45–S55

Sengupta S, Vonesch JL, Waltzinger C, Zheng H, Wasylyk B (2000) Negative cross-talk between p53 and the glucocorticoid receptor and its role in neuroblastoma cells. EMBO J 19:6051–6064

Shackney SE, McCormack GW, Cuchural GJ, Jr (1978) Growth rate patterns of solid tumors and their relation to responsiveness to therapy: an analytical review. Ann Intern Med 89:107–121

Sherlock P, Hartmann WH (1962) Adrenal steroids and the pattern of metastases of breast cancer. Jama 181:313–317

Sinclair WK (1968) Cyclic x-ray responses in mammalian cells in vitro. Radiat Res 33:620–643

Song G, Ouyang G, Bao S (2005) The activation of Akt/PKB signaling pathway and cell survival. J Cell Mol Med 9:59–71

Sorensen HT, Mellemkjaer L, Nielsen GL, Baron JA, Olsen JH, Karagas MR (2004) Skin cancers and non-hodgkin lymphoma among users of systemic glucocorticoids: a population-based cohort study. J Natl Cancer Inst 96:709–711

Srikanth S, Franklin CC, Duke RC, Kraft RS (1999) Human DU145 prostate cancer cells overexpressing mitogen-activated protein kinase phosphatase-1 are resistant to Fas ligand-induced mitochondrial perturbations and cellular apoptosis. Mol Cell Biochem 199:169–178

Steinmetz R, Wagoner HA, Zeng P, Hammond JR, Hannon TS, Meyers JL, Pescovitz OH (2004) Mechanisms regulating the constitutive activation of the extracellular signal-regulated kinase (ERK) signaling pathway in ovarian cancer and the effect of ribonucleic acid interference for ERK1/2 on cancer cell proliferation. Mol Endocrinol 18:2570–2582

Stevenson BR, Siliciano JD, Mooseker MS, Goodenough DA (1986) Identification of ZO-1: a high molecular weight polypeptide associated with the tight junction (zonula occludens) in a variety of epithelia. J Cell Biol 103:755–766

Sui M, Chen F, Fan W (2006) Glucocorticoids interfere with therapeutic efficacy of paclitaxel against human breast and ovarian xenograft tumors. Int J Cancer 119:712–717

Tao YS, Edwards RA, Tubb B, Wang S, Bryan J, McCrea PD (1996) beta-Catenin associates with the actin-bundling protein fascin in a noncadherin complex. J Cell Biol 134:1271–1281

Teicher BA, Herman TS, Holden SA, Wang YY, Pfeffer MR, Crawford JW, Frei EIIIrd, (1990) Tumor resistance to alkylating agents conferred by mechanisms operative only in vivo. Science 247:1457–1461

Terasima T, Tolmach LJ (1961) Changes in x-ray sensitivity of HeLa cells during the division cycle. Nature 190:1210–1211

Topper YJ, Freeman CS (1980) Multiple hormone interactions in the developmental biology of the mammary gland. Physiol Rev 60:1049–1106

Tormey DC, Gray R, Gilchrist K, Grage T, Carbone PP, Wolter J, Woll JE, Cummings FJ (1990) Adjuvant chemohormonal therapy with cyclophosphamide, methotrexate, 5-fluorouracil, and prednisone (CMFP) or CMFP plus tamoxifen compared with CMF for premenopausal breast cancer patients. An Eastern Cooperative Oncology Group trial. Cancer 65:200–206

Valeriote F, van Putten L (1975) Proliferation-dependent cytotoxicity of anticancer agents: a review. Cancer Res 35:2619–2630

van Noort M, Meeldijk J, van der Zee R, Destree O, Clevers H (2002) Wnt signaling controls the phosphorylation status of beta-catenin. J Biol Chem 277:17901–17905

Vander Heiden MG, Plas DR, Rathmell JC, Fox CJ, Harris MH, Thompson CB (2001) Growth factors can influence cell growth and survival through effects on glucose metabolism. Mol Cell Biol 21:5899–5912

Wang JM, Chao JR, Chen W, Kuo ML, Yen JJ, Yang-Yen HF (1999) The antiapoptotic gene mcl-1 is up-regulated by the phosphatidylinositol 3-kinase/Akt signaling pathway through a transcription factor complex containing CREB. Mol Cell Biol 19:6195–6206

Wang R, Zhang L, Zhang X, Moreno J, Luo X, Tondravi M, Shi Y (2001) Differential regulation of the expression of CD95 ligand, receptor activator of nuclear factor-kappa B ligand (RANKL), TNF-related apoptosis-inducing ligand (TRAIL), and TNF-alpha during T cell activation. J Immunol 166:1983–1990

Webster JC, Huber RM, Hanson RL, Collier PM, Haws TF, Mills JK, Burn TC, Allegretto EA (2002) Dexamethasone and tumor necrosis factor-alpha act together to induce the cellular inhibitor of apoptosis-2 gene and prevent apoptosis in a variety of cell types. Endocrinology 143:3866–3874

Wellington JS, Moon HD (1961) Effect of hydrocortisone on human cells in tissue culture. Proc Soc Exp Biol Med 107:556–559

Wen LP, Madani K, Fahrni JA, Duncan SR, Rosen GD (1997) Dexamethasone inhibits lung epithelial cell apoptosis induced by IFN- gamma and Fas. Am J Physiol 273:L921–L929

Wolf J, Spear P, Yesner R, Patno ME (1960) Nitrogen mustard and the steroid hormones in the treatment of inoperable bronchogenic carcinoma. Am J Med 29:1008–1016

Wolff JE, Jurgens H (1994) Dexamethasone induced partial resistance to methotrexate in C6-glioma cells. Anticancer Res 14:1585–1588

Wolff JE, Denecke J, Jurgens H (1996) Dexamethasone induces partial resistance to cisplatinum in C6 glioma cells. Anticancer Res 16:805–809
Wong V, Ching D, McCrea PD, Firestone GL (1999) Glucocorticoid down-regulation of fascin protein expression is required for the steroid-induced formation of tight junctions and cell-cell interactions in rat mammary epithelial tumor cells. J Biol Chem 274:5443–5453
Woo PL, Cha HH, Singer KL, Firestone GL (1996) Antagonistic regulation of tight junction dynamics by glucocorticoids and transforming growth factor-beta in mouse mammary epithelial cells. J Biol Chem 271:404–412
Woodgett JR (1990) Molecular cloning and expression of glycogen synthase kinase-3/factor A. Embo J 9:2431–2438
Wu W, Chaudhuri S, Brickley DR, Pang D, Karrison T, Conzen SD (2004) Microarray analysis reveals glucocorticoid-regulated survival genes that are associated with inhibition of apoptosis in breast epithelial cells. Cancer Res 64:1757–1764
Wu W, Pew T, Zou M, Pang D, Conzen SD (2005) Glucocorticoid receptor-induced MKP-1 expression inhibits paclitaxel-associated MAP kinase activation and contributes to breast cancer cell survival. J Biol Chem 280:4117–4124
Wu W, Zou M, Brickley DR, Pew T, Conzen SD (2006) Glucocorticoid receptor activation signals through FOXO3a in breast cancer cells. Mol Endocrinol 20(10):2304–2314
Yamamoto M, Fukuda K, Miura N, Suzuki R, Kido T, Komatsu Y (1998) Inhibition by dexamethasone of transforming growth factor beta1-induced apoptosis in rat hepatoma cells: a possible association with Bcl-xL induction. Hepatology 27:959–966
Yamamoto T, Nishiguchi M, Inoue N, Goto HG, Kudawara I, Ueda T, Yoshikawa H, Tanigaki Y, Nishizawa Y (2002) Inhibition of murine osteosarcoma cell proliferation by glucocorticoid. Anticancer Res 22:4151–4156
Yang Y, Mercep M, Ware CF, Ashwell JD (1995) Fas and activation-induced Fas ligand mediate apoptosis of T cell hybridomas: inhibition of Fas ligand expression by retinoic acid and glucocorticoids. J Exp Med 181:1673–1682
Yang Y, Fang S, Jensen JP, Weissman AM, Ashwell JD (2000) Ubiquitin protein ligase activity of IAPs and their degradation in proteasomes in response to apoptotic stimuli. Science 288:874–877
Yerramasetti R, Gollapudi S, Gupta S (2002) Rifampicin inhibits CD95-mediated apoptosis of Jurkat T cells via glucocorticoid receptors by modifying the expression of molecules regulating apoptosis. J Clin Immunol 22:37–47
Zettl KS, Sjaastad MD, Riskin PM, Parry G, Machen TE, Firestone GL (1992) Glucocorticoid-induced formation of tight junctions in mouse mammary epithelial cells in vitro. Proc Natl Acad Sci U S A 89:9069–9073
Zhang C, Kolb A, Mattern J, Gassler N, Wenger T, Herzer K, Debatin KM, Buchler M, Friess H, Rittgen W, Edler L, Herr I (2005) Dexamethasone desensitizes hepatocellular and colorectal tumours toward cytotoxic therapy. Cancer Lett 242(1):104–111
Zhang C, Beckermann B, Kallifatidis G, Liu Z, Rittgen W, Edler L, Büchler P, Debatin K-M, Büchler MW, Friess H, Herr I (2006a) Corticosteroids induce chemotherapy resistance in the majority of tumour cells from bone, brain, breast, cervix, melanoma and neuroblastoma. Int J Oncol 29(5):1295–1301
Zhang C, Kolb A, Buechler P, Cato ACB, Mattern J, Rittgen W, Edler L, Debatin K-M, Buechler M, Friess H, Herr I (2006b) Corticosteroid co-treatment induces resistance to chemotherapy in surgical resections, xenografts and established cell lines of pancreatic cancer. BMC Cancer 6:61
Zhang C, Marme A, Wenger T, Gutwein P, Edler L, Rittgen W, Debatin K-M, Altevogt P, Mattern J, Herr I (2006c) Glucocorticoid-mediated inhibition of chemotherapy in ovarian carcinomas. Int J Oncology 2:551–557
Zhang C, Mattern J, Haferkamp A, Pfitzenmaier J, Hohenfellner M, Rittgen W, Edler L, Debatin KM, Groene E, Herr I (2006d) Corticosteroid-induced chemotherapy resistance in urological cancers. Cancer Biol Ther 5:59–64
Zhang C, Wenger T, Mattern J, Ilea S, Frey C, Gutwein P, Altevogt P, Bodenmuller W, Gassler N, Schnabel PA, Dienemann H, Marme A, Hohenfellner M, Haferkamp A, Pfitzenmaier J, Grone

HJ, Kolb A, Buchler P, Buchler M, Friess H, Rittgen W, Edler L, Debatin KM, Krammer PH, Rutz HP, Herr I (2007) Clinical and mechanistic aspects of glucocorticoid-induced chemotherapy resistance in the majority of solid tumors. Cancer Biol Ther 6:278–287

Zibera C, Gibelli N, Butti G, Pedrazzoli P, Carbone M, Magrassi L, Robustelli della Cuna G (1992) Proliferative effect of dexamethasone on a human glioblastoma cell line (HU 197) is mediated by glucocorticoid receptors. Anticancer Res 12:1571–1574

Zipp F, Wendling U, Beyer M, Grieger U, Waiczies S, Wagenknecht B, Haas J, Weller M (2000) Dual effect of glucocorticoids on apoptosis of human autoreactive and foreign antigen-specific T cells. J Neuroimmunol 110:214–222

Targeting Death-Receptors in Radiation Therapy

Olivier Niemöller and Claus Belka

Abstract The development of apoptosis resistance is a crucial step during the pathogenesis of malignant tumors. Thus, any treatment approach overcoming apoptosis resistance may be a valuable tool in oncology. Although a variety of treatments induce apoptosis, only very few specifically trigger programmed cell death. In this regard, the class of apoptosis inducing ligands may turn out to have a considerable potential in oncology. TNF-α-related apoptosis-inducing ligand (TRAIL/Apo2L) is the most promising candidate, either alone or in combination with established cancer therapies, since it induces apoptosis in a wide range of malignant cells while sparing most normal tissues.

Since death-receptor induced apoptosis is mainly mediated via nonmitochondrial death pathways, it is possible to induce apoptosis in cancer cell systems which mainly harbor defects within the mitochondrial death cascades.

Even more so it has been shown that conventional DNA damaging approaches reduced the killing threshold for receptor induced apoptosis, making TRAIL an ideal candidate for combined approaches. Thus, combined treatments might offer the chance to enhance therapeutic efficiency and overcome resistance. In combination, additive or synergistic apoptotic responses and substantially enhanced clonogenic cell kill has been documented. Furthermore, in several settings it has been shown that combined modality teatments were effective in malignant cells, which are highly resistant to either treatment, alone. Ionizing radiation is one of the most effective modalities in oncology. Thus, it is reasonable to test, how far combinations of TRAIL with ionizing radiation may increase the efficacy. Indeed, the combination of TRAIL with ionizing radiation in several in vitro settings as well as xenograft models resulted in highly increased rates of cell kill and long-term tumor control. No increase in the rate and severity of side effects has been documented, indicating that the combination really increases the therapeutic ratio. It is important to note that TRAIL and TRAIL receptor agonistic antibodies, either as single

O. Niemöller and C. Belka (✉)
Department of Radiation Oncology, Ludwig-Maximilians Universität Munich,
Marchioninistr. 15, 81377 Munich, Germany
e-mail: claus.belka@med.uni-muenchen.de

Results Probl Cell Differ, DOI 10.1007/400_2009_17

agent or in combination with cytotoxic drugs, were safe in human phase I trials. Ongoing phase II trials will hopefully document the clinical efficacy of this treatment approach.

1 Introduction

1.1 Apoptosis in Cancer Onset and Therapy

Resistance towards apoptosis induction plays a pivotal role in the onset of cancer progression and in the failure of treatment. In this regard different models of carcinogenesis have provided evidence that apoptosis resistance routinely occurs during the multistep pathogenesis of cancer (for review see (Vogelstein and Kinzler 2004). Although other means of cell kill exist, most anticancer strategies (e.g., Radiation Therapy and Chemotherapy) exert their effects to a relevant amount by inducing apoptosis in malignant cells (Debatin et al. 2002; Herr and Debatin 2001; Johnstone et al. 2002). Thus alterations of apoptosis signaling may strongly affect therapeutic efficacy (Goh et al. 1995). Understanding the regulation of apoptosis in normal and malignant cells is therefore of importance, when attempting to improve the treatment gain.

1.2 Apoptotic Pathways

In principle, two main apoptotic pathways are distinguished (for review see (Danial and Korsmeyer 2004)), both culminating in a common downstream signal, namely the activation of cysteine–aspartate-specific proteases, called caspases.

The most important signaling step during chemo- and radiation-induced apoptotic cell death is mediated via the so called "intrinsic" or mitochondrial pathway of apoptosis (Debatin et al. 2002; Herr and Debatin 2001; Norbury and Zhivotovsky 2004). The crucial step is the release of proapoptotic substances from the mitochondria, triggering caspase activation. In opposition, the "extrinsic" pathway is initially activated by an autocatalytic, receptor mediated caspase activation. Although paradigmatically distinct, both pathways are highly interconnected at different levels.

1.2.1 The Mitochondrial (Intrinsic) Apoptotic Pathway

Signaling of the mitochondrial pathway culminates in permeabilization of the outer mitochondrial membrane, which leads to release of proapoptotic molecules, including cytochrome c into the cytosol (Liu et al. 1996). Cytosolic cytochrome c binds Apoptotic protease-activating Factor (Apaf-1), d-ADP and pro-caspase-9 to form a multiprotein complex, called the apotosome. Through the apotosome pro-caspase-9 is activated by auto-cleavage and active caspase-9 is formed. Caspase-9 in turn cleaves effector caspases, thus leading to execution of the apoptotic program. Permeabilization of the mitochondrial membrane is tightly regulated, mainly by proteins of the Bcl-2-Family (Belka and Budach 2002; Yang et al. 1997), consisting of proapoptotic (Bax, Bak, Noxa, PUMA) and antiapoptotic (Bcl-2, $BclX_L$) members.

Irradiation and most chemotherapeutics target the intrinsic pathway of apoptosis through their DNA-damaging and cell cycle interfering effects, leading to transcriptional or conformational activation of proapoptotic and to the inhibition of antiapoptotic members of the Bcl-2 family. Although different mechanisms, which connect DNA-damage to the mitochondrial death machinery have been identified (for review see (Norbury and Zhivotovsky 2004)), the p53 cascade may serve as paradigm. After induction of p53, the molecule increases the transcription rate of several proapoptotic members of the Bcl-2-family, namely Bax, Bak, Puma, and Noxa (Miyashita and Reed 1995; Nakano and Vousden 2001; Oda et al. 2000). Mutations in the p53 tumor suppressor gene are commonly found in many forms of human cancer and might be associated with poor prognosis (Goh et al. 1995) and decreased sensitivity to chemo and radiotherapy (Lowe et al. 1993, 1994).

1.2.2 The Death-Receptor (Extrinsic) Apoptotic Pathway

Signaling of the extrinsic pathway occurs through cell surface receptors, belonging to the TNF-Superfamily (TNFR1, CD95/Apo-1/Fas, TRAIL-R1/DR4 and TRAIL-R2/DR5) and their respective ligands (TNF-α, CD95L/Apo1L/FasL, TNF-α-related apoptosis-inducing ligand: TRAIL/Apo2L). Although the receptors are clearly distinct, stimulation of the different death-receptors results in similar intracellular death processes.

Since TRAIL has been shown to have the highest potential as therapeutic agent most of the data in the following paragraphs will be focused on TRAIL and the respective TRAIL receptors.

To date five TRAIL receptors are known: The proapoptotic receptors TRAIL-R1/DR4 (Pan et al. 1997b), TRAIL-R2/DR5 (Pan et al. 1997a; Schneider et al. 1997) and the truncated forms TRAIL-R3/DcR1 (DegliEsposti et al. 1997; Pan et al. 1997a) and TRAIL-R4/DcR2 (Marsters et al. 1997), the latter two lacking a functional intracellular death domain (DD). The function of the soluble receptor osteoprotegerin is unknown, since it does not bind TRAIL at physiological temperatures. TRAIL-R3/DcR1 and TRAIL-R4/DcR2 act as decoy receptors, possibly through interfering with TRAIL-signaling by competitive binding of TRAIL (DcR1) and hetero-multimerization with DR5 (Merino et al. 2006; Pan et al. 1997a; Sheridan et al. 1997).

Proapoptotic signaling occurs only though DR4 and DR5, both containing a specific intracellular domain, commonly called DD. Binding of TRAIL leads to receptor trimerization and recruitment of Fas associated death domain (FADD) to the DD. FADD binds pro-caspase-8 or pro-caspase-10, to form the death-inducing signaling complex (DISC), resulting in cleavage of pro-caspase-8 or pro-caspase-10

(Sprick et al. 2002). Active caspase-8 or 10 in turn process pro-caspase-3, triggering the cleavage of multiple proteins, finally resulting in an apoptotic phenotype (for review see (Falschlehner et al. 2007).

Although, initially thought to be a simple cascade process, it turned out that apoptosis induction via death-receptors is a tightly regulated and complex signaling process.

Although being a simplification, two principally different reaction patterns may be distinguished: In Type I cells, activation of death-receptors alone is sufficient to induce apoptosis. In contrast, in Type II cells adequate death induction relies on the amplification of the apoptotic signal via activation of the mitochondrial pathway (Scaffidi et al. 1998). The interconnection of both pathways is mainly mediated through cleavage of the proapoptotic Bid molecule by caspase-8 (Li et al. 1998; Luo et al. 1998; Wang et al. 1996). The resulting truncated Bid (t-bid) inactivates antiapoptotic Bcl-2-family members, thereby releasing the proapoptotic members Bax and Bak to initiate the efflux of cytochrome c and other proapoptotic factors from the mitochondria. Furthermore, a direct interaction of t-Bid with Bax precedes the insertion of Bax into the outer mitochondrial membrane (Eskes et al. 2000). Subsequently the activation of pro-caspase-9 and all downstream molecules takes place. Although initially described for CD95 triggered cells death, the Type I, Type II reaction pattern is also relevant during TRAIL induced cell death (Rudner et al. 2005).

In addition to the Bcl-2 regulated interplay on the mitochondrial death level, multiple negative regulators of TRAIL induced apoptosis have been identified. c-FLIP for example competes with pro-caspase-8 for binding to the DISC. Another class of negative regulators of apoptosis consists of the inhibitor of apoptosis proteins (IAPs). IAPs directly bind to caspases, thereby inhibiting the proteolytic activation (Suliman et al. 2001). The observation, that IAPs in turn are counteracted by Smac/DIABOLO upon release from the mitochondria represents a perfect example for the tight and complex regulatory network of TRAIL induced apoptosis.

Moreover, NFκB is a possible candidate for the regulation of TRAIL induced apoptosis, since it was shown that the RelA/p65 subunit of NFκB inhibits the expression of DR4 and DR5 and caspase-8. Furthermore, RelA/p63 positively regulates the expression of cIAP1 and cIAP2, thereby acting as a survival factor (Chen et al. 2003; Yoshida et al. 2001).

2 TNF-α and Fas Ligand in Cancer Therapy

Since TNF-α and CD95 were highly effective inducers of apoptosis, the initial attempt to implement death-receptor agonists in cancer therapy was made with TNF-α or CD95L/Apo1L/FasL.

Although TNF-α and Fas-Ligand induce apoptosis in a variety of malignant cells, severe side effects in vivo prohibited a real clinical application. Triggering apoptosis by activation of the CD95 system for example causes extensive liver damage already found in murine test systems (Ni et al. 1994; Ogasawara et al. 1993; Yin et al. 1999).

3 TRAIL in Cancer Therapy

3.1 Rationale

As stated above, malignant cell systems are characterized by alterations of intrinsic apoptotic pathways. In contrast, apoptosis mediated through the activation of death-receptors is not necessarily dependent on the mitochondrial release of cytochrome c. Thus, therapeutic targeting of death-receptors might induce cell death in cells with established resistance within mitochondrial death pathways. In addition, in situations of partial resistance the interaction of different apoptosis triggers may result in synergistic death induction. Indeed, profound synergistic effects in combined treatment with DNA-damaging approaches had been observed, especially in malignant cells resistant to conventional treatment.

Although apoptosis induction is relevant in a multitude of cell systems one of the crucial observations regarding TRAIL as an anticancer approach was the fact that the agent seemingly was specific for malignant cells (Ashkenazi et al. 1999; Walczak et al. 1999).

Moreover, in contrast to former death-receptor based approaches, administration of TRAIL and agonistic TRAIL receptor antibodies in vivo seems to be well tolerated (Herbst et al. 2006; Pacey et al. 2005; Tolcher et al. 2007).

3.2 TRAIL as Single Agent and in Combination with Cytotoxic Agents

As already stated a growing body of evidence exists concerning the clinical use of TRAIL and TRAIL-receptor agonistic antibodies, either as a single agent or in combination with cytotoxic drugs or so called targeted therapies. A detailed description is beyond the scope of this work. An overview of published and ongoing studies is found in a recent review by Ashkenazi (Ashkenazi 2008).

3.3 TRAIL in Combination with Radiotherapy

Based on the hypothesis that death-receptor agonists and ionizing radiation induce apoptotic cell death through distinct and possibly synergistic mechanisms, the mode of action of these two therapeutic approaches was investigated (Table 1).

The fact that radiation and death ligands act via distinct death pathways has been described using Bcl-2 and $BclX_L$ as marker molecules. Whereas radiation induced apoptosis was abrogated in Bcl-2 or $BclX_L$ over-expressing cells, CD95 induced apoptosis was only mildly influenced (Belka et al. 2000). In accordance, it was shown that caspase-8-deficiency lead to resistance to CD95L-, but not to radiation-induced apoptosis (Belka et al. 2000).

Although, most observations were initially made using CD95 as a trigger, subsequent studies proved that similar reaction patterns are responsible for TRAIL induced apoptosis. Using Bcl-2 over-expressing Jurkat-T-cells under tet–control, dose–response studies revealed the complex behavior of TRAIL induced cell death in relation to the expression level of Bcl-2 (Rudner et al. 2005). Whereas after low dose stimulation, Bcl-2 over-expression rendered the cells apoptosis-resistant, high doses of TRAIL readily killed all cells. In parallel, Bcl-2 over-expression shifts the onset of apoptosis to later time points.

These data shown conclusively is that, depending on the trigger dose and the level of Bcl-2 being present, apoptosis induction may rely on direct receptor cascades or a combination of both, direct caspase activation and mitochondrial feedback. Interestingly, the expression of a caspase-9 dominant negative mutant in Jurkat-T did not render those cells TRAIL resistant, indicating that death-receptor induced apoptosis might be indeed independent of mitochondrial caspase activation (Verbrugge et al. 2008).

Nevertheless both pathways are interconnected, as activation of caspase-8 and subsequent processing of Bid occurs in response to irradiation in wild type Jurkat-T-cells, but not in Bcl-2 over-expressing cells, leading to the conclusion that caspase-8 activation might not only occur in response to death-receptor stimulation, but also in response to irradiation, downstream of mitochondrial permeabilization.

On the other hand, processing of the Bid to its proapoptotic form occurred irrespective of Bcl-2 when apoptosis was induced with high doses of TRAIL (Belka et al. 2001). In this experimental setting, subsequent transactivation of the mitochondrial pathway was ascribed to Bid-cleavage (Luo et al. 1998).

3.3.1 Combined Treatment In Vitro

These findings demonstrated that induction of apoptosis by TRAIL indeed occurred via distinct but interconnected pathways, making a combination a rational approach. Consequently, the effect of combined treatment with ionizing radiation and death-receptor stimulation was studied.

First data on the efficacy of ionizing radiation and TRAIL were provided by Chinnaiyan and coworkers using breast cancer cell lines. Synergy of apoptosis induction following TRAIL-exposure after irradiation was demonstrated (Chinnaiyan et al. 2000).

Interestingly, combined treatment was found to exert synergistic effects in cells with low intrinsic sensitivity to radiation-induced apoptosis and less than synergistic effects in cells sensitive to radiation alone. Furthermore, the biomathematical degree of interaction was not linked to the Bcl-2 status, but to the p53 status. In more detail, synergism was observed in cells expressing wild type p53, whereas in p53 mutants only additive effects occurred (Chinnaiyan et al. 2000).

In a more mechanistically oriented approach, employing Jurkat-T-cells rendered apoptosis resistant by over-expression of Bcl-2, Belka and coworkers provided evidence that the combination of TRAIL and radiation was highly efficient in situations of mitochondrial death resistance. In detail, at higher dose levels the combination turned out to be of similar efficacy compared to death induction in Bcl-2 negative cells. Of notice, a detailed mathematical analysis revealed that combined treatment exerted additive effects in wild type Jurkat-T-cells, whereas cell kill was synergistic in Bcl-2 over-expressing cells. Importantly these effects were found at low concentrations of TRAIL and clinically relevant radiation doses.

After the initial finding that radiation and TRAIL may act synergistically, these findings were confirmed in a multitude of different cell systems.

Marini and coworkers exposed cell lines derived from breast-, lung-, colorectal- and head and neck cancer to ionizing radiation (10 Gy) followed by treatment with TRAIL at low concentrations (0.1 ng ml^{-1}), either immediately, or 12 h after irradiation (Marini et al. 2005). Treatment with TRAIL alone induced apoptosis to a low extent, as measured by Hoechst staining, while irradiation had substantial apoptosis inducing effects. Isobologramm analysis confirmed that a combination of irradiation, followed by immediate treatment with TRAIL induced apoptosis in a synergistic manner in three out of six cell lines. However when TRAIL was applied 12 h after irradiation, synergistic effects were observed in five of six cell lines, indicating that preirradiation is crucial for effective sensitization (Marini et al. 2005).

Up till now there are multiple sets of data describing synergistic or sensitizing effects of combined treatments with ionizing radiation and TRAIL in multiple malignant cell lines. Synergistic effects had been found in Jurkat-T-cells (Belka et al. 2001), prostate cancer cells (Shankar et al. 2004), breast cancer cells (Chinnaiyan et al. 2000; Shankar et al. 2004a), and human glioma cells (Nagane et al. 2007).

Sensitizing effects of combined treatment in cells, normally exhibiting low radiation sensitivity, were found in erythroleukemic cells (Di Pietro et al. 2001), gastric cancer cells, lung cancer – and prostate cancer cell lines (Hamasu et al. 2005), and cervical cancer cells (Maduro et al. 2008).

Meanwhile, the relevance of sequential treatment to induce strong apoptotic effects was confirmed in different malignant cell models, e. g. breast cancer cells (Shankar et al. 2004a), prostate cancer cells (Shankar et al. 2004b), lung -, colorectal - and head and neck-cancer cells (Marini et al. 2005), T-lymphoblastic leukemia- (Rezacova et al. 2008) and erythroleukemic cells (Di Pietro et al. 2001).

Although these results are an important observation, an increased rate of apoptosis does not necessarily mean that a given combination will be of clinical value. Since other complex death mechanisms exist, there is no one by one correlation of apoptosis induction and final eradication of clonogenic tumor cells (Aldridge et al. 1995). Thus, it is necessary to prove, that a given treatment approach finally results in increased and durable tumor cell eradication. In order to do so, colony forming assays comprise the initial experimental in vitro set up.

In this regard, several lines of evidence show that the increased apoptosis induction after combined treatment with ionizing radiation and TRAIL, translates into a substantial eradication of malignant clones (Belka et al. 2001; Fiveash et al. 2008; Maduro et al. 2008; Nagane et al. 2007; Shankar et al. 2004b, 2005; Wissink et al. 2006).

3.3.2 Combined Treatment In Vivo

In addition to colony formation assays, animal studies allow a much more precise efficacy estimation of a given anticancer strategy, since effects of tumor oxygenation, drug distribution, tumor–stroma interactions are integrated within the experimental set up.

One of the first works elucidating the in vivo activity of TRAIL in combination with ionizing radiation in an animal setting was carried out by Chinnaiyan and coworkers (Chinnaiyan et al. 2000). They demonstrated high efficiency of combined treatment in NIH III nude mice carrying MFC7 breast carcinoma xenografts. Radiation was administered in three fractions of 5 Gy. Multimeric-polyhistidine-tagged TRAIL (5 mg kg^{-1}) was given intraperitoneally for 7 days. Within 15 days after single agent treatment with TRAIL or ionizing radiation, the mean tumor volume increased by 70 and 150%, respectively. In contrast, upon combined treatment tumors regressed more than 50% in the same time. These findings were confirmed by diffusion-weighted MRI and histopathological analysis. No apparent toxicity was described.

In 2004 Shankar and coworkers assessed the combination of ionizing radiation and TRAIL in two xenograft models.

In the fist model, they found that sequential treatment with irradiation, followed by TRAIL, completely eradicated breast cancer xenografts in atymic nude mice and enhanced survival, without toxicity to normal tissues. Sequential treatment was superior to concomitant treatment, probably through radiation induced up-regulation of DR5 in a p53 dependent manner (Shankar et al. 2004a).

Similar findings were demonstrated in prostate cancer bearing mice (Shankar et al. 2004b). In vitro, combined treatment yielded synergistic apoptotic responses. TRAIL exposure following irradiation was superior to concurrent or reverse treatment in respect to induction of apoptosis and activation of caspase-3 and caspase-8, indicating improved efficacy of TRAIL induced caspase-8 activity after irradiation. In vivo, treatment with 5 Gy, followed by TRAIL administration after 24 h, weekly for 3 weeks, synergistically inhibited tumor growth and prolonged survival in Balb c nude mice bearing prostate cancer xenografts. Tumor tissues from treated mice showed increased apoptosis and activity of caspase-3 and caspase-7, as compared to single treatment. Furthermore, expression of DR5, Bax and Bak was increased following irradiation in vivo.

Similar results were described in SCID mice, bearing gastric cancer xenografts after treatment with TRAIL and ionizing radiation (Takahashi et al. 2008).

4 Monoclonal Agonistic Death-Receptor Antibodies

In addition to different preparations of recombinant TRAIL, agonistic monoclonal antibodies targeting either DR4 or DR5 had been developed (Ichikawa et al. 2001). Phase I and phase II clinical studies have shown low toxicity and some evidence of tumoricidal activity (Chow et al. 2006; Georgakis et al. 2005; Kanzler et al. 2005; Pacey et al. 2005; Pukac et al. 2005; Tolcher et al. 2007).

Xenograft and in vitro experiments, combining agonistic antibodies directed against DR4 or DR5 with ionizing radiation, yielded similar results as former trials with recombinant TRAIL. In a colorectal xenograft model, median tumor doubling time was increased from about 5 days for untreated and irradiated animals, to 27.6 and 71 days, after application of HGS-ETR1, a monoclonal DR4-agonistic antibody, alone and combined treatment, respectively (Marini et al. 2006).

Similar in vivo activity of combined treatment was demonstrated in malignant glioma-xenograft bearing nude mice with a death-receptor agonistic antibody (Fiveash et al. 2008).

5 Normal Tissue Toxicity

Since TRAIL was found to have a wide therapeutic ratio as single agent, as well as in combination with cytotoxic drugs, experiments were performed to investigate normal tissue toxicity after combined treatment with ionizing radiation. After combined treatment of human hepatocytes, fibroblasts, small muscle cells and epithelial cells from prostate, kidney, breast, small airways and umbilical cord with 10 Gy, 12 h prior to TRAIL exposure (1 ng ml^{-1}), no induction of apoptosis was observed, as assessed by morphological criteria after 48 h (Marini et al. 2005).

Similar results have been observed in erythroblasts and erythroleukemic cells after combined treatment. Ionizing radiation had a sensitizing effect on TRAIL-induced apoptosis in erythroleukemic cells but not in normal erythroblasts (Di Pietro et al. 2001). In normal human astrocytes, TRAIL (100 ng ml^{-1}) in combination with ionizing radiation (10 and 20 Gy) had no effect on viability, while in malignant T98G cells enhanced cytotoxicity was observed (Nagane et al. 2007).

In normal prostate cells, concurrent treatment induced slightly more apoptosis than either treatment alone. In these experiments, single treatment and combined treatment was found to induce roughly tenfold higher rate of apoptosis in malignant cells, compared to normal cells (Shankar et al. 2004b).

Little is still known about toxicity of combined treatment in vivo. To date increased toxicity has not been reported. In breast cancer bearing mice no apparent normal tissue toxicity was observed after combined treatment with irradiation and TRAIL (Shankar et al. 2004a).

First clinical studies combining TRAIL or monoclonal agonistic TRAIL-receptor antibodies with DNA-damaging drugs, did not find markedly increased normal tissue toxicity (Chow et al. 2006; Mom et al. 2006).

6 Mechanistic Aspects

Although a few basic aspects of the underlying mechanisms have been described above, several important putative mechanisms will be described in the following paragraph (Fig. 1).

One of the earliest observations was the fact that DNA-damage induces an up-regulation of CD95 type death-receptors (Fuchs et al. 1997; Williams et al. 1997).

Subsequently in several reports a close link between up-regulation of the TRAIL-receptors and an increased sensitivity was described.

Up-regulation of DR5, in a p53-dependent or p53-independent manner, after ionizing radiation or administration of chemotherapeutic drugs has been observed in different cell models (Guan et al. 2001; Hamasu et al. 2005; Wu et al. 1997).

In human glioma cells (Nagane et al. 2007), prostate cancer cell lines (Shankar et al. 2004b), breast cancer cells (Chinnaiyan et al. 2000), cervical cancer cells (Maduro et al. 2008) and gastric cancer cells, lung cancer cells and prostate cancer cells (Hamasu et al. 2005) increased levels of DR5 after irradiation had been observed. Moreover, it was found that DR5 was up-regulated in response to ionizing radiation in malignant, but not in normal prostate cells (Shankar et al. 2004b). In this study, up-regulation of DR5 was shown to be dependent on p53 since p53-siRNA markedly reduced radiation induced up-regulation of DR5.

Up-regulation of DR4 in response to ionizing radiation has been described as well in hematological malignancies (Di Pietro et al. 2001; Rezacova et al. 2008).

Nevertheless, the amount of DR4 and DR5 is not strictly associated with sensitivity to TRAIL induced apoptosis (Marini et al. 2005; Wagner et al. 2007). Marini and coworkers for example, provided evidence that increased TRAIL-receptor levels after irradiation were detectable in four of six cell lines, but no one by one correlation between sensitization and up-regulation was detectable (Marini et al. 2005).

Similarly, several studies revealed that TRAIL-receptor levels did not tightly correlate with sensitization by ionizing radiation (Chuntharapai et al. 2001; Ganten et al. 2004; Lacour et al. 2003; Singh et al. 2003).

In addition to up-regulation of the respective receptors, apoptosis threshold shift via up-regulation of proapoptotic members of the Bcl-2 family may constitute a putative mechanism of action (Chinnaiyan et al. 2000; Rudner et al. 2005). In one setting, the interaction of TRAIL and radiation was found to depend on the expression level of the proapoptotic Bcl-2 family member Bax. Lack of Bax was associated with a loss of interaction in several cell systems, including prostate cancer and colorectal cancer (von Haefen et al. 2004).

In addition to the analysis of death signals above, other factors explaining the interaction have been described. Signaling from both pathways culminates in activation of effector caspases, thereby resulting in increased levels of active caspase-3 and caspase-7.

In this regard it has been shown, that both apoptosis pathways participated in apoptosis induction following combined treatment. Radiation induced apoptosis involved Bax and caspase-3, while TRAIL induced apoptosis involved caspase-8 or caspase-10 and the adapter protein FADD which links caspase-8 to the death-receptor (Marini et al. 2003).

Recently, while trying to further elucidate the underlying mechanisms, Verbrugge and coworkers showed, that irradiation preceding application of TRAIL allowed bypassing of the mitochondrial pathway, rendering Type II cells into Type I cells (Verbrugge et al. 2008). In addition they demonstrated that TRAIL- and CD95-L

Table 1 Studies of combined treatment with TRAIL or agonistic TRAIL receptor antibodies and ionizing radiation

Description	Cell type	Agonist	Efficacy	Clonogenic cell kill	DR-up-regulation	Normal tissue toxicity	Citation
In vitro and xenograft study	Breast cancer cell lines	Polyhistidine-tagged TRAIL	Synergy in vitro, dependent on p53	Not assessed	DR5-up-regulation	No apparent toxicities	Chinnaiyan et al.(2000)
In vitro Study	Jurkat-T-cells	Recombinant human TRAIL	Enhanced cell kill	Enhanced clonogenic cell kill	DR5-up-regulation in MOLT4 cells	Not assessed	Gong and Almasan (2000)
In vitro study	Jurkat-T-cells	Recombinant human TRAIL	Synergy in Bcl-2 over-expressing cells	Synergistic clono-genic cell kill	Not assessed	Not assessed	Belka et al. (2001)
In vitro study	Jurkat-T-cells	Recombinant human TRAIL	Synergistic cell kill in TRAIL resistant cell clones	Not assessed	No up-regulation of DR4/DR5	Not assessed	Kim et al. (2001)
In vitro study	Erythroleukemic cells, erythro-blasts	His6-tagged TRAIL	Specific sensitization of erythroleukemic cells	Not assessed	DR4-up-regulation	Moderate cell kill through single treatment, no additive effects of combined treatment	Di Pietro et al. (2001)
In vitro and xenograft study	Breast cancer cell lines	Anti-DR5 antibody	High efficacy of combined treatment	Not assessed	Not assessed	Not assessed	Buchsbaum et al. (2003)
In vitro study	Renal cell carci-noma cell lines	Recombinant human TRAIL	Sensitization in one of eight cell lines	Not assessed	Not assessed	Not assessed	Ramp et al. (2003)
In vitro and xenograft study	Breast cancer cell lines	Recombinant human TRAIL	Synergistic cell kill in vitro, complete eradication of xenografts	Not assessed	p53-dependent DR5-up-regulation	No apparent toxicities	Shankar et al. (2004a)

(continued)

Table 1 (continued)

Description	Cell type	Agonist	Efficacy	Clonogenic cell kill	DR-up-regulation	Normal tissue toxicity	Citation
In vitro and xenograft study	Prostate cancer cell lines	Recombinant human TRAIL	Synergistic cell kill, sensitization of TRAIL-resistant cells. Eradication of xenografts	Enhanced clonogenic cell kill (data not published)	p53-dependent DR5-up-regulation	Moderate increased apoptosis in normal prostate cells	Shankar et al. (2004b)
In vitro study	Breast-, lung-, colorectal-, head and neck-cancer cell lines. Multiple normal tissue cultures	Recombinant human TRAIL	Additive and synergistic induction of apoptosis	Not assessed	DR5-up-regulation	No normal tissue toxicity	Marini et al. (2005)
In vitro study	Gastric cancer-, lung cancer- and prostate cancer cell lines	Recombinant human TRAIL	Enhanced induction of apoptosis	Not assessed	DR5-up-regulation in some cell types dependent on the cell red ox state	Not assessed	Hamasu et al. (2005)
In vitro and xenograft study	Colorectal cancer cell lines	DR5- and DR4-agonistic antibodies	Enhanced induction of apoptosis in vitro. Strongly prolonged tumor doubling time in vivo	Not assessed	Not assessed	No apparent toxicity	Marini et al. (2006)
In vitro study	Human glioma cells	Recombinant human TRAIL	Synergistically enhanced apoptosis	Enhanced clonogenic cell kill	DR5-up-regulation	Not assessed	Nagane et al. (2007)
In vitro study	Human melanoma cell lines	Recombinant human TRAIL	Enhanced induction of apoptosis	Not assessed	DR5- and DR4-up-regulation	Not assessed	Ivanov et al. (2007)

In vitro and xenograft study	Jurkat-T-cells	Isoleucine zipped TRAIL	Synergistically enhanced apoptosis. Delayed tumor growth in vivo	Enhanced clonogenic cell kill	Not assessed	No apparent toxicity in vivo	Wissink et al. (2006)
In vitro study	T-lymphoblastic leukemia cells	Recombinant human TRAIL	Sensitization of TRAIL resistant cells	Decreased proliferation	DR5-up-regulation	Not assessed	Rezacova et al. (2008)
In vitro study	Cervical cancer cells	Recombinant human TRAIL and DR5- and DR4-agonistic antibodies	More than additive enhanced apoptosis	Enhanced clonogenic cell kill	DR5- and DR4-up-regulation.	Not assessed	Maduro et al. (2008)
Xenograft study	Gastric adenocarcinoma cell lines	Recombinant human TRAIL	Synergistic growth inhibition in vivo	Not assessed	DR5-up-regulation in vivo	No apparent toxicity	Takahashi et al. (2008)

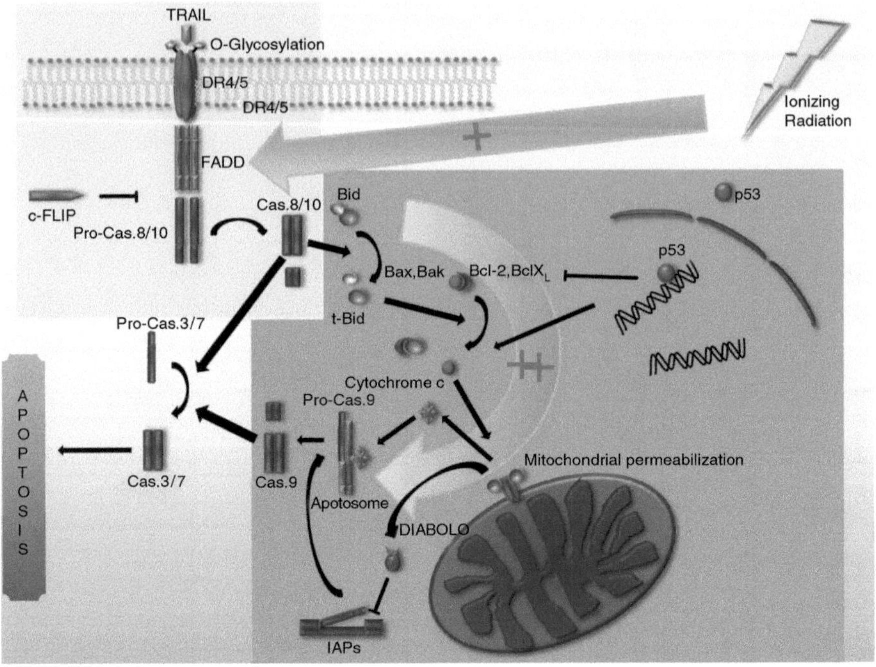

Fig. 1 Interaction of death-receptor induced and irradiation induced apoptosis: Death-receptor ligation induces recruitment of FADD and Caspase-8 or -10 and subsequent autoactivation. Active caspase-8 and -10 activate caspse-3. Amplification of the death-signal is mediated by caspase-8-dependent cleavage of Bid. t-bid releases proapoptotic Bax from inhibition by Bcl-2 and BclXL. Bax leads to the release of cytochrome c from the mitochondria. Cytochrome c, dATP and Apaf-1 form the apotosome which activates caspase-9. Caspase-9 subsequently activates caspase-3 and -7, thus the apoptotic signal culminates in activation of effector caspases-3 and -7. Caspase activation is further facilitated by the antagonizing effect of Smac/DIABOLO on caspase inhibition by IAPs. Irradiation mediated mitochondrial permeabilization, can be mediated by p53-dependent up-regulation of proapoptotic – and downregulation of antiapoptotic Bcl-2-proteins. Possible mechanisms of increased induction of apoptosis in combined treatment are: I. Increased activation of effector caspases by coactivation of the intrinsic pathway. II. Increased expression of death-receptors leads to increased activation of caspase-8. III. Enhanced efficacy of FADD and caspse-8-recruciment to death-receptors following irradiation occurs

induced apoptosis, after irradiation or treatment with the DNA-damaging agent etoposide, was independent of the known components of the intrinsic apoptosis pathway like Bax/Bak and caspase-9, alteration of the NFκB pathway or levels of c-FLIP. Upon combined treatment, no release of cytochrome c was observed and a stable knock down of Bax and Bak did not affect efficacy of combined treatment, indicating that the sensitizing effect occurs independent of the mitochondrial pathway. Moreover, the sensitivity of Jurkat-T-cells expressing a dominant negative caspase-9 to combined treatment was not diminished. Consequently, besides activation of the mitochondrial amplification loop, another mechanism of sensitization can be proposed.

Caspase-3 processing upon TRAIL- and CD95-Ligand treatment was found to be significantly enhanced in irradiated cells, leading to the conclusion that pretreatment with ionizing radiation improves the capacity of death-receptors to activate effector caspases. To exclude decreased levels of IAPs and hence decreased inhibition of caspases, genome-wide mRNA expression of IAP genes and protein levels of IAPs, following ionizing radiation, were determined. No relevant decrease in expression was detected. The expression of a Smac mutant, constitutively localized in the cytoplasm to antagonize IAPs, also had only minimal impact on the sensitizing effects of pre-treatment with ionizing radiation. Furthermore, only a moderate increase in DR5 and CD95-R1 expression levels was found. DR4 and decoy receptors were not expressed. So the conclusion was drawn that alteration of TRAIL sensitivity after irradiation, might occur at the receptor level. Finally, improved caspase-8 activation and improved DISC formation following irradiation was found. A fourfold FADD to TRAIL ratio, from 2 h after irradiation on, and higher amounts of caspase-8 and -10 in the DISC were observed, indicating an improved capacity of TRAIL-R2 to form a DISC including an improved recruitment of initiator caspases.

Similar results were obtained by Lacour and coworkers in 2003 for the combination of DNA-damaging substances with TRAIL. An increased recruitment of FADD and pro-caspase to form the DISC was observed (Lacour et al. 2003).

Moreover, decoy receptors may play some role in altering sensitivity to TRAIL induced apoptosis, although the presence of mRNA encoding decoy receptors did not correspond to sensitivity to TRAIL induced apoptosis (Griffith et al. 1998) and the decreased expression of DcR1 and DcR2 did not correlate with sensitivity of colon cancer cells to cytotoxic drugs (Lacour et al. 2001).

7 Death-Receptors and Radiation Induced Toxicities

Treatment of cancer is always limited by the side effects of a given approach to treatment. For a long time, radiation induced side effects were considered to be solely stochastic processes, not susceptible by any active means. Currently, it has become clear that radiation induced side effects result at least partially from active reactions of cells and tissues, making them indeed susceptible to interventions. In case of cell death research, several lines of evidence underline the importance of apoptosis induction as a basic trigger for radiation induced side effects (for review see (Kolesnick and Fuks 2003).

Although, the main focus of research on death ligands in radiation oncology is directed towards the use of death ligands as active inducers of cell death in tumors, several observations support the interpretation that the TRAIL -, and CD95-system are also strongly involved during the pathogenesis of radiation induced side effects. In this regard the phenotypes of several null mice systems provide clear evidence for this interpretation.

After preparing a seemingly normal DR5 knock out mouse Finnberg and cow-orkers showed that in response to ionizing radiation, tissues from brain, spleen and payer plaques, derived from these mice exhibit a reduced rate of apoptosis, compared to wild type mice. In tissue derived from the colon, the rate of apoptosis did not differ between the two models, indicating a role of DR5 in radiation induced apoptosis of some normal tissues in this mouse model (Finnberg et al. 2005).

Interestingly, silencing of DR5 in cancer cells in the colon significantly enhanced tumor growth in vivo (Wang and El-Deiry 2004). Moreover, DR5-deficient mice show enhanced metastasis of squamous carcinoma cell to the lymph nodes (Grosse-Wilde et al. 2008) and likewise, DR5-deficient lymphomas exhibit increased metastatic potential (Finnberg et al. 2008), indicating a possible role of TRAIL in immune surveillance. Following irradiation with sublethal doses, DR5-null mice developed signs of bronchopneumonia and tumorigenesis and showed decreased long-term survival (Finnberg et al. 2008).

As CD95/Fas and its ligand is involved in inflammatory responses in the lung (Grassme et al. 2000), it was tested in how far the CD95-system may be involved in the pneumonitic response after irradiation of the lung. Heinzelmann and coworkers demonstrated that induction of pneumonitis by irradiation is attenuated in CD95-knockout mice and in CD95L-knockout mice. Irradiated knockout mice, in contrast to wild-type mice, showed no increase in breathing frequency, airway resistance and histopathological changes after single irradiation of one lung with 12.5 Gy (Heinzelmann et al. 2006).

Thus, also the main research focus still aims at increasing the cell kill rate, a variety of data also supports the interpretation that specific death signaling is involved in the pathogenesis of radiation induced side effects.

8 Future Directions

All data available indicate that TRAIL in combination with ionizing radiation is highly effective regarding apoptosis induction and clonogenic cell kill. Based on in vitro studies and animal data, there are no hints pointing to an increased risk of side effects whenever TRAIL is combined with radiation. Thus, it seems reasonable to assume that this approach finally will result in a substantial therapeutic gain. However only phase III trials will be able to ultimately determine the clinical value of TRAIL. Beside several open questions regarding mechanistical aspects, it is currently problematic to integrate TRAIL into already highly complex treatment protocols, combining radiation, chemotherapy and surgery. Although many aspects of dual interactions (radiation plus TRAIL or chemotherapy plus TRAIL) have been elucidated, the understanding in how far three and even more partners interact is limited (Fiveash et al. 2008).

From what is known currently, TRAIL in combination with radiation would be worth being tested in bronchial cancers, rectal cancer or head and neck cancer.

Regarding the putative involvement of CD95 or TRAIL in the physiological modulation of early and late effects of radiation it is currently too early to conclude in how far these steps turn out to be a realistic prophylactic and/or therapeutic target.

Nevertheless the available data strongly suggest that radiation induced side effects are actively regulated and death cascades are clearly involved in these processes.

References

Aldridge DR, Arends MJ, Radford R (1995) Increasing the susceptibility of the rat 208f fibroblast cell-line to radiation-induced apoptosis does not alter its clonogenic survival dose – response. Br J Cancer 71:571–577

Ashkenazi A (2008) Directing cancer cells to self-destruct with pro-apoptotic receptor agonists. Nat Rev Drug Discov 7:1001–1012

Ashkenazi A, Pai RC, Fong S, Leung S, Lawrence DA, Masters SA, Blackie C, Chang L, McMurtrey AE, Hebert A et al (1999) Safety and antitumor activity of recombinant soluble Apo2 ligand. J Clin Invest 104:155–162

Belka C, Budach W (2002) Anti-apoptotic Bcl-2 proteins: structure, function and relevance for radiation biology. Int J Radiat Biol 78:643–658

Belka C, Rudner J, Wesselborg S, Stepczynska A, Marini P, Lepple-Wienhues A, Faltin H, Bamberg M, Budach W, Schulze-Osthoff K (2000) Differential role of caspase-8 and BID activation during radiation- and CD95-induced apoptosis. Oncogene 19:1181–1190

Belka C, Schmid B, Marini P, Durand E, Rudner J, Faltin H, Bamberg M, Schulze-Osthoff K, Budach W (2001) Sensitization of resistant lymphoma cells to irradiation-induced apoptosis by the death ligand TRAIL. Oncogene 20:2190–2196

Buchsbaum DJ, Zhou T, Grizzle WE, Oliver PG, Hammond CJ, Zhang S, Carpenter M, LoBuglio AF (2003) Antitumor efficacy of TRA-8 anti-DR5 monoclonal antibody alone or in combination with chemotherapy and/or radiation therapy in a human breast cancer model. Clin Cancer Res 9:3731–3741

Chen XF, Kandasamy K, Srivastava RK (2003) Differential roles of RelA (p65) and c-Rel subunits of nuclear factor kappa B in tumor necrosis factor-related apoptosis-inducing ligand signaling. Cancer Res 63:1059–1066

Chinnaiyan AM, Prasad U, Shankar S, Hamstra DA, Shanaiah M, Chenevert TL, Ross BD, Rehemtulla A (2000) Combined effect of tumor necrosis factor-related apoptosis-inducing ligand and ionizing radiation in breast cancer therapy. Proc Natl Acad Sci U S A 97:1754–1759

Chow LQ, Eckhardt SG, Gustafson DL, O'Bryant C, Hariharan S, Diab S, Fox NL, Corey A, Padavic K, Brown M et al (2006) HGS-ETR1, an antibody targeting TRAIL-R1, in combination with paclitaxel and carboplatin in patients with advanced solid malignancies: results of a phase I and PK study. J Clin Oncol 24:103S

Chuntharapai A, Dodge K, Grimmer K, Schroeder K, Marsters SA, Koeppen H, Ashkenazi A, Kim KJ (2001) Isotype-dependent inhibition of tumor growth in vivo by monoclonal antibodies to death receptor 4. J Immunol 166:4891–4898

Danial NN, Korsmeyer SJ (2004) Cell death: critical control points. Cell 116:205–219

Debatin KM, Poncet D, Kroemer G (2002) Chemotherapy: targeting the mitochondrial cell death pathway. Oncogene 21:8786–8803

DegliEsposti MA, Smolak PJ, Walczak H, Waugh J, Huang CP, DuBose RF, Goodwin RG, Smith CA (1997) Cloning and characterization of TRAIL-R3, a novel member of the emerging TRAIL receptor family. J Exp Med 186:1165–1170

Di Pietro R, Secchiero P, Rana R, Gibellini D, Visani G, Bemis K, Zamai L, Miscia S, Zauli G (2001) Ionizing radiation sensitizes erythroleukemic cells but not normal erythroblasts to tumor necrosis factor-related apoptosis-inducing ligand (TRAIL)-mediated cytotoxicity by selective up-regulation of TRAIL-R1. Blood 97:2596–2603

Eskes R, Desagher S, Antonsson B, Martinou JC (2000) Bid induces the oligomerization and insertion of Bax into the outer mitochondrial membrane. Mol Cell Biol 20:929–935

Falschlehner C, Emmerich CH, Gerlach B, Walczak H (2007) TRAIL signalling: decisions between life and death. Int J Biochem Cell Biol 39:1462–1475

Finnberg N, Gruber JJ, Fei PW, Rudolph D, Bric A, Kim SH, Burns TF, Ajuha H, Page R, Wu GS et al (2005) DR5 knockout mice are compromised in radiation-induced apoptosis. Mol Cell Biol 25:2000–2013

Finnberg N, Klein-Szanto AJP, El-Deiry WS (2008) TRAU-R deficiency in mice promotes susceptibility to chronic inflammation and tumorigenesis. J Clin Invest 118:111–123

Fiveash JB, Gillespie GY, Oliver PG, Zhou T, Belenky ML, Buchsbaum DJ (2008) Enhancement of glioma radiotherapy and chemotherapy response with targeted antibody therapy against death receptor 5. Int J Radiat Oncol Biol Phys 71:507-516

Fuchs EJ, Paik JE, Engstrom LE, Bedi A (1997) CD95 (Fas/APO-1)-dependence of T cell apoptosis induced by DNA damage. Blood 90: 225

Ganten TM, Haas TL, Sykora J, Stahl H, Sprick MR, Fas SC, Krueger A, Weigand MA, Grosse-Wilde A, Stremmel W et al (2004) Enhanced caspase-8 recruitment to and activation at the DISC is critical for sensitisation of human hepatocellular carcinoma cells to TRAIL-induced apoptosis by chemotherapeutic drugs. Cell Death Differ 11:S86–S96

Georgakis GV, Li Y, Humphreys R, Andreeff M, O'Brien S, Younes M, Carbone A, Albert V, Younes A (2005) Activity of selective fully human agonistic antibodies to the TRAIL death receptors TRAIL-R1 and TRAIL-R2 in primary and cultured lymphoma cells: induction of apoptosis and enhancement of doxorubicin- and bortezomib-induced cell death. Br J Haematol 130:501–510

Goh HS, Yao J, Smith DR (1995) P53 point mutation and survival in colorectal-cancer patients. Cancer Res 55:5217–5221

Gong BD, Almasan A (2000) Apo2 ligand/TNF-related apoptosis-inducing ligand and death receptor 5 mediate the apoptotic signaling induced by ionizing radiation in leukemic cells. Cancer Res 60:5754–5760

Grassme H, Kirschnek S, Riethmueller J, Riehle A, von Kurthy G, Lang F, Weller M, Gulbins E (2000) CD95/CD95 ligand interactions on epithelial cells in host defense to Pseudomonas aeruginosa. Science 290:527–530

Griffith TS, Chin WA, Jackson GC, Lynch DH, Kubin MZ (1998) Intracellular regulation of TRAIL-induced apoptosis in human melanoma cells. J Immunol 161:2833–2840

Grosse-Wilde A, Voloshanenko O, Bailey SL, Longton GM, Schaefer U, Csernok AI, Schutz G, Greiner EF, Kemp CJ, Walczak H (2008) TRAIL-R deficiency in mice enhances lymph node metastasis without affecting primary tumor development. J Clin Invest 118:100–110

Guan BX, Yue P, Clayman GL, Sun SY (2001) Evidence that the death receptor DR4 is a DNA damage-inducible, p53-regulated gene. J Cell Physiol 188:98–105

Hamasu T, Inanami O, Asanuma T, Kuwabara M (2005) Enhanced induction of apoptosis by combined treatment of human carcinoma cells with X rays and death receptor agonists. J Radiat Res 46:103–110

Heinzelmann F, Jendrossek V, Lauber K, Nowak K, Eldh T, Boras R, Handrick R, Henkel M, Martin C, Uhlig S et al (2006) Irradiation-induced pneumonitis mediated by the CD95/CD95-ligand system. J Natl Cancer Inst 98:1248–1251

Herbst RS, Mendolson DS, Ebbinghaus S, Gordon MS, O'Dwyer P, Lieberman G, Ing J, Kurzrock R, Novotny W, Eckhardt G (2006) A phase I safety and pharmacokinetic (PK) study of recombinant Apo2L/TRAIL, an apoptosis-inducing protein in patients with advanced cancer. J Clin Oncol 24:124S

Herr I, Debatin KM (2001) Cellular stress response and apoptosis in cancer therapy. Blood 98:2603–2614

Ichikawa K, Liu WM, Zhao LM, Wang Z, Liu D, Ohtsuka T, Zhang HG, Mountz JD, Koopman WJ, Kimberly RP et al (2001) Tumoricidal activity of a novel anti-human DR5 monoclonal antibody without hepatocyte cytotoxicity. Nat Med 7:954–960

Ivanov N, Zhou HN, Hei TK (2007) Sequential treatment by ionizing radiation and sodium arsenite dramatically accelerates TRAIL-mediated apoptosis of human melanoma cells. Cancer Res 67:5397–5407

Johnstone RW, Ruefli AA, Lowe SW (2002) Apoptosis: a link between cancer genetics and chemotherapy. Cell 108:153–164

Kanzler S, Trarbach T, Heinemann V, Kohne CH, Sneller V, Bieber F, Galle PR, Seeber S (2005) Results of a phase 2 study of HGS-ETR1, a fully human agonistic monoclonal antibody to TRAIL Receptor 1, in subjects with relapsed or refractory colorectal cancer (CRC). EJC Suppl 3:178–178

Kim MR, Lee JY, Park MT, Chun YJ, Jang YJ, Kang CM, Kim HS, Cho CK, Lee YS, Jeong HY et al (2001) Ionizing radiation can overcome resistance to TRAIL in TRAIL-resistant cancer cells. FEBS Lett 505:179–184

Kolesnick R, Fuks Z (2003) Radiation and ceramide-induced apoptosis. Oncogene 22:5897–5906

Lacour S, Hammann A, Wotawa A, Corcos L, Solary E, Dimanche-Boitrel MT (2001) Anticancer agents sensitize tumor cells to tumor necrosis factor-related apoptosis-inducing ligand-mediated caspase-8 activation and apoptosis. Cancer Res 61:1645–1651

Lacour S, Micheau O, Hammann A, Drouineaud V, Tschopp J, Solary E, Dimanche-Boitrel MT (2003) Chemotherapy enhances TNF-related apoptosis-inducing ligand DISC assembly in HT29 human colon cancer cells. Oncogene 22:1807–1816

Li HL, Zhu H, Xu CJ, Yuan JY (1998) Cleavage of BID by caspase 8 mediates the mitochondrial damage in the Fas pathway of apoptosis. Cell 94:491–501

Liu XS, Kim CN, Yang J, Jemmerson R, Wang XD (1996) Induction of apoptotic program in cell-free extracts: requirement for dATP and cytochrome c. Cell 86:147–157

Lowe SW, Ruley HE, Jacks T, Housman DE (1993) P53-dependent apoptosis modulates the cytotoxicity of anticancer agents. Cell 74:957–967

Lowe SW, Bodis S, McClatchey A, Remington L, Ruley HE, Fisher DE, Housman DE, Jacks T (1994) P53 status and the efficacy of cancer-therapy in-vivo. Science 266:807–810

Luo X, Budihardjo I, Zou H, Slaughter C, Wang XD (1998) Bid, a Bcl2 interacting protein, mediates cytochrome c release from mitochondria in response to activation of cell surface death receptors. Cell 94:481–490

Maduro JH, de Vries EGE, Meersma GJ, Hougardy BMT, van der Zee AGJ, De Jong S (2008) Targeting pro-apoptotic trail receptors sensitizes HeLa cervical cancer cells to irradiation-induced apoptosis. Int J Radiat Oncol Biol Phys 72:543–552

Marini P, Jendrossek V, Durand E, Gruber C, Budach W, Belka C (2003) Molecular requirements for the combined effects of TRAIL and ionising radiation. Radiother Oncol 68:189–198

Marini P, Schmid A, Jendrossek V, Faltin H, Daniel PT, Budach W, Belka C (2005) Irradiation specifically sensitises solid tumour cell lines to TRAIL mediated apoptosis. BMC Cancer 5:5

Marini P, Denzinger S, Schiller D, Kauder S, Welz S, Humphreys R, Daniel PT, Jendrossek V, Budach W, Belka C (2006) Combined treatment of colorectal tumours with agonistic TRAIL receptor antibodies HGS-ETR1 and HGS-ETR2 and radiotherapy: enhanced effects in vitro and dose-dependent growth delay in vivo. Oncogene 25:5145–5154

Marsters SA, Sheridan JP, Pitti RM, Huang A, Skubatch M, Baldwin D, Yuan J, Gurney A, Goddard AD, Godowski P et al (1997) A novel receptor for Apo2L/TRAIL contains a truncated death domain. Current Biology 7:1003–1006

Merino D, Lalaoui N, Morizot A, Schneider P, Solary E, Micheau O (2006) Differential inhibition of TRAIL-mediated DR5-DISC formation by decoy receptors 1 and 2. Mol Cell Biol 26:7046–7055

Miyashita T, Reed JC (1995) Tumor-suppressor p53 is a direct transcriptional activator of the human bax gene. Cell 80:293–299

Mom CH, Sleijfer S, Gietema JA, Fox NL, Piganeau C, Lo L, Uges DRA, Loos W, de Vries EGE, Verweij J (2006) Mapatumumab, a fully human agonistic monoclonal antibody that targets TRAIL-R1, in combination with gemcitabine and cisplatin: a phase 1 study in patients with advanced solid malignancies. EJC Suppl 4:202

Nagane M, Cavenee WK, Shiokawa Y (2007) Synergistic cytotoxicity through the activation of multiple apoptosis pathways in human glioma cells induced by combined treatment with ionizing radiation and tumor necrosis factor-related apoptosis-inducing ligand. J Neurosurg 106:407–416

Nakano K, Vousden KH (2001) PUMA, a novel proapoptotic gene, is induced by p53. Mol Cell 7:683–694

Ni R, Tomita Y, Matsuda K, Ichihara A, Ishimura K, Ogasawara J, Nagata S (1994) FAS-mediated apoptosis in primary cultured mouse hepatocytes. Exp Cell Res 215:332–337

Norbury CJ, Zhivotovsky B (2004) DNA damage-induced apoptosis. Oncogene 23:2797–2808

Oda E, Ohki R, Murasawa H, Nemoto J, Shibue T, Yamashita T, Tokino T, Taniguchi T, Tanaka N (2000) Nora, a BH3-only member of the Bcl-2 family and candidate mediator of p53-induced apoptosis. Science 288:1053–1058

Ogasawara J, Watanabefukunaga R, Adachi M, Matsuzawa A, Kasugai T, Kitamura Y, Itoh N, Suda T, Nagata S (1993) Lethal effect of the anti-Fas antibody in mice. Nature 364:806–809

Pacey S, Plummer RE, Attard G, Bale C, Calvert AH, Blagden S, Fox NL, Corey A, de Bono JS (2005) Phase I and pharmacokinetic study of HGS-ETR2, a human monoclonal antibody to TRAIL R2, in patients with advanced solid malignancies. J Clin Oncol 23: 205S

Pan GH, Ni J, Wei YF, Yu GL, Gentz R, Dixit M (1997a) An antagonist decoy receptor and a death domain-containing receptor for TRAIL. Science 277:815–818

Pan GH, Orourke K, Chinnaiyan AM, Gentz R, Ebner R, Ni J, Dixit M (1997b) The receptor for the cytotoxic ligand TRAIL. Science 276:111–113

Pukac L, Kanakaraj P, Humphreys R, Alderson R, Bloom M, Sung C, Riccobene T, Johnson R, Fiscella M, Mahoney A et al (2005) HGS-ETR1, a fully human TRAIL-receptor 1 monoclonal antibody, induces cell death in multiple tumour types in vitro and in vivo. Br J Cancer 92:1430–1441

Ramp U, Caliskan E, Mahotka C, Krieg A, Heikaus S, Gabbert HE, Gerharz CD (2003) Apoptosis induction in renal cell carcinoma by TRAIL and gamma-radiation is impaired by deficient caspase-9 cleavage. Br J Cancer 88:1800–1807

Rezacova M, Vavrova J, Vokurkova D (2008) Ionizing radiation sensitizes leukemic MOLT-4 cells to TRAIL-induced apoptosis. Acta Medica (Hradec Kralove) 51:101–105

Rudner J, Jendrossek V, Lauber K, Daniel PT, Wesselborg S, Belka C (2005) Type I and type II reactions in TRAIL-induced apoptosis – results from dose-response studies. Oncogene 24:130–140

Scaffidi C, Fulda S, Srinivasan A, Friesen C, Li F, Tomaselli KJ, Debatin KM, Krammer PH, Peter ME (1998) Two CD95 (APO-1/Fas) signaling pathways. EMBO J 17:1675–1687

Schneider P, Bodmer JL, Thome M, Hofmann K, Holler N, Tschopp J (1997) Characterization of two receptors for TRAIL. FEBS Lett 416:329–334

Shankar S, Singh TR, Chen XF, Thakkar H, Firnin J, Srivastava RK (2004a) The sequential treatment with ionizing radiation followed by TRAIL/Apo-2L reduces tumor growth and induces apoptosis of breast tumor xenografts in nude mice. Int J Oncol 24:1133–1140

Shankar S, Singh TR, Srivastava RK (2004b) Ionizing radiation enhances the therapeutic potential of TRAIL in prostate cancer in vitro and in vivo: intracellular mechanisms. Prostate 61:35–49

Shankar S, Chen XF, Srivastava RK (2005) Effects of sequential treatments with chemotherapeutic drugs followed by TRAIL on prostate cancer in vitro and in vivo. Prostate 62:165–186

Sheridan JP, Marsters SA, Pitti RM, Gurney A, Skubatch M, Baldwin D, Ramakrishnan L, Gray CL, Baker K, Wood WI et al (1997) Control of TRAIL-induced apoptosis by a family of signaling and decoy receptors. Science 277:818–821

Singh TR, Shankar S, Chen WF, Asim M, Srivastava RK (2003) Synergistic interactions of chemotherapeutic drugs and tumor necrosis factor-related apoptosis-inducing ligand/Apo-2 ligand on apoptosis and on regression of breast carcinoma in vivo. Cancer Res 63:5390–5400

Sprick MR, Rieser E, Stahl H, Grosse-Wilde A, Weigand MA, Walczak H (2002) Caspase-10 is recruited to and activated at the native TRAIL and CD95 death-inducing signalling complexes in a FADD-dependent manner but can not functionally substitute caspase-8. EMBO J 21:4520–4530

Suliman A, Lam A, Datta R, Srivastava RK (2001) Intracellular mechanisms of TRAIL: apoptosis through mitochondrial-dependent and -independent pathways. Oncogene 20:2122–2133

Takahashi M, Yasui H, Ogura A, Asanuma T, Kubota N, Tsujitan M, Kuwabara M, Inanami O (2008) X irradiation combined with TNF alpha-related apoptosis-inducing ligand (TRAIL) reduces hypoxic regions of human gastric adenocarcinoma xenografts in SCID mice. J Radiat Res 49:153–161

Tolcher AW, Mita M, Meropol NJ, von Mehren M, Patnaik A, Padavic K, Hill M, Mays T, McCoy T, Fox NL et al (2007) Phase I pharmacokinetic and biologic correlative study of mapatumu-

mab, a fully human monoclonal antibody with agonist activity to tumor necrosis factor-related apoptosis-inducing ligand receptor-1. J Clin Oncol 25:1390–1395
Verbrugge I, de Vries E, Tait SWG, Wissink EHJ, Walczak H, Verheij M, Borst J (2008) Ionizing radiation modulates the TRAIL death-inducing signaling complex, allowing bypass of the mitochondrial apoptosis pathway. Oncogene 27:574–584
Vogelstein B, Kinzler KW (2004) Cancer genes and the pathways they control. Nat Med 10:789–799
von Haefen C, Gillissen B, Hemmati PG, Wendt J, Guner D, Mrozek A, Belka C, Dorken B, Daniel PT (2004) Multidomain Bcl-2 homolog Bax but not Bak mediates synergistic induction of apoptosis by TRAIL and 5-FU through the mitochondrial apoptosis pathway. Oncogene 23:8320–8332
Wagner KW, Punnoose EA, Januario T, Lawrence DA, Pitti RM, Lancaster K, Lee D, von Goetz M, Yee SF, Totpal K et al (2007) Death-receptor O-glycosylation controls tumor-cell sensitivity to the proapoptotic ligand Apo2L/TRAIL. Nat Med 13:1070–1077
Walczak H, Miller RE, Ariail K, Gliniak B, Griffith TS, Kubin M, Chin W, Jones J, Woodward A, Le T et al (1999) Tumoricidal activity of tumor necrosis factor related apoptosis-inducing ligand in vivo. Nat Med 5:157–163
Wang SL, El-Deiry WS (2004) Inducible silencing of KILLER/DR5 in vivo promotes bioluminescent colon tumor xenograft growth and confers resistance to chemotherapeutic agent 5-fluorouracil. Cancer Res 64:6666–6672
Wang K, Yin XM, Chao DT, Milliman CL, Korsmeyer SJ (1996) BID: a novel BH3 domain-only death agonist. Genes Dev 10:2859–2869
Williams BA, Makrigiannis AP, Blay J, Hoskin DW (1997) Treatment of the P815 murine mastocytoma with cisplatin or etoposide up-regulates cell-surface Fas (CD95) expression and increases sensitivity to anti-Fas antibody-mediated cytotoxicity and to lysis by anti-CD3-activated killer-T cells. Int J Cancer 73:416–423
Wissink EHJ, Verbrugge I, Vink SR, Schader MB, Schaefer U, Walczak H, Borst J, Verheij M (2006) TRAIL enhances efficacy of radiotherapy in a p53 mutant, Bcl-2 overexpressing lymphoid malignancy. Radiother Oncol 80:214–222
Wu GS, Burns TF, McDonald ER, Jiang W, Meng R, Krantz D, Kao G, Gan DD, Zhou JY, Muschel R et al (1997) KILLER/DR5 is a DNA damage-inducible p53-regulated death receptor gene. Nat Genet 17:141–143
Yang J, Liu XS, Bhalla K, Kim CN, Ibrado AM, Cai JY, Peng TI, Jones DP, Wang XD (1997) Prevention of apoptosis by Bcl-2: release of cytochrome c from mitochondria blocked. Science 275:1129–1132
Yin XM, Wang K, Gross A, Zhao YG, Zinkel S, Klocke B, Roth KA, Korsmeyer SJ (1999) Bid-deficient mice are resistant to Fas-induced hepatocellular apoptosis. Nature 400:886–891
Yoshida T, Maeda A, Tani N, Sakai T (2001) Promoter structure and transcription initiation sites of the human death receptor 5/TRAIL-R2 gene. FEBS Lett 507:381–385

Death Ligands Designed to Kill: Development and Application of Targeted Cancer Therapeutics Based on Proapoptotic TNF Family Ligands

Jeannette Gerspach, Harald Wajant, and Klaus Pfizenmaier

Abstract The identification of molecular markers associated with cancer development or progression, opened a new era in the development of therapeutics. The successful introduction of a few low molecular weight chemicals and recombinant protein therapeutics with targeted actions into clinical practice have raised great expectations to broadly improve cancer therapy with respect to both overall clinical responses and tolerability. Targeting the apoptotic machinery of malignant cells is an attractive concept to combat cancer, which is currently exploited for the proapoptotic members of the TNF ligand family at various stages of preclinical and clinical development. This review summarizes recent progress in this rapidly progressing field of "biologic" therapies targeting the death receptors of TNF, CD95L, and TRAIL by means of its cognate protein ligands, receptor specific antibodies, and gene therapeutic approaches. Preclinical data on newly derived variants and fusion proteins based on these death ligands, designed to act in a tumor restricted manner, thereby preventing a systemic, potentially harmful action, will also be discussed.

1 Introduction

In the past decade, cancer therapy has changed tremendously. Although for many malignancies traditional cytotoxic chemotherapy attacking rapidly dividing cells still is the treatment of choice, targeted therapies gain increasing importance for cancer treatment. Targeted cancer therapies generally are better tolerated although not free of adverse effects. They are represented by monoclonal antibodies and small molecule inhibitors, targeting extracellular components or interfering with

J. Gerspach (✉) and K. Pfizenmaier (✉)
Institute of Cell Biology and Immunology, University of Stuttgart, Stuttgart, Germany
e-mail: jeanette.gerspach@izi.uni-stuttgart.de
e-mail: klaus.pfizenmaier@izi.uni-stuttgart.de

H. Wajant
Department of Molecular Internal Medicine, Medical Clinic and Polyclinic II,
University of Wuerzburg, Wuerzburg, Germany

Results Probl Cell Differ, doi:10.1007/400_2008_22

mediators of intracellular signaling pathways, respectively. The anti-CD20 antibody Rituximab for treatment of non-Hodgkin's lymphoma (NHL) and the tyrosine kinase inhibitor Imatinib for treatment of chronic myeloid leukemia (CML) are lucid examples of the superior clinical efficacy of targeted therapies (reviewed in Gerber 2008). Cytokines of the tumor necrosis factor (TNF) family, especially the death inducing ligands TNF, CD95L, and TNF-related apoptosis-inducing ligand (TRAIL), also hold great promise as novel targeted therapeutics. As mediators of the immune system, most TNF family members have a role in immune homeostasis and defense against pathogens, while in addition, TNF, CD95L, and TRAIL, in particular, play an important role in tumor surveillance (see below). The broad use of genetically engineered derivatives of these cytokines as effective targeted therapeutics will depend on success in restricting their actions to the tumor tissue to eliminate unwanted side effects. In fact, systemic application of TNF and CD95 agonists is presently not possible due to overt generalized toxicity.

The death ligands, like almost all TNF family members, come in two flavors, as genuine transmembrane proteins with full signaling capability and as soluble mediators, derived by proteolytic processing and differing from the membrane ligand in their receptor activating potential (Bodmer et al. 2002; Wajant et al. 2003). Thus the membrane form of TNF, CD95L, and TRAIL efficiently activate their cognate receptors, that is, both TNF receptors (TNF-R1 and TNF-R2), CD95, and TRAILR1 and TRAILR2, respectively. In contrast, the soluble ligands, despite high affinity binding, show a receptor selective reduced signaling capacity. Accordingly, soluble TNF (sTNF) fails to properly activate TNFR2, soluble TRAIL (sTRAIL) displays very limited activity on TRAILR2 and, soluble CD95L (sCD95L) is practically unable to trigger CD95 (Grell et al. 1995; Wajant et al. 2001; Schneider et al. 1998). TNF ligands assemble into homotrimers by virtue of their common TNF homology domain (THD), which is in addition responsible but not necessarily sufficient for receptor binding (Bodmer et al. 2002; Berg et al. 2007). Modifications introduced by genetic engineering, in particular generation of death ligand fusion proteins, can lead to the conversion of sTNF, sCD95L, and sTRAIL into fully signaling competent ligands. This conversion is either mediated by secondary multimerization or by cell surface immobilization (see below).

Several copies of an extracellularly located conserved cysteine-rich domain (CRD) characterize the TNF receptor family members, from which the subgroup of eight death receptors can be distinguished by their common intracellular protein interaction domain, the death domain (DD) (reviewed in Hehlgans and Pfeffer 2005; Bodmer et al. 2002). The prototype death receptors are TNFR1 (DR1), CD95 (DR2), and two of the TRAILRs, TRAILR1 and TRAILR2 (DR4 and DR5). Besides apoptosis induction, these death receptors might also transduce non-apoptotic signals leading to the activation of, for example, NFkB (Park et al. 2005; Kreuz et al. 2004; Wajant 2004). In addition, for TRAIL and CD95L, signaling deficient, so-called decoy receptors are known (DcR1 to 3 and OPG).

Here, we summarize recent progress in the rapidly progressing field of "biologic" therapies targeting death receptors and the apoptotic machinery of tumor cells. We focus on the three death ligands, TNF, CD95L, and TRAIL, and newly designed molecules derived thereof, presently evaluated in preclinical and clinical trials for usefulness, safety, and efficacy as protein therapeutic and in gene therapy approaches alike. To complete the current picture, we also briefly review the antibody based strategies, targeting death receptors.

2 Tumor Necrosis Factor

Now over 30 years ago, TNF has been identified as the factor responsible for bacterially induced hemorrhagic necrosis of tumors, thus being the name giving member of the present TNF superfamily (Carswell et al. 1975). Because of its severe systemic side effects culminating in septic shock syndrome and multiorgan failure, TNF's potent antitumor activity still is not amenable to a general application in cancer therapy. This is based on the fact that the maximum tolerated dose of TNF is 30–545 $\mu g\ m^{-2}$ depending on route of injection. This is about 10–50 times less than the effective dose in animal tumor models (Lejeune et al. 2006; Eggermont et al. 2003). Presently, only in the sophisticated clinical setting of isolated limb perfusion (ILP), high dose TNF treatment is successfully and safely employed in combination with the chemotherapeutic melphalan for treatment of regionally advanced melanomas and soft tissue sarcomas. Although here discussed primarily as a death inducer, TNF is a pleiotropic cytokine with far-reaching effects (Pfeffer 2003). Actually its antitumor activities themselves are diverse and apparently to a lesser extent due to direct cytotoxic action on the tumor cells. Rather, TNF's ability to selectively damage tumor-associated endothelium and to induce clot formation leading to vessel obstruction is now recognized to play an important role (Nawroth et al. 1988 and reviewed in Mocellin et al. 2007; Lejeune et al. 2006). In addition, already low doses of TNF can enhance blood vessel permeability, an effect from which combination therapy with therapeutics directly toxic to the tumor can greatly profit (Kristensen et al. 1996; Seynhaeve et al. 2007 and reviewed in Lejeune et al. 2006). For this process the disruption of VE-cadherin complexes at vascular endothelial cell junctions seems to be responsible (Menon et al. 2006). Furthermore, there is increasing evidence for TNF playing an active role in both innate and adaptive antitumor immune responses (reviewed in Mocellin et al. 2007). Thus, besides critical requirement of TNF signaling for effective priming, proliferation, and recruitment of tumor-specific T-cells, an essential role of TNF in tumor immune surveillance has been described (Calzascia et al. 2007). For TNF-mediated tumor rejection, activation of TNFR2 might be as important as TNFR1 stimulation. This is gathered from reports describing that (1) TNFR2 expressed on host innate immune cells is sufficient to mediate the antitumor effect of TNF, which is dependent on nitric oxide production (Zhao et al. 2007), (2) TNFR2 seems to suppress metastasis development, whereas TNFR1 rather has promoting activity (Guillem and Sampsel 2006), (3) TNFR2 seems to provide an important costimulatory signal for optimal T-cell activation (Kim and Teh, 2001, 2004; Kim et al. 2006b; Aspalter et al. 2003, 2007), (4) endothelial TNFR2 is essential for TNF-induced leukocyte-endothelial cell interaction (Chandrasekharan et al. 2007).

For preclinical analysis of human TNF-based therapeutics in murine tumor models, it has to be considered that human TNF binds murine TNFR1, but not murine TNFR2, which excludes a possible potentiation of antitumor responses related to TNFR2 activation (Ranges et al. 1989; Ameloot et al. 2001).

2.1 TNF Fusion Proteins

During the last two decades, a number of TNF fusion proteins have been developed, which are composed of a targeting domain and the TNF moiety connected either by chemical conjugation or genetic fusion. In general, targeting is achieved by antibodies, natural ligands, or derivatives thereof. Target structures typically represent cell surface markers either expressed by tumor cells or tumor stroma cells, for example epidermal growth factor receptor (EGFR)-1, carcino-embryonic antigen (CEA), tumor associated glycoprotein (TAG) 72, transferrin receptor, the ganglioside GD2, fibroblast activation protein (FAP), and vascular markers like CD13 (aminopeptidase N) or B-FN, a fibronectin (FN) isoform containing the ED-B domain. Most of these fusion proteins retain normal TNF activity on target negative cells, but display 10- to 1,000-fold higher activity on target positive cells in vitro and possess enhanced anti-tumor activity in vivo. Noteworthy, antigen-dependent cell surface immobilization can be accompanied by conversion of soluble to membrane-like TNF activity, resulting in efficient activation of both TNF receptors (reviewed in Wajant et al. 2005). For some of these TNF fusion proteins, clinical investigations are already underway or pending. An example is NGR-muTNF, murine TNF fused to the C-terminus of a cyclic CNGRCG peptide, which targets a CD13 (Aminopeptidase N) isoform expressed on angiogenic blood vessels (Curnis et al. 2000). On L-M cells, cytotoxicity of this TNF variant was comparable to that of wild-type TNF as was systemic toxicity upon i.p. injection in tumor bearing mice (LD50: 45 μg NGR-muTNF; 60 μg murine TNF). Likewise, NGR-muTNF displayed similar serum half life and tumor/blood ratios as murine TNF. The latter one was attributed to the fact that endothelial cells represent a minor component of tumor mass. In contrast, inhibition of tumor growth by NGR-muTNF was 12–15 times more efficient compared to murine TNF. Interestingly, the authors observed a bell-shaped dose response curve for NGR-muTNF with antitumor activities between approx. 0.01–1 ng and 100 ng–10 μg, which could be accounted to TNF-induced shedding of TNFR2, neutralizing intermediate levels of circulating TNF. In support of this reasoning, NGR-huTNF did not show a bell-shaped response because murine TNFR2 does not bind human TNF (Curnis et al. 2002). Treatment of B16F1 tumor bearing mice with the chemotherapeutic Melphalan in combination with NGR-muTNF, but not murine TNF, showed increased inhibition of tumor growth compared to single agents. More important, NGR-muTNF as well as NGR-huTNF strongly enhanced antitumor activity of Melphalan, Doxorubicin, Cisplatin, Paclitaxel, and Gemcitabine at subnanogram doses (about six orders of magnitude lower than the LD50) with no increase in systemic toxicity (Curnis et al. 2002). In this process IFNγ appears critically involved (Sacchi et al. 2004). Recently, Crippa et al. (2008) showed synergistic damage of tumor vessels when NGR-TNF was combined with similar subnanomolar doses of endothelial-monocyte activating polypeptide (EMAP) II, a soluble tumor derived cytokine known to sensitize tumor neovasculature to TNF. These successful preclinical studies paved the way for presently ongoing clinical studies (phase I/II) investigating NGR-hTNF as single agent or in combination with chemotherapeutics in patients with solid tumors and different metastatic cancers.

Another presently investigated TNF fusion protein is the L19-TNF, a fusion protein of 134 kDa composed of murine TNF fused to the C-terminus of the scFv fragment "L19" which recognizes the angiogenesis marker B-FN. In vitro, L19-mTNF was five- to sixfold more active on B-FN expressing L-M cells than murine TNF. In vivo biodistribution experiments revealed stable and high level accumulation in B-FN positive tumors, with specific and selective targeting of tumor vasculature. Interestingly, although molecular size of L19-mTNF resembled that of a normal IgG (approximately 150 kDa), serum α and β phase half-lives of the TNF fusion protein were about twofold and 27-fold shorter, respectively. Histological analysis of different organs after intravenous (i.v.) injection of up to 1 pmol g^{-1} revealed no morphologic side effects. In contrast, bolus i.v. injection of 1 pmol g^{-1} into F9 tumor bearing mice induced cell death in tumor as well as tumor endothelial cells and resulted in at least fourfold higher antitumor activity compared to murine TNF. Combined treatment with L19-mTNF and L19 fusion proteins of IL2 or IL12 showed superior tumoricidal effects in comparison to each single agent alone, with complete responses and no (L19-IL2) or only a moderate increase (IL12-L19) in systemic toxicity (Borsi et al. 2003; Halin et al. 2003). In addition, combination therapy with Melphalan resulted in a considerably higher synergistic antitumor activity compared with murine TNF or a control fusion protein in combination with Melphalan in three different syngeneic murine tumor models (Borsi et al. 2003). Importantly, treatment of s.c. established tumors with L19-muTNF and Melphalan led to complete responses in case of C51 tumors or an increase in complete response rate up to 83% in case of WEHI-164 tumors, whereas the same treatment could only retard WEHI-164 tumors when growing in immunocompromised mice. Cured mice rejected challenges with syngeneic tumor cells of different histologic origin with long-lasting antitumor immunity (Balza et al. 2006). Further analysis of mice treated with the combination regimen showed (1) that both $CD4^+$ and $CD8^+$ T-cells, but not NK cells, were responsible for tumor rejection and $CD4^+$ cells seemed to act as classical helper cells, (2) a mixed Th1/Th2-type response is induced in which IFNγ seemed to play a central role as combination therapy failed to cure tumor bearing IFNγ−/− mice, (3) each single agent alone as well as both combined led to a down-regulation of $CD4^+$ $CD25^+$ Treg cells in tumor draining lymph nodes, which recently have been shown to potentially inhibit TNF action by shedding off large amounts of TNFR2 (van Mierlo et al. 2008), and (4) spleen cells of cured mice displayed strong CTL activity whereas those of nonresponders did not (Mortara et al. 2007). At present, phase I clinical studies in patients with colorectal cancer are underway with the L19-TNF fusion protein.

The fusion protein scFvMEL/TNF with a molecular mass of about 135 kDa, comprised of human TNF fused to the antibody fragment scFvMEL, is specific for the melanoma antigen gp240, which is present in most human melanomas and some types of breast cancer. On gp240 positive cells, scFvMEL/TNF showed up to more than 250-fold higher apoptotic activity than human TNF. Analysis of organ distribution 72 h after injection revealed highest accumulation in tumor tissue with tumor/blood ratio of about 8 although accumulation in kidney, spleen, and liver was also noted (Liu et al. 2004). Plasma half-live was triphasic with 0.38, 3.9, and 17.6 h for the α, β, and γ phases. The maximal tolerated dose (MTD) upon five daily i.v. injections was 4 mg kg^{-1}. Doses above were accompanied by lethal toxicity. At doses up to 3 mg kg^{-1} no changes in gross pathology, clinical chemistry, or hematologic parameters were observed. Treatment of nude mice bearing established A375GFP tumors with scFvMEL/TNF (five daily i.v. injections, 2.5 mg kg^{-1}) resulted in complete regression in all treated mice, whereas in all control treated groups no regression occurred (Liu et al. 2004). Clinical investigation of this TNF fusion protein is pending.
Bauer and colleagues developed a humanized FAP specific IgG1-TNFderivative in which the constant regions CH2 and CH3 of the IgG were replaced by human TNF. This TNF fusion protein revealed an apparent molecular mass of about 150 kDa indicative of a dimeric assembly, whereas a corresponding fusion protein with a scFv fragment antibody domain assembled into trimers (Bauer et al. 2004). The FAP-specific IgG1-TNF fusion protein showed reduced TNFR1 binding and cytotoxic activity compared to human TNF or the trimeric scFv-TNF variant. Likewise, reduced systemic TNF activity was observed upon i.v. injection into mice. Here up to 150 μg of TNF-equivalent doses were applicable without lethal outcome. In contrast, 30 μg of huTNF or an equivalent dose of the corresponding scFv-TNF fusion protein induced lethality in all mice tested. Furthermore, when used in a TNF-equivalent dose of 100 μg, the FAP-specific IgG1-TNF construct induced a much stronger tumor growth delay of xenotransplanted HT1080-FAP tumors compared to human TNF or the scFv-TNF variant each applied at the corresponding MTD (Bauer et al. 2006).
Recently, two new formats of TNF fusion proteins have been developed, namely a TNF prodrug and single chain TNF (Gerspach et al. 2006a,b, Krippner-Heidenreich et al. 2008). Besides a specific targeting domain and the TNF module, TNF prodrugs additionally are endowed with an inhibitory TNF receptor fragment linked to the core molecule via a protease-sensitive linker. The linker is composed of consensus cleavage sites for matrix metalloproteinases such as MMP-2 or the urokinase-type plasminogen activator (uPA) or combinations thereof. Both proteases are reported to be upregulated in a broad variety of tumors (Egeblad and Werb 2002; Stefanidakis and Koivunen 2006; Pillay et al. 2007). On target antigen negative, but MMP-2 and uPA positive cells, TNF prodrugs remained intact and thus possessed only marginal activity compared to TNF. In contrast, upon binding on target antigen positive cells, TNF prodrugs were efficiently processed by the respective proteases and converted into highly active TNF variants with membrane-TNF-like activity (Gerspach et al. 2006a, β). Thus, by keeping TNF's bioactivity shielded until reaching the site of target antigen expression and encountering

tumor-associated proteases, TNF prodrugs represent a new class of cancer therapeutics with a highly antitumoral potential and double safety features.

The single chain TNF (scTNF) is another novel TNF fusion protein comprised of three human TNF monomers covalently linked via two glycine serine linkers. Cell binding studies revealed similar maximum binding capacities of TNF and scTNF and somewhat higher binding affinities of the scTNF for both TNFR1 and TNFR2. TNF and scTNF showed comparable bioactivities in cell culture experiments, but the latter displayed strongly increased thermal and serum stability upon incubation at 37°C in human serum (Krippner-Heidenreich et al. 2008). Interestingly, systemic toxicity of scTNF in a D-galactosamine-sensitized liver injury model was significantly lower compared to TNF. Antitumor experiments with CFS-1 tumor bearing C3H/HeN mice or nude mice implanted with HCT116 tumor fragments revealed enhanced or at least similar antitumor activity. Thus, by offering a wider therapeutic window compared to TNF, the scTNF bears great potential for development of new TNF-based cancer therapeutics, including improved targeted TNF fusion proteins (Krippner-Heidenreich et al. 2008).

2.2 TNF Gene Therapy Approaches

Development of gene therapy approaches for cancer treatment has received great attention as alternatives to conventional pharmaceutic therapy. Viruses are the most commonly used vectors for clinical gene therapy. They are excellent vectors to deliver foreign DNA with high efficiency. Depending on the viral system, dividing and also non-dividing cells can be transfected and short-term as well as long-term expression can be achieved. However, clinical trials uncovered all the pitfalls and drawbacks, namely acute and delayed immune responses, insertion mutagenesis as well as dissemination of a viral vector into the environment via excreta from treated patients, which all are serious safety concerns. (Seth 2005; Gao et al. 2007b; Schenk-Braat et al. 2007). Alternatively, foreign DNA can be delivered by nonviral physical or chemical approaches. One of the main disadvantages of the latter strategy is that this is in general much less efficient than viral vectors, thereby limiting therapeutic applications.

TNF gene therapy is expected to act locally thereby reducing systemic side effects. However, early studies showed that upon intratumoral injection of an adenoviral TNF construct, TNF protein expressed in the tumor can leak into the vasculature, thereby inducing severe systemic side effects (Okada et al. 2003; Marr et al. 1997). On the other hand, earlier studies already indicated that gene therapy with a TNF mutant exclusively expressed on cell membrane can prevent leakage, yet displayed similar antitumor activities as wild-type TNF (Marr et al. 1997). Nevertheless, for cytokine gene therapy in general and TNF in particular, sustained high-level expression by constitutive vectors seemed not to be a safely applicable strategy. Further investigations that aimed at cytokine treatment in the tumor vicinity with moderate but therapeutically effective doses were more successful. For example in a human glioma xenograft model, TNF gene therapy driven by the stress-inducible promoter of gadd 153 combined with tumor-localized hyperthermia using magnetite cationic liposomes, both injected intratumorally, displayed superior tumoricidal activity compared to single therapy, while liver damage and TNF leakage into circulation were not observed (Ito et al. 2001). In another example, a jet-injected, nonviral TNF vector under the control of the human multidrug resistance gene (mdr1) promoter induced significant tumor growth reduction of established HCT116 tumors when combined with hyperthermia and Doxorubicin (Walther et al. 2007). The mdr1 promoter contains drug-inducible as well as heat-inducible elements, thus Doxorubicin and hyperthermia jointly induce TNF expression, which in turn is known to enhance antitumoral effects of chemo- and hyperthermia (Walther et al. 2000, 2007).

Another promising nonviral based approach uses surface-shielded ligand-polycation/DNA complexes for tumor-targeted gene delivery. Plasmid-DNA coding for murine TNF can be condensed by the polycation polyethylenimine (PEI), known to confer efficient transfection and to facilitate escape of DNA from the endosomal compartment. In experiments described by Kircheis et al., targeted delivery and receptor mediated cellular uptake of TNF encoding DNA/PEI complexes were achieved by transferrin chemically coupled to PEI, making use of frequent overexpression of transferrin receptors on a variety of tumor cells (Kircheis et al. 2002b). The high load of transferrin in these complexes additionally conferred surface shielding in order to overcome unspecific uptake by liver or spleen and to reduce nonspecific interactions with blood components, extracellular matrix, and nontarget cells (Kircheis et al. 2002b). Injection of these complexes resulted in predominant TNF expression in the tumor, no significant increase in TNF serum levels, and no obvious TNF-related systemic toxicities. Moreover, upon repeated i.v. injections significantly higher tumor responses as compared to the untargeted complexes or control complexes containing the empty or β galactosidase encoding vector were observed in neuroblastoma, melanoma, and fibrosarcoma tumor models (Kircheis et al. 2002a). Targeted PEI/DNA polyplexes surface shielded by covalent linkage of polyethylene glycol (PEG) likewise were well tolerated upon repeated systemic application and displayed antitumor activities (Kursa et al. 2003). Thus, with both polyplex variants, it was shown that in animal models it is possible to reach tumor targeted action of TNF upon systemically applied polyplex mediated TNF gene therapy.

Similar results were obtained with TNF gene therapy transcriptionally regulated by telomerase gene promoters with polyplex formulations based on polypropylenimine dendrimer. Repeated i.v. treatment of epidermoid carcinoma-, cervix carcinoma-, or colorectal carcinoma-bearing mice with polyplexed TNF vector was significantly more potent than TNF treatment alone or when dendrimers and TNF were administered subsequently, although the cationic dendrimer carrier by itself exhibited transgene-independent antitumor activity ranging from growth retardation to complete tumor regression (Dufes et al. 2005). Until now clinical trials with general cationic molecule-based systems did not succeed due to low delivery efficiency and toxicity (Gao et al. 2007b). It remains to be seen if surface shielded, transferrin receptor targeted DNA complexes will perform better.

To date, the most successful TNF gene therapy approach appears to be the TNFerade vector, currently being in a pivotal phase II/III clinical investigation. TNFerade is a radiation-inducible radio-sensitizing second generation E1-, partial E3-, and E4-deleted adenoviral vector carrying the transgene encoding human TNF downstream of the radiation- and chemo-inducible Egr-1 promoter. The Egr-1 promoter allows optimized spatial and temporal gene expression following ionizing radiation, which increases intratumoral TNF levels about 12-fold (Rasmussen et al. 2002). Preclinical studies showed that upon direct intratumoral injection of Ad.Egr-TNF and fractionated IR radio-resistant human epidermoid carcinoma, prostate and glioma xenografts were radio-sensitized. Histopathological analysis revealed extensive intratumoral vessel thrombosis and tumor necrosis sparing adjacent normal tissue vessels (Mezhir et al. 2006). Activation of the Egr-1 promoter by ionizing radiation is mediated by DNA-damage- and radical oxygen intermediate (ROI)-induced signaling to the CArG sequence contained within the promoter. Thus, different chemotherapeutics like Doxorubicin and Temozolomide were also able to induce TNF expression, thereby enhancing the tumoricidal effects of these drugs without increasing systemic toxicity. Recently, Yamini and colleagues showed that survival upon treatment of malignant glioma bearing mice with temozolomide and ionizing radiation was significantly enhanced by combination with TNFerade (Yamini et al. 2007). Additionally, for radiation or chemotherapy intolerant patients, resveratrol, a polyphenol with anti-inflammatory, immunomodeling, and chemopreventive properties, might be an alternative agent. Resveratrol caused antitumor responses in Ad.Egr-TNF-infected human or rat tumor xenograft models comparable to that of radio- or chemotherapy (Bickenbach et al. 2008). Taken together, in this approach radio- or chemotherapy displays dual function as it not only acts as anticancer therapeutic by itself, but also induces its own enhancer. Presently, delivery is limited to intratumoral administration and thus constrained to accessible sites of local disease. Studies from MacGill and colleagues, however, provide evidence that local treatment of a solid tumor with TNFerade and radiation may also reduce metastatic disease by a host-dependent response (MacGill et al. 2007). Furthermore, tumor-selective delivery of systemically applied TNF encoding adenoviral vector might be achievable by PEGylation using the enhanced permeability and retention (EPR) effect related to PEG or by capsid modification targeting $\alpha v\beta 3/5$ integrin receptors together with control of expression by the potentially tumor selective MUC-1 promoter (Gao et al. 2007a; Murugesan et al. 2007).

2.3 Alternative Approaches to Increase Efficacy of TNF

Cyt-6091(Aurimune, formerly known as PT-cAu-TNF) is a nanotherapeutic prepared by covalent coupling of TNF to PEGylated colloidal gold. This in turn is a sole of about 30 nm nanoparticles of Au^0 coated with 20 kDa PEG via disulfide bonds. Covalent binding of human TNF by itself is pH-dependent and might occur via its internal cysteine residue. Approximately 400 molecules of TNF are carried by one nanoparticle (Paciotti et al. 2004). Within 6 h upon injection of Cyt-6091 in TNF-sensitive MC38 colon carcinoma bearing C57BL/6 mice TNF accumulation in the tumor and TNF levels in blood were about 10- and 20-fold higher, respectively, than for conventional TNF. In contrast, colloidal gold TNF lacking PEG (cAu-TNF) revealed only about 2.5-fold TNF accumulation and reduced TNF blood levels compared to TNF (Paciotti et al. 2004). These differences are related to a strong uptake in liver and spleen of cAu-TNF, but not Cyt-6091. Nevertheless, compared to native TNF intravenously injected cAu-TNF displayed increased safety and efficacy in tumor therapy experiments, as 24 μg TNF equivalents, which were 100% lethal in case of native TNF, induced significant tumor reduction without causing death. Efficacy of Cyt-6091 was even higher as a dose of 7.5 μg induced maximal tumor response also without mediating lethal side effects (Paciotti et al. 2004). Further analysis showed that Cyt-6091 can significantly increase hyperthermia induced tumor responses by concomitant reduction in tumor perfusion and without increasing toxicity as compared to combination with native TNF (Visaria et al. 2006, 2007). Moreover, video microscopy in TNF-sensitive and -resistant tumor bearing mice revealed that cAu-TNF as well as TNF induced blood vessel permeability, which was rapidly, selectively, and significantly increased in areas of neovasculature compared to those of normal vasculature (Farma et al. 2007). First satisfying data from seven patients of a clinical phase I study in patients with advanced cancer demonstrate, upon systemic administration of 50, 100, or 150 μg m^{-2} no dose limiting hypotensive response, blood chemistries and urinalysis not significantly different following treatment and serum half life between 117 and 145 min. In addition, a tenfold increase in the number of gold nanoparticles in tumor tissue compared to adjacent healthy tissue could be observed (Libutti et al. 2007).

PEGylation of TNF represents a further approach to increase efficacy of TNF in treatment of cancer. PEGylation, the covalent coupling of PEG to therapeutic proteins, has been accepted as a standard technique in industrial settings and has become the dominant protein drug modification/delivery system for the biotech industry. Thus, several commercially produced PEGylated drugs are already available, for example, PEGASYS® (PEG-Interferon alpha-2a) and PEGIntron® (PEG-Interferon alpha-2b) for treatment of chronic hepatitis B and/or C. Modification by PEG can improve therapeutic properties of peptides, proteins, small molecules, or oligonucleotides, generally due to prolonged half-life, higher stability, water solubility, lower immunogenicity, and antigenicity and sometimes facilitating specific targeting to tissues or cells (Pasut and Veronese, 2007). Already about 15 years ago, Mayumi and co-workers demonstrated that by covalent PEG modification of TNF, an up to 100-fold higher antitumor activity can be reached in MethA tumor-bearing mice without observation of systemic toxicity (Tsutsumi et al. 1994, 1995). This initial modification of TNF was done by a randomized process in which all lysine residues of TNF are potential target sites for PEGylation, generating a mixture of differently modified TNF molecules. Later on a lysine-deficient TNF variant was developed thereby enabling N-terminal, site-specific PEGylation. This mono-

PEGylated TNF had about ten times higher antitumor activity in the MethA fibrosarcoma model with about three times less toxicity to mice than wild-type TNF (Yamamoto et al. 2003). Using phage display an improved lysine-deficient TNF mutant with superior in vitro bioactivity and higher affinities for both TNF receptors was selected. This improved lysine deficient TNF mutant displayed about tenfold higher antitumor activity compared with wild-type TNF. After N-terminal PEGylation in vitro, bioactivity was about threefold higher compared to unmodified TNF. Furthermore, induction of tumor responses was about 30-fold higher and systemic toxicity about twofold lower compared to wild-type TNF, thereby increasing the width of therapeutic window about 60-fold (Shibata et al. 2004). Recently, a transferrin-PEG-TNF was generated, which is composed of randomly PEGylated TNF and transferrin coupled via the PEG chains. Treatment of S-180 tumor-bearing mice resulted in about fivefold higher tumor responses compared with wild-type TNF (Jiang et al. 2007). Potentially, higher antitumor activity might be revealed by using lysine-deficient TNF-mutant mono-PEGylated with branched PEG and transferrin coupled to all side chains of PEG.

By developing a chemically coupled anti-CEA/anti-TNF bispecific antibody (BsAb), Azria and colleagues were able to target TNF to the tumor. This antibody construct was shown to increase TNF concentration in the CEA-positive human colorectal carcinoma T380 tumor xenograft model up to eightfold (Robert et al. 1996). In two CEA positive xenograft tumor models (colon cancer, LS174T and pancreatic cancer, BxPC-3), the use of BsAb in combination with TNF and radiotherapy (RT) significantly enhanced median tumor growth delay compared with treatment only with TNF and RT, whereby no body weight loss was observed (Azria et al. 2004). Recently, the authors showed that treatment of CEA transfected mouse tumor cells growing in CEA transgenic immunocompetent mice with a combination of RT plus TNF plus BsAb resulted in complete tumor regression in 50% of the treated animals compared to about 20% for RT with or without TNF (Larbouret et al. 2007).

Increasing efficacy of TNF is also achievable by prevention of systemic toxicity by chemical compounds. For example, tumor-bearing mice systemically treated with TNF/IFNγ in combination with the broad-spectrum metalloproteinase inhibitor BB-94 were protected against TNF/IFNγ-induced mortality, but antitumor activity was preserved (Wielockx et al. 2001). Substitution of BB-94 by specific MMP inhibitors seems not to be possible as judged by analysis in different MMP-deficient mice (Van Roy et al. 2007a). However, adenoviral vector mediated delivery of human tissue inhibitors of matrix metalloproteinase (hTIMP)-1 and hTIMP-2 gene to the liver led to significantly faster regression of tumors upon TNF/IFNγ treatment (Van Roy et al. 2007b). Interestingly by upregulation of HSP70 in different organs, zinc was also able to prevent TNF-induced lethality without affecting antitumor activity (Van Molle et al. 2007).

Several approaches are also aiming at increasing antineoplastic activity of TNF by employing TNF sensitizers like, for example, inhibitors of the NFkB or PKB/Akt pathway or of nitric oxide synthase (reviewed in Mocellin et al. 2007).

3 CD95L (FasL, APO-1L, CD178)

CD95L physiologically plays a critical role in tumor immune surveillance, and its natural antitumor activity is apparent, for example, through tumor specific CD8+ T-cells, which utilize CD95L as one effector molecule to directly kill tumor cells (Wajant 2006; Seki et al. 2002; Dobrzanski et al. 2004). In addition, an interference of CD95L with the development of solid tumors in the liver has been recently described (Park et al. 2008). However, systemic CD95 activation by agonistic antibodies or secondarily crosslinked CD95L results in death of treated animals due to massive hepatotoxicity (Ogasawara et al. 1993; Schneider et al. 1998). Therefore, systemic application of CD95 agonists was not taken into consideration for clinical application, whereas various agonistic CD95L variants are presently exploited for local treatment strategies (see below). Interestingly, mice deleted for both TNF receptors were resistant to death and fulminant liver injury induced by agonistic anti-CD95 antibody (Costelli et al. 2003). Although CD95L can induce tumor regression, it has been proposed that tumors may suppress antitumor responses and thus evade immune mediated destruction by expressing themselves this death ligand on their cell surface. This phenomenon has been termed "tumor counterattack" and is thought to operate via CD95L-mediated apoptosis induction in tumor specific effector T-cells. The tumor counterattack differs from other mechanisms of induced resistance, also known for TNF and TRAIL, where the parallel induction of antiapoptotic signaling pathways, predominantly via NFkB, may lead to a net state of apoptosis resistance, enhanced tumor growth, or even promotion of metastasis (Igney and Krammer, 2005; Whiteside, 2007; Balkwill, 2006). For CD95L, Wada and colleagues provide evidence that the expression level is a critical parameter that determines whether tumor cell expressed CD95L leads to tumor regression or immune escape. Unexpectedly, their data suggest that high level CD95L expression led to tumor rejection, whereas low level expression enhanced tumor growth (Wada et al. 2007). Thus, therapy approaches that aim at delivering relatively high amounts of CD95 agonists in a tumor specific manner should be able to primarily induce tumor regression while sparing normal tissue from toxic action.

3.1 CD95L Fusion Proteins

As mentioned earlier, soluble trimeric CD95L despite showing specific binding to its cognate receptor is poorly able to activate CD95 and might even act as an antagonist for membrane CD95L. Yet apoptotic signaling of soluble trimeric CD95L can be fully reconstituted, for example, upon secondary crosslinking by antibody or membrane-assisted presentation. This latter feature of soluble CD95L is presently exploited for the development of systemically applicable therapeutics with tumor targeted activity. Such fusion proteins comprising CD95L and a tumor specific targeting module have to meet following criteria: (1) fusion of the respective components have to occur without formation of aggregates of latently active ligand trimers, as already hexameric CD95L is able to activate CD95 (Holler et al. 2003); (2) upon membrane binding, the fusion protein should be presented such that a signaling competent ligand complex assembles, mimicking the naturally expressed membrane form of CD95L. While the latter process seems to be less demanding, as it has been reported to occur spontaneously upon interaction of soluble CD95L with extracelluar matrix proteins (Aoki et al. 2001), the generation and maintenance of strictly trimeric, that is, a priori signaling deficient, CD95L fusion proteins appears more challenging. This is based on observations that especially when antibody fragments are used as targeting modules, higher order complexes may occur due to an intrinsic tendency of aggregation (for example, see Watermann et al. 2007). Nevertheless, several fusion proteins have been developed which met both criteria. One of them, sc40-FasL, is composed N-terminally of a human FAP-specific scFv fragment and C-terminally of the extracellular domain (ECD) of CD95L linked via a Flag peptide containing linker. This construct displayed highly increased cytotoxicity on FAP-expressing cells, which results from FAP-dependent cell surface immobilization, leading to conversion into a fully active CD95L moiety. The sc40-FasL prevented outgrowth of FAP-expressing, but not target antigen negative HT1080 tumor xenografts (Samel et al. 2003). Remarkably, systemic application of up to 90 μg of the trimeric fusion protein into mice (corresponding to ~10× LD50 of conventional oligomerized CD95L) resulted in no detectable toxicity, verifying lack of nontargeted activity of this CD95L fusion protein and thus fulfilling the criterion of safe systemic applicability. By genetically coupling CD95L (ECD) to an N-terminally located scFv fragment derived of the CD20 specific chimeric monoclonal Ab Rituximab, Bremer and coworkers generated scFvRit:sFasL (Bremer et al. 2008). Rituximab is an approved biotherapeutic for the treatment of NHL and mediates its effect by antibody induced cell-mediated cytotoxicity (ADCC), complement-mediated cytotoxicity as well as direct induction of apoptosis upon crosslinking of CD20 on B cells. scFvRit:sFasL displayed CD20-dependent dual apoptotic activity via both functional moieties. Of note, this fusion protein displayed superior activity compared to the combined application of reagents targeting CD20 and CD95 individually, that is, Rituximab and an agonistic anti-CD95 antibody. In vitro, strikingly low picomolar concentrations of scFvRit:sFasL were sufficient to induce apoptosis in B-cell lines as well as malignant B-cells from NHL or chronic B-lymphocytic leukemia (B-CLL) patients, yet normal B-cells and peripheral blood lymphocytes remained unaffected. In addition, upon intraocular injection into mice, no obvious systemic toxicity was observed. Thus, scFvRit:sFasL might be a novel potential therapeutic for treatment of B-cell malignancies (Bremer et al. 2008). The scFvCD7:sFasLfusion protein is another example for a CD95L fusion protein with antigen dependent activity, in this case targeting the T cell leukemia-associated antigen CD7. Incubation of CD7 expressing T-cell leukemic cell lines or patient derived material resulted in induction of apoptosis, while CD7 negative cells or activated CD7 positive T-cells were not or moderately affected (Bremer et al. 2006). Similar to the fusion proteins discussed so far, targeting CD95L by other means, as exemplified by use of the ECD of CD40 or RANK, targeting CD40L and RANKL, respectively, likewise resulted in fusion proteins displaying cell surface antigen-restricted activity (Assohou-Luty et al. 2006). Interestingly, a CD40-CD95Lfusion protein was shown to possess dual activity such that the CD40 domain, on the one hand, interferes with CD40L-CD40 signaling and, on the other hand, mediates cell surface presentation of CD95L and reconstitution of its full, thus membrane CD95L-like activity. Similar to the above described scFv-CD95L fusion proteins, the homotrimeric CD40-CD95L fusion protein exerted no apparent toxicity at doses up to 100 μg per mouse upon i.v. injection.

If the target antigen of choice is not strictly tumor specific, that is, strongly overexpressed on tumor cells, but also found on normal tissues, CD95L fusion proteins as described above may not sufficiently meet the criterion of tumor restricted activity. To cope with this situation and to introduce another level of safety for all CD95L fusion proteins independent of the target antigen exploited, CD95L prodrugs have been developed (Watermann et al. 2007). These are structurally related to the TNF prodrugs (see above), making use of the CD95 receptor extracellular domain as an inhibitor of CD95L. Different from the TNF prodrugs, in the CD95L prodrugs the inhibitory domain, thus the ECD of CD95, was N-terminally located followed by a Tenascin C trimerization domain, a MMP-sensitive linker, a FAP targeting scFv fragment, a Flag peptide, and the ECD of CD95L at the C-terminus. Such CD95L prodrugs existed in a hexameric configuration, potentially due to intermolecular ligand–receptor interactions between two trimeric molecules. Despite this hexameric and thus potentially signal competent assembly, the prodrug remained inactive unless processed at the membrane of target antigen expressing, MMP positive tumor cells. In vivo analysis showed that a FAP-specific CD95L prodrug lacking acute systemic toxicity prevented growth of FAP positive, but not FAP negative tumors upon repeated application (Watermann et al. 2007).

Although tumor localized action of CD95 agonists seems to be preferable to prevent systemic toxicity under certain conditions, treatment with highly active nontargeted CD95L variants might be conceivable. The so-called Mega-FasL (MFL) is composed of the ECD of human CD95L fused to the dimer-forming collagen domain of adiponectin (ACRP30), forcing a hexameric structure, that is, two closely coupled CD95L trimers, resulting in a highly active apoptosis inducing agent (Holler et al. 2003; Greaney et al. 2006). MFL efficiently induces apoptosis in various cancer cell lines as well as primary tumor cells with up to 100-fold higher activity compared to secondarily crosslinked soluble CD95L (Holler et al. 2003; Greaney et al. 2006; Etter et al. 2007). Combination with cisplatin synergistically enhanced apoptotic activity (Etter et al. 2007). Moreover, MFL diminished ascites production and peritoneal spread in the orthotopic HOC79 model of human ovarian carcinoma upon repeated intraperitoneous injections. Surprisingly, at a dose of 20 $\mu g\ kg^{-1}$ per injection, reversible liver toxicity was observed, but no severe regional or systemic toxicity (Etter et al. 2007). Thus, MFL displays high potential for intracavitary cancer treatment. Currently a phase I dose-escalation study starting with 2.5 $\mu g\ m^{-2}$ i.v. bolus injections in patients with untreatable advanced or refractory solid tumors is conducted. Another example of a highly active CD95L variant is composed of a Flag-tagged isoleucine zipper motif fused to the N-terminus of the ECD of CD95L, FIZ-shFasL (Shiraishi et al. 2004). On Jurkat cells, FIZ-shFasL was similarly active as, for example, MFL (Holler et al. 2003), but surprisingly, the version without the Flag peptide lost about 90% of activity. Nevertheless, upon systemic injection in rats, FIZ-shFasL displayed dose-dependent toxicity at doses of 0.3 mg kg^{-1} and higher and thus might be more suitable for local cancer treatment (Shiraishi et al. 2004). Of note, this fusion protein contains the stalk region of CD95L, which recently has been shown to have a ligand specific capability to facilitate the formation of secondary aggregated trimers (Berg et al. 2007).

In another approach, a CD95L-Fc fusion protein in a complex with protein A/ palmitate was used to enhance therapeutic efficacy of adoptive T-cell transfer against established tumors. The protein A was chemically derivatized with palmitate and thereby served to anchor the CD95L-Fc on the cell surface by cell membrane integration of the palmitate moiety. Intratumoral injection of these CD95L-Fc/ protein A–palmitate complexes led to elimination of tumor infiltrating regulatory T-cells (Treg), thereby strongly enhancing antitumor responses mediated by adoptively transferred tumor reactive $CD8^+$ T-cells. (Chen et al. 2007). Thus targeted depletion of Tregs by appropriately designed CD95L might confer advantage over systemic depletion of CD4+/CD25+ Tregs, which has been shown to greatly improve adoptive therapy, but might potentiate autoimmune complications (Danese and Rutella 2007).

3.2 *CD95L Gene Therapy*

Intracellular CD95L expression by $AdGFPFasL_{TET}$, an adenoviral vector expressing a modified murine CD95L gene N-terminally fused to the green fluorescent protein (GFP), was able to overcome CD95L-resistance of prostate cancer and head and neck squamous cell carcinoma (HNSCC) cells seen upon treatment with agonistic anti-CD95 Ab (Hyer et al. 2000; ElOjeimy et al. 2006). Thus, at least in these cases effective CD95L gene delivery to tumor cells might be superior to application of protein-based CD95 agonists. Although systemic delivery of a CD95L encoding adenovirus dose-dependently caused death due to massive hepatic apoptosis (Muruve et al. 1997), intratumoral injection resulted in considerable antitumor responses. For example, adenoviral vector encoded murine CD95L induced strong tumor regression in CD95 positive renal epithelial (Renca) as well as CD95 negative colon carcinoma CT26 tumors. While in the first model tumor destruction seemed to be due to a direct apoptosis inducing effect of CD95L, in the second model it was mediated by infiltrating neutrophils and monocytes independent of T-, B-, and NK-cells (Arai et al. 1997). Using a tissue specific promoter (smooth muscle cell-specific SM22α) and an noncleavable version of CD95L (CD95LΔQP), an adenoviral vector was developed, which upon intratumoral injection induced specific regression of smooth muscle derived tumors, but showed no effect on tumors of other origin (Aoki et al. 2000). The replication-incompetent AdGFPFasL, intratumorally injected, likewise led to significantly reduced tumor growth of SCC-14a (HNSCC) xenografts and prolonged survival of mice compared to treatment with AdGFP control vector. No signs of systemic toxicity were recorded (ElOjeimy et al. 2006). Although for adenoviral vectors expressing the CD95L gene under tissue-specific regulation i.v. application was achievable without causing death of experimental animals, therapeutic studies for this route of administration were not performed (Aoki et al. 2000; Rubinchik et al. 2001).

3.3 Agonistic CD95 Specific Antibodies

Since agonistic receptor specific antibodies mimic the action of the respective ligand, they also can represent potential biotherapeutics. In mice it has been shown that agonistic CD95 specific antibodies can exert different, partly puzzling outcomes upon application. While the hamster anti-mouse CD95 monoclonal antibody Jo2 leads to rapid liver damage and death, the hamster anti-mouse CD95 monoclonal antibody RK8 is only moderately hepatotoxic, but nevertheless induces thymocyte apoptosis and thymic atrophy (Ogasawara et al. 1993; Nishimura et al. 1997). Similar to RK8, the cross-species CD95-specific antibody HEF7A also fails to cause hepatic injury (Ichikawa et al. 2000) despite activating CD95 on thymocytes. It seems that the antibodies RK8 and HEF7A discriminate between so-called type-I and -II cells, being independent and dependent, respectively, of mitochondrial contribution to apoptosis, with these two antibodies apparently inducing apoptosis preferentially in type-I cells/tissues. Furthermore, Xu and colleagues provide evidence that cytotoxicity of antibodies can be modulated by different types of Fc receptors independent of classical ADCC with Jo2 and HEF7A depending on FcγRIIB and FcγRI/FcγRIII, respectively (Xu et al. 2003). Recently, HEF7A has been shown to reduce growth of MMN9 melanoma tumor xenografts upon repeated intratumoral injection without any signs of overt hepatotoxicity (Hiramoto et al. 2006). An attempt to target antibody-mediated activation of CD95 was done by the development of bispecific antibodies either in single chain scFv format or as chemically hybridized F(ab′)2 fragments. Two bispecific antibodies targeting CD95 and the neuronal glial antigen-2 or Tenascin C, respectively, were effective in activation of CD95 in an antigen dependent manner (Herrmann et al. 2008). However, these results need to be confirmed in in vivo models.

4 TNF-Related Apoptosis-Inducing Ligand (TRAIL, APO-2L)

The TRAIL/TRAILR pathway is critically involved in antitumorimmune responses in the body (Wajant 2006; Grosse-Wilde et al. 2008; Finnberg et al. 2008; Zerafa et al. 2005). With respect to therapeutic applications, a major advantage of TRAIL in contrast to TNF and CD95L is that it induces apoptosis in various tumor cells, but is apparently nontoxic to most normal cells and tissues (Ashkenazi 2002; Ashkenazi et al. 1999; Kelley and Ashkenazi 2004). The initial controversial debate about toxic actions of TRAIL against normal cells, especially hepatocytes, but also neuronal tissues now appears to be resolved to the point that aggregated TRAIL variants, like His-TRAIL or crosslinked Flag-TRAIL display toxicity towards normal cells, whereas strictly trimeric TRAIL displays tumor selective apoptosis induction (reviewed in Koschny et al. 2007c). Unfortunately, although not harmful to normal cells, trimeric versions of soluble TRAIL also display low activity on many tumor cells (Koschny et al. 2007c). Further, there are cumulative reports that the majority of primary tumor cells may be TRAIL resistant, despite expressing the necessary TRAIL death receptors (Koschny et al. 2007b; Dyer et al. 2007; Todaro et al. 2008). Thus, it is likely that for effective therapy some form of sensitization for TRAIL-induced apoptosis will be required. Concurrently this faces the potential problem that under such combined treatment options normal cells might also be sensitized. Indeed, there are some initial reports showing that combinations of TRAIL with, for example, Cisplatin, HDAC inhibitors or high dose Bortezomib sensitized hepatocytes, lymphocytes, or liver explants to TRAIL-induced cell death, although for combination with several other chemotherapeutics, for example, 5-FU or Gemcitabine toxicity to normal cells was not observed (Koschny et al. 2007a, c; Meurette et al. 2006; Volkmann et al. 2007; Ganten et al. 2006). It has to be mentioned that overall results of preclinical in vivo studies carried out in mice are not easily translated in man as the repertoire of TRAIL receptors differs in both species. Thus, in mice only one death receptor more closely related to human TRAILR2/DR5 than to TRAILR1/DR4 exists. In men, there are three decoy receptors TRAILR3/DcR1, TRAILR4/DcR2, and the soluble receptor osteoprotegerin (OPG), which is the physiologic inhibitor of RANK ligand, another member of the TNF superfamily. Mice possess OPG and the decoy receptors mDcTRAIL-R1/mDcR1 and mDcTRAIL-R2/mDcR2, but DcR1 only binds mouse but not human TRAIL (Yagita et al. 2004). Many preclinical studies evaluating the potential toxicity of TRAIL to normal cells were performed in vitro. It is common knowledge that phenotype and functions of primary tissues may change dramatically with time in culture. In particular, for primary human hepatocytes an extrapolation to in vivo sensitivity might be difficult if not impossible to make. Therefore, available results have to be handled with care (Ganten et al. 2006). Thus, presently, neither the question of an effective TRAIL sensitizing regimen nor the problem of potential dose limiting side effects associated with these combined treatment strategies can be answered to satisfaction and await further results from ongoing preclinical and clinical trials. In addition, the TRAIL reagents presently under clinical evaluation are characterized by a rather short in vivo half-life and thus require high dose or repeated applications to potentially reach effective doses. Altogether, the presently published knowledge about TRAIL's clinical performance points to the need of a further optimization of TRAIL reagents, aiming at an improvement of specific activity and in vivo stability, and options to combine TRAIL function with tumor targeting as pursued for its sister molecules TNF and CD95L.

4.1 Soluble TRAIL (sTRAIL)

Initial preclinical studies have been conducted to demonstrate the antitumoral activity of soluble TRAIL as a monotherapeutic (reviewed in Wajant et al. 2005). More recent preclinical investigations concentrated on examination of combination therapies to identify novel treatment options for TRAIL, with enhanced tumoricidal activity compared to monotherapy. For example, combination of TRAIL with Doxorubicin, Etoposide, Paclitaxel, or Camptothecin synergistically reduced tumor growth of prostate cancer xenografts and combination with Rituximab cooperated to shrink or attenuate growth of various NHL tumor xenografts in SCID mice (Shankar et al. 2005; Daniel et al. 2007). Combined treatment of TRAIL with gemcitabine, the standard therapeutic in pancreas carcinoma, did result in considerably reduced tumor growth of patient-derived pancreatic adenocarcinoma xenografts (Hylander et al. 2005). Interestingly, curcumin, a major constituent of the yellow spice turmeric presently under clinical investigation, can convert TRAIL-resistant pancreas xenografts to yield TRAIL antitumor activity (Shankar et al. 2008). Concurrently, different TRAIL combination therapies were also analyzed in a variety of primary tumor as well as normal cells, which showed increased tumor cell death induction with no toxic effect to most normal cell types (Koschny et al. 2007c). Soluble TRAIL (rhApo2L/TRAIL) is currently tested in phase I/II clinical trials in combination with Camptosar®/Erbitux® chemotherapy in subjects with previously treated, refractory metastatic colorectal cancer, in combination with Rituximab in subjects with low grade NHL and in combination with chemotherapy +/– Bevacizumab in subjects with previously untreated NSCLC. So far, promising intermediate results from six patients intravenously treated with 4 mg kg^{-1} d^{-1} in combination with Rituximab revealed safe application and evidence of activity in subjects with low grade NHL that has relapsed following previous Rituximab-containing therapy (Yee et al. 2007).

4.2 TRAIL Fusion Proteins

As pointed out earlier, the ongoing clinical studies may reveal that TRAIL therapy will require an improvement of tumor selective action due to possible toxicity to normal cells when combined with sensitizing drugs. Therefore, tumor targeting approaches which aim to restrict TRAIL's cytotoxic activity to the cancerous tissue could bear advantages. Independent thereof, a further increase in the specific activity of TRAIL, for example, by targeting approaches and/or engineering functional features, would be favorable. The latter aspect is based on the observation that the specific activity of soluble recombinant TRAIL can be substantially increased by oligomerization, resembling similar features of TNF and CD95L. For example, it has been shown that efficient activation of TRAILR2/DR5 is only reached by natural membrane TRAIL, crosslinked soluble, or cell surface immobilized TRAIL

(Wajant et al. 2001). In analogy to CD95L and TNF, cell surface targeted TRAIL fusion proteins should acquire, upon target binding, a signal capability mimicking membrane TRAIL and thus are potentially superior to the presently available TRAIL reagents. In addition, TRAIL fusion proteins are of higher molecular mass compared to conventional soluble TRAIL and let expect longer half-life (sTRAIL is between 3 and 31 min (Kelley et al. 2001)) and tissue distribution.

MBOS4-TRAIL(95–281), a FAP-specific minibody fused to the N-terminus of soluble TRAIL, displayed about 30 times higher cytotoxicity on FAP expressing vs. antigen negative HT1080 cells. Importantly, already at low concentrations, MBOS4-TRAIL induced FAP-dependent induction of cell death predominantly via TRAILR2, pointing to membrane TRAIL-like activity of the fusion protein upon targeting (Wajant et al. 2001). Helfrich and co-workers generated a variety of similar TRAIL fusion proteins, which were composed of soluble TRAIL fused to single chain scFv antibody fragments with specificity for the pancarcinoma-associated antigen EGP2/EpCAM, human CD7, a T cell surface antigen also found on T-cell leukemia and lymphoma, CD19 expressed on B-cells and on the majority of B-CLL, NHL as well as acute B-lymphoblastic leukemia (B-ALL) cells, and epidermal growth factor receptor (EGFR) overexpressed by many epithelial-derived cancers. The corresponding fusion proteins scFvC54:sTRAIL, scFvCD7:sTRAIL, scFvCD19:sTRAIL, and scFv425:sTRAIL, respectively, have been shown to exist in soluble homotrimeric forms either directly by size exclusion chromatography and/or indirectly by displaying no cytotoxicity to Jurkat cells in which apoptosis can only be induced by TRAILR2/DR5 activation and which therefore are resistant against homotrimeric TRAIL. All fusion proteins displayed target-dependent activity in the absence of crosslinker on Jurkat cells, ectopically expressing the respective antigen as well as against other target positive cells, pointing to target-restricted membrane TRAIL-like activity. Of interest, cell surface immobilized TRAIL variants also significantly induced apoptosis to antigen negative bystander cells (Bremer et al. 2004, 2005a, b; Stieglmaier et al. 2008). More important, scFvCD7:sTRAIL induced apoptosis in blood cells of two of three T-ALL patients, while it failed to kill CD7-positive T- and NK-cells of freshly isolated leukocytes. Similarly, activated T-cells and human umbilical vein endothelial cells (HUVECs), either resting or activated, were not affected by cell surface presented scFvCD7:sTRAIL (Bremer et al. 2005a). The CD19 targeted TRAIL likewise induced CD19-dependent apoptosis induction in most patient-derived mononuclear cells of B-ALL and B-CLL. Valproic acid, a histone deacetylase inhibitor (HDACi), synergistically increased apoptotic activity in B-CLL, even in cells from a patient resistant to incubation with only scFvCD19:sTRAIL. In either case normal blood cells were left unaffected. Furthermore, NOD/SCID mice intravenously injected with tumor cells of a human pre B-ALL cell line and treated on day 3, 5, and 7 with 4 μg of scFvCD19:sTRAIL showed significantly prolonged survival, with about 80% of mice still alive on day 112, while all control-treated mice lived not longer than 56 days (Stieglmaier et al. 2008). In case of the EGFR-specific TRAIL fusion protein scFv425:sTRAIL, the scFv antibody domain not only mediates tumor localized binding and action of TRAIL, but also interferes with EGFR signaling.

Thus, the scFv part of this fusion protein induced rapid dephosphorylation of EGFR accompanied by inhibition of the MAPK and PI3K pathways, decreased expression of the anti-apoptotic caspase 8 homologue cFLIP-L, and decreased phosphorylation of the proapoptotic molecule Bad. In combination with Irissa, an EGFR-tyrosine kinase inhibitor, synergistic apoptosis induction could be observed (Bremer et al. 2005b). By developing an approach for treatment of gliomas, it was shown that the scFv425:sTRAIL producing CHO-K1 cells could be encapsulated in alginate microcapsules and that produced TRAIL fusion protein is able to inhibit outgrowth of A172 spheroids (Kuijlen et al. 2006). Unfortunately, intracerebrally and intratumorally infused scFv425:sTRAIL fusion protein by the convection enhanced delivery method (CED) did not result in tumor response of xenografted SW948 tumors growing in the cerebrum (Kuijlen et al. 2007). As already shown for CD95L fusion proteins, the antibody-based targeting module of TRAIL fusion proteins can be replaced by other specific targeting domains (Assohou-Luty et al. 2006)
Recently, two integrin targeting approaches for TRAIL have been published (Cao et al. 2008; Tarrus et al. 2008). RGD-L-TRAIL, a variant targeting αVβ3/αVβ5 integrins expressed by angiogenic vessels and tumor cells, was generated by fusion of the ACDCRGDCFC peptide to TRAIL. RGD-L-TRAIL showed specific binding and enhanced apoptosis inducing activity on microvascular endothelial cells and αVβ3/αVβ5 integrin-positive cancer cells. Importantly, RGD-L-TRAIL suppressed tumor growth of Colo-205 xenografts more efficiently than TRAIL. Co-treatment with irinotecan hydrochloride (CPT-11) resulted in further increased antitumor activity also in TRAIL-resistant HT-29 tumor xenografts (Cao et al. 2008). In a second report, RGD peptides were coupled to avidin via PEG linkers to mediate specific binding to αVβ3 integrins expressed on HUVECs. His-tagged TRAIL (114–281), already cytotoxic to Jurkat cells, was biotinylated either via lysine or methionine residues, resulting in biotin-TRAIL variants. Upon target binding of RGDPEG-avidin/biotinylated TRAILcomplexes, a significantly increased apoptotic activity in Jurkat cells was observed (Tarrus et al. 2008).

4.3 TRAIL Gene Therapy

In the last few years, preclinical studies assessing TRAIL gene therapy strategies either as monotherapy or as part of a combination therapy have been performed, which all demonstrate principle feasibility of such approaches. For example, combination therapy with an immune stimulating agent (CpG oligodeoxynucleotide) was shown to generate a significant tumor response, while full-length murine TRAIL encoding recombinant adenoviral based therapy alone was only minimally active (VanOosten and Griffith, 2007). To restrict gene expression to tumor tissue, the human telomerase reverse transcriptase (hTERT) promoter is increasingly used for gene therapy approaches, as it is active in most human cancers but quiescent in most somatic cells (Shay and Bacchetti 1997; Wirth et al. 2005). Based on this fact the adenoviral vector Ad/gTRAILhas been constructed encoding a fusion protein of

GFP and full length human TRAIL, which was regulated by an additionally hTERT promoter-controlled transactivator. In vitro, this adenoviral vector induced transgene expression and apoptosis in a variety of human breast, lung, colon, and pancreatic cancer cell lines, while no or minimal toxicity in different normal human primary cells like mammary epithelial cells and fibroblasts was observed. Moreover, repeated intratumoral administration of Ad/gTRAIL effectively suppressed the growth of human tumor xenografts derived from human breast, colon, and pancreatic cancer cell lines. Treatment was well tolerated and liver toxicity was not observed (Lin et al. 2002a, β; Katz et al. 2003). Further modifications of this vector resulted in Ad/TRAIL-F/RGD in which the GFP was omitted and a RGD peptide inserted in the capsid fiber protein. The RGD peptide mediates the use of an alternative receptor during cell entry, allowing efficient transduction in cells resistant to conventional adenoviral vectors (Jee et al. 2002). In tumor xenograft models of NSCLC and esophageal cancer, repeated intratumoral injections in combination with radiotherapy were well tolerated and resulted in significant tumor responses and significantly prolonged survival compared to control treated groups (Jacob et al. 2004, 2005; Zhang et al. 2005; Chang et al. 2006). Importantly, repeated i.v. injections in combination with Gemcitabine treatment similarly were well tolerated and resulted in significant growth inhibition of pancreas tumors growing in the liver (Jacob et al. 2005). When this vector was co-applied with an oncolytic adenovirusinto breast cancer tumors growing subcutaneously in immunocompetent mice, there was significant tumor growth reduction compared to all control groups. In addition, the oncolytic vector led to a strong reduction of spontaneously arising liver metastasis, which could be further reduced by the combination treatment (Guo et al. 2006). Recently, an oncolytic adenovector was constructed carrying the full length human TRAIL and the viral E1A both under control of a modified hTERT promoter. This vector was superior in cell death induction in different cancer cell lines compared to Ad/TRAIL-F/RGD and the oncolytic vector without TRAIL gene. Moreover, established NSCLC xenografts showed complete regression after repeated intratumoral injections and no treatment-associated toxicity was observed (Dong et al. 2006). Although clinical studies with an oncolytic virus revealed replication, but not induction of apoptosis or necrosis in tumor cells, recent improvements in viral oncolytic therapy together with the fact that the majority of adenovectors remain in a small tumor area around the needle track after intratumoral injection point to a feasible option for death ligand armed oncolytic viral vectors for local cancer therapy (Wadler et al. 2003; Davis and Fang, 2005; Jacob et al. 2004). Very recently, an adeno-associated virus (AAV) containing the TRAIL gene under control of the hTERT promoter (AAV-hTERT-TRAIL) was reported, which likewise displayed cancer-specific cytotoxicity and significantly suppressed the growth of xenograft tumors upon intratumoral administration (Wang et al. 2008).

Combining cell and gene therapy, Carlo-Stella and colleagues employed $CD34^+$ cell stransduced with an adenoviral vector encoding full length TRAIL. These cells not only potently induced apoptosis in a variety of tumor cells in vitro, but also led to strong tumor responses, resulting in significantly prolonged survival of mice bearing tumor xenografts. Interestingly, the authors also observed extensive damage

of the tumor vasculature associated with hemorrhagic necrosis induced by these cells (Carlo-Stella et al. 2007).
Radiotherapy together with an in vivo electroporated plasmid carrying the TRAIL gene controlled by the radio-responsive promoter Egr-1has also been shown to significantly reduce tumor growth of glioma xenografts compared to control groups (Tsurushima et al. 2007). There are also gene therapy approaches using exclusively the soluble form of TRAIL. For example, a vector carrying the cDNA for a fusion protein of hFlex(the soluble form of the hematopoietic growth factor Flt3L) and TRAIL(95–281) with a leucine zipper in between revealed serum levels of up to 40 $\mu g\ ml^{-1}$ of the TRAIL fusion protein upon i.v. hydrodynamic-based gene delivery (Wu et al. 2001). Two injections beginning two days after tumor inoculation resulted in significant tumor growth inhibition of MDA-231 human mammary adenocarcinomas compared with control treated groups. In this construct, hFlex not only served for efficient secretion due to its signal peptide, but was also shown to retain immune stimulating function by strongly enhancing expansion of dendritic cells. Although hydrodynamic-based gene delivery is known to lead to high level gene expression, particularly in the liver, no apparent toxicity was observed (Wu et al. 2001). Similarly, Kim and co-workers constructed an adenoviral vector encoding TRAIL(114–281) N-terminally fused to an isoleucine zipper sequence, Ad-stTRAIL. This vector not only efficiently induced apoptosis in several tumor cell lines, but also led to strong tumor regression of glioma xenografts, without apparent toxicity upon intratumoral injection (Kim et al. 2006a).

Although all these preclinical studies point to promising therapeutic approaches, further improvements especially concerning safety will most likely be necessary to enter a clinical evaluation. But as exemplified for TNFerade (see Sect. 1.1.2) it appears possible that some of these approaches will succeed.

4.4 TRAILR Specific Antibodies

In contrast to therapeutics based on the soluble ligand TRAIL and genetic variants thereof, agonistic antibodies with specificity for either TRAILR1/DR4 or TRAILR2/DR5 provide the advantage that their agonistic activity cannot be deviated and neutralized by naturally occurring TRAIL decoy receptors. Receptor specific antibodies enable differentiated receptor stimulation, if required. Thus, it is consistent that in addition to sTRAIL also agonistic TRAIL death receptor-specific antibodies showed promising antitumor responses in mono- as well as combination therapy (reviewed in Koschny et al. 2007c and Buchsbaum et al. 2006; Shanker et al. 2008; Smith et al. 2007). Interestingly, using a mouse TRAILR2/DR5 specific antibody, Uno et al. were able to eradicate pre-established primary tumors even when composed of up to 90% tumor cells as well as pre-established multiorgan metastasis in a triple monoclonal antibody combination therapy together with anti-CD137 and anti-CD40 antibodies (Uno et al. 2006). So far, several agonistic TRAIL receptor specific antibodies are under clinical investigation with some preliminary results already available. For example, the fully human TRAILR1/DR4 specific monoclonal antibody Mapatumumab (TRM-1, HGS-ETR1) revealed safe and feasible i.v. administration of 10 mg kg^{-1} every 14 or 21 days, but no objective single agent activity from phase I/II studies in patients with advanced solid disease and relapsed or refractory NSCLC (Tolcher et al. 2007; Greco et al. 2008). Currently, this reagent is in phase II in combination with bortezomib in subjects with relapsed or refractory multiple myeloma and, in combination with paclitaxel and carboplatin, as first-line therapy in patients with NSCLC. Lexatumumab (HGS-ETR2) is a fully human monoclonal antibody to TRAILR2. This reagent also has been shown to be safe and well tolerated at doses up to 10 mg kg^{-1} every 21 days associated with sustained stable disease in several patients (Plummer et al. 2007). Currently, a phase I study is conducted in treating young patients with solid tumors or lymphoma that have relapsed or not responded to first line treatment. First data of AMG 655, a fully human pro-apoptotic TRAILR2 agonist, from studies in adult patients with advanced solid tumors revealed safe application of up to 20 mg kg every two weeks with 1/16 partial response and 4/16 stable disease reported (LoRusso et al. 2007). AMG 655 now is investigated in combination with gemcitabine for treatment of metastatic pancreatic cancer and with mFOLFOX6 and Bevacizumab for metastatic colorectal cancer (phase I/II). Another fully human anti-TRAILR2 agonistic monoclonal antibody is Apomab. This antibody likewise seems to be well tolerated with a minor response seen in a colorectal cancer patient at this interim time point (Camidge et al. 2007). Apomab is currently in phase I/II studies in patients with advanced chondrosarcoma, in patients with previously untreated, advanced-stage NSCLC, in combination with Rituximab in NHL patients with progressing disease following previous Rituximab therapy, and in combination with Cetuximab and Irinotecan chemotherapy in patients with previously treated metastatic colorectal cancer. Two further TRAILR2-specific antibodies CS-1008 (the humanized version of TRA-8) and LBY135 (a chimeric mouse/human derivative) join these studies in being in phase I/II patients with advanced solid malignancies and lymphoma/untreated and unresectable pancreatic cancer and advanced solid tumors, respectively, either as single agent or in combination with gemcitabine or capecitabine, but early results are not yet available.

Ongoing clinical investigations will show if apparent advantages of, for example, long serum half-life and activating Fc-mediated killing mechanisms might not only increase their antitumor effect compared to sTRAIL, but also enhance their toxicity especially when combined with chemotherapeutic drugs.

5 Conclusions

Over 20 years of clinical experience with TNF and the more recently initiated clinical investigations on CD95L and TRAIL's use as cancer therapeutics have taught us basic rules about the applicability of these multifunctional, pleiotropic cytokines and revealed their great potential as cancer therapeutics, but, foremost, the necessity of restricting action to the tumor area to prevent untolerable, in part life threatening side effects. In this regard, TRAIL stands as a superior candidate; here an improvement in pharmacokinetic properties and signaling potential appears as the prevailing task. Another salient feature already evident from the preclinical and clinical results with the death ligands is the requirement of combination therapy, with reagents targeting intracellular signal pathways regulating apoptosis sensitivity of tumor cells to obtain significant clinical responses. While the clinical exploitation of recombinant proteins, whether based on death ligands or antibodies, stands on firm grounds and will eventually identify suitable therapeutics capable of broad application, the clinical application and success of gene therapeutic approaches, though in part conceptionally very elegant, is less foreseeable despite enormous research efforts in this direction. With new concepts of targeted therapies steadily emerging and a full pipeline of new molecules at a preclinical level waiting to enter clinical evaluation, it will require several years more of clinical studies to reveal which of the various competing strategies and reagents based on targeting death receptors will live up to its expectations.

Acknowledgements Original work of the authors referred to in this review was supported by Deutsche Krebshilfe, grant no. 107551 and 106235 to H.W. and K.P. as well as by a career fellowship award by the Peter and Traudl Engelhorn Foundation to J.G.

References

Ameloot P, Fiers W, De Bleser P, Ware CF, Vandenabeele P, Brouckaert P (2001) Identification of tumor necrosis factor (TNF) amino acids crucial for binding to the murine p75 TNF receptor and construction of receptor-selective mutants. J Biol Chem 276:37426–37430

Aoki K, Akyurek LM, San H, Leung K, Parmacek MS, Nabel EG, Nabel GJ (2000) Restricted expression of an adenoviral vector encoding Fas ligand (CD95L) enhances safety for cancer gene therapy. Mol Ther 1:555–565

Aoki K, Kurooka M, Chen JJ, Petryniak J, Nabel EG, Nabel GJ (2001) Extracellular matrix interacts with soluble CD95L: retention and enhancement of cytotoxicity. Nat Immunol 2:333–337

Arai H, Gordon D, Nabel EG, Nabel GJ (1997) Gene transfer of Fas ligand induces tumor regression in vivo. Proc Natl Acad Sci USA 94:13862–13867

Ashkenazi A (2002) Targeting death and decoy receptors of the tumour-necrosis factor superfamily. Nat Rev Cancer 2:420–430

Ashkenazi A, Pai RC, Fong S, Leung S, Lawrence DA, Marsters SA, Blackie C, Chang L, McMurtrey AE, Hebert A, DeForge L, Koumenis IL, Lewis D, Harris L, Bussiere J, Koeppen H, Shahrokh Z, Schwall RH (1999) Safety and antitumor activity of recombinant soluble Apo2 ligand. J Clin Invest 104:155–162

Aspalter RM, Eibl MM, Wolf HM (2003) Regulation of TCR-mediated T cell activation by TNF-RII. J Leukoc Biol 74:572–582

Aspalter RM, Eibl MM, Wolf HM (2007) Defective T-cell activation caused by impairment of the TNF receptor 2 costimulatory pathway in common variable immunodeficiency. J Allergy Clin Immunol 120:1193–1200

Assohou-Luty C, Gerspach J, Siegmund D, Muller N, Huard B, Tiegs G, Pfizenmaier K, Wajant H (2006) A CD40-CD95L fusion protein interferes with CD40L-induced prosurvival signaling and allows membrane CD40L-restricted activation of CD95. J Mol Med 84:785–797

Azria D, Larbouret C, Garambois V, Gourgou S, Martineau P, Robert B, Dubois JB, Pelegrin A (2004) A bispecific antibody to enhance radiotherapy by tumor necrosis factor-alpha in human CEA-expressing digestive tumors. Int J Radiat Oncol Biol Phys 58:580–588

Balkwill F (2006) TNF-alpha in promotion and progression of cancer. Cancer Metastasis Rev 25:409–416

Balza E, Mortara L, Sassi F, Monteghirfo S, Carnemolla B, Castellani P, Neri D, Accolla RS, Zardi L, Borsi L (2006) Targeted delivery of tumor necrosis factor-alpha to tumor vessels induces a therapeutic T cell-mediated immune response that protects the host against syngeneic tumors of different histologic origin. Clin Cancer Res 12:2575–2582

Bauer S, Adrian N, Williamson B, Panousis C, Fadle N, Smerd J, Fettah I, Scott AM, Pfreundschuh M, Renner C (2004) Targeted bioactivity of membrane-anchored TNF by an antibody-derived TNF fusion protein. J Immunol 172:3930–3939

Bauer S, Adrian N, Fischer E, Kleber S, Stenner F, Wadle A, Fadle N, Zoellner A, Bernhardt R, Knuth A, Old LJ, Renner C (2006) Structure-activity profiles of Ab-derived TNF fusion proteins. J Immunol 177:2423–2430

Berg D, Lehne M, Muller N, Siegmund D, Munkel S, Sebald W, Pfizenmaier K, Wajant H (2007) Enforced covalent trimerization increases the activity of the TNF ligand family members TRAIL and CD95L. Cell Death Differ 14:2021–2034

Bickenbach KA, Veerapong J, Shao MY, Mauceri HJ, Posner MC, Kron SJ, Weichselbaum RR (2008) Resveratrol is an effective inducer of CArG-driven TNF-alpha gene therapy. Cancer Gene Ther 15:133–139

Bodmer JL, Schneider P, Tschopp J (2002) The molecular architecture of the TNF superfamily. Trends Biochem Sci 27:19–26

Borsi L, Balza E, Carnemolla B, Sassi F, Castellani P, Berndt A, Kosmehl H, Biro A, Siri A, Orecchia P, Grassi J, Neri D, Zardi L (2003) Selective targeted delivery of TNFalpha to tumor blood vessels. Blood 102:4384–4392

Bremer E, Kuijlen J, Samplonius D, Walczak H, de Leij L, Helfrich W (2004) Target cell-restricted and -enhanced apoptosis induction by a scFv:sTRAIL fusion protein with specificity for the pancarcinoma-associated antigen EGP2. Int J Cancer 109:281–290

Bremer E, Samplonius DF, Peipp M, van Genne L, Kroesen BJ, Fey GH, Gramatzki M, de Leij LF, Helfrich W (2005a) Target cell-restricted apoptosis induction of acute leukemic T cells by a recombinant tumor necrosis factor-related apoptosis-inducing ligand fusion protein with specificity for human CD7. Cancer Res 65:3380–3388

Bremer E, Samplonius DF, van Genne L, Dijkstra MH, Kroesen BJ, de Leij LF, Helfrich W (2005b) Simultaneous inhibition of epidermal growth factor receptor (EGFR) signaling and enhanced activation of tumor necrosis factor-related apoptosis-inducing ligand (TRAIL) receptor-mediated apoptosis induction by an scFv:sTRAIL fusion protein with specificity for human EGFR. J Biol Chem 280:10025–10033

Bremer E, ten Cate B, Samplonius DF, de Leij LF, Helfrich W (2006) CD7-restricted activation of Fas-mediated apoptosis: a novel therapeutic approach for acute T-cell leukemia. Blood 107:2863–2870

Bremer E, ten Cate B, Samplonius DF, Mueller N, Wajant H, Stel AJ, Chamuleau M, van de Loosdrecht AA, Stieglmaier J, Fey GH, Helfrich W (2008) Superior activity of fusion protein scFvRit:sFasL over cotreatment with rituximab and Fas agonists. Cancer Res 68:597–604

Buchsbaum DJ, Zhou T, Lobuglio AF (2006) TRAIL receptor-targeted therapy. Future Oncol 2:493–508

Calzascia T, Pellegrini M, Hall H, Sabbagh L, Ono N, Elford AR, Mak TW, Ohashi PS (2007) TNF-alpha is critical for antitumor but not antiviral T cell immunity in mice. J Clin Invest 117:3833–3845

Camidge D, Herbst RS, Gordon M, Eckhardt S, Kurzroc R, Durbin B, Ing J, Ling J, Sager J, Mendelson D (2007) A phase I safety and pharmacokinetic study of apomab, a human DR5 agonist antibody, in patients with advanced cancer. J Clin Oncol (ASCO Meeting Abstracts) 25:3582

Cao L, Du P, Jiang SH, Jin GH, Huang QL, Hua ZC (2008) Enhancement of antitumor properties of TRAIL by targeted delivery to the tumor neovasculature. Mol Cancer Ther 7:851–861
Carlo-Stella C, Lavazza C, Locatelli A, Vigano L, Gianni AM, Gianni L (2007) Targeting TRAIL agonistic receptors for cancer therapy. Clin Cancer Res 13:2313–2317
Carswell EA, Old LJ, Kassel RL, Green S, Fiore N, Williamson B (1975) An endotoxin-induced serum factor that causes necrosis of tumors. Proc Natl Acad Sci USA 72:3666–3670
Chandrasekharan UM, Siemionow M, Unsal M, Yang L, Poptic E, Bohn J, Ozer K, Zhou Z, Howe PH, Penn M, DiCorleto PE (2007) Tumor necrosis factor alpha (TNF-alpha) receptor-II is required for TNF-alpha-induced leukocyte-endothelial interaction in vivo. Blood 109:1938–1944
Chang JY, Zhang X, Komaki R, Cheung R, Fang B (2006) Tumor-specific apoptotic gene targeting overcomes radiation resistance in esophageal adenocarcinoma. Int J Radiat Oncol Biol Phys 64:1482–1494
Chen A, Liu S, Park D, Kang Y, Zheng G (2007) Depleting intratumoral CD4+CD25+ regulatory T cells via FasL protein transfer enhances the therapeutic efficacy of adoptive T cell transfer. Cancer Res 67:1291–1298
Costelli P, Aoki P, Zingaro B, Carbo N, Reffo P, Lopez-Soriano FJ, Bonelli G, Argiles JM, Baccino FM (2003) Mice lacking TNFalpha receptors 1 and 2 are resistant to death and fulminant liver injury induced by agonistic anti-Fas antibody. Cell Death Differ 10:997–1004
Crippa L, Gasparri A, Sacchi A, Ferrero E, Curnis F, Corti A (2008) Synergistic damage of tumor vessels with ultra low-dose endothelial-monocyte activating polypeptide-II and neovasculature-targeted tumor necrosis factor-alpha. Cancer Res 68:1154–1161
Curnis F, Sacchi A, Borgna L, Magni F, Gasparri A, Corti A (2000) Enhancement of tumor necrosis factor alpha antitumor immunotherapeutic properties by targeted delivery to aminopeptidase N (CD13). Nat Biotechnol 18:1185–1190
Curnis F, Sacchi A, Corti A (2002) Improving chemotherapeutic drug penetration in tumors by vascular targeting and barrier alteration. J Clin Invest 110:475–482
Danese S, Rutella S (2007) The Janus face of CD4+CD25+ regulatory T cells in cancer and autoimmunity. Curr Med Chem 14:649–666
Daniel D, Yang B, Lawrence DA, Totpal K, Balter I, Lee WP, Gogineni A, Cole MJ, Yee SF, Ross S, Ashkenazi A (2007) Cooperation of the proapoptotic receptor agonist rhApo2L/TRAIL with the CD20 antibody rituximab against non-Hodgkin lymphoma xenografts. Blood 110:4037–4046
Davis JJ, Fang B (2005) Oncolytic virotherapy for cancer treatment: challenges and solutions. J Gene Med 7:1380–1389
Dobrzanski MJ, Reome JB, Hollenbaugh JA, Hylind JC, Dutton RW (2004) Effector cell-derived lymphotoxin alpha and Fas ligand, but not perforin, promote Tc1 and Tc2 effector cell-mediated tumor therapy in established pulmonary metastases. Cancer Res 64:406–414
Dong F, Wang L, Davis JJ, Hu W, Zhang L, Guo W, Teraishi F, Ji L, Fang B (2006) Eliminating established tumor in nu/nu nude mice by a tumor necrosis factor-alpha-related apoptosis-inducing ligand-armed oncolytic adenovirus. Clin Cancer Res 12:5224–5230
Dufes C, Keith WN, Bilsland A, Proutski I, Uchegbu IF, Schatzlein AG (2005) Synthetic anticancer gene medicine exploits intrinsic antitumor activity of cationic vector to cure established tumors. Cancer Res 65:8079–8084
Dyer MJ, MacFarlane M, Cohen GM (2007) Barriers to effective TRAIL-targeted therapy of malignancy. J Clin Oncol 25:4505–4506
Egeblad M, Werb Z (2002) New functions for the matrix metalloproteinases in cancer progression. Nat Rev Cancer 2:161–174
Eggermont AM, de Wilt JH, Ten Hagen TL (2003) Current uses of isolated limb perfusion in the clinic and a model system for new strategies. Lancet Oncol 4:429–437
ElOjeimy S, McKillop JC, El Zawahry AM, Holman DH, Liu X, Schwartz DA, Day TA, Dong JY, Norris JS (2006) FasL gene therapy: a new therapeutic modality for head and neck cancer. Cancer Gene Ther 13:739–745
Etter AL, Bassi I, Germain S, Delaloye JF, Tschopp J, Sordat B, Dupuis M (2007) The combination of chemotherapy and intraperitoneal MegaFas Ligand improves treatment of ovarian carcinoma. Gynecol Oncol 107:14–21

Farma JM, Puhlmann M, Soriano PA, Cox D, Paciotti GF, Tamarkin L, Alexander HR (2007) Direct evidence for rapid and selective induction of tumor neovascular permeability by tumor necrosis factor and a novel derivative, colloidal gold bound tumor necrosis factor. Int J Cancer 120:2474–2480

Finnberg N, Klein-Szanto AJ, El Deiry WS (2008) TRAIL-R deficiency in mice promotes susceptibility to chronic inflammation and tumorigenesis. J Clin Invest 118:111–123

Ganten TM, Koschny R, Sykora J, Schulze-Bergkamen H, Buchler P, Haas TL, Schader MB, Untergasser A, Stremmel W, Walczak H (2006) Preclinical differentiation between apparently safe and potentially hepatotoxic applications of TRAIL either alone or in combination with chemotherapeutic drugs. Clin Cancer Res 12:2640–2646

Gao JQ, Eto Y, Yoshioka Y, Sekiguchi F, Kurachi S, Morishige T, Yao X, Watanabe H, Asavatanabodee R, Sakurai F, Mizuguchi H, Okada Y, Mukai Y, Tsutsumi Y, Mayumi T, Okada N, Nakagawa S (2007a) Effective tumor targeted gene transfer using PEGylated adenovirus vector via systemic administration. J Control Release 122:102–110

Gao X, Kim KS, Liu D (2007b) Nonviral gene delivery: what we know and what is next. AAPS J 9:E92–E104

Gerber DE (2008) Targeted therapies: a new generation of cancer treatments. Am Fam Physician 77:311–319

Gerspach J, Muller D, Munkel S, Selchow O, Nemeth J, Noack M, Petrul H, Menrad A, Wajant H, Pfizenmaier K (2006a) Restoration of membrane TNF-like activity by cell surface targeting and matrix metalloproteinase-mediated processing of a TNF prodrug. Cell Death Differ 13:273–284

Gerspach J, Nemeth J, Munkel S, Wajant H, Pfizenmaier K (2006b) Target-selective activation of a TNF prodrug by urokinase-type plasminogen activator (uPA) mediated proteolytic processing at the cell surface. Cancer Immunol Immunother 55:1590–1600

Greaney P, Nahimana A, Lagopoulos L, Etter AL, Aubry D, Attinger A, Beltraminelli N, Huni B, Bassi I, Sordat B, Demotz S, Dupuis M, Duchosal MA (2006) A Fas agonist induces high levels of apoptosis in haematological malignancies. Leuk Res 30:415–426

Greco FA, Bonomi P, Crawford J, Kelly K, Oh Y, Halpern W, Lo L, Gallant G, Klein J (2008) Phase 2 study of mapatumumab, a fully human agonistic monoclonal antibody which targets and activates the TRAIL receptor-1, in patients with advanced non-small cell lung cancer. Lung Cancer 61(1):82–90

Grell M, Douni E, Wajant H, Löhden M, Clauss M, Maxeiner B, Georgopoulos S, Lesslauer W, Kollias G, Pfizenmaier K, Scheurich P (1995) The transmembrane form of tumor necrosis factor is the prime activating ligand of the 80 kDa tumor necrosis factor receptor. Cell 83(5):793–802

Grosse-Wilde A, Voloshanenko O, Bailey SL, Longton GM, Schaefer U, Csernok AI, Schutz G, Greiner EF, Kemp CJ, Walczak H (2008) TRAIL-R deficiency in mice enhances lymph node metastasis without affecting primary tumor development. J Clin Invest 118:100–110

Guillem EB, Sampsel JW (2006) Antitumor-associated antigens IgGs: dual positive and negative potential effects for cancer therapy. Adv Exp Med Biol 587:341–374

Guo W, Zhu H, Zhang L, Davis J, Teraishi F, Roth JA, Stephens C, Fueyo J, Jiang H, Conrad C, Fang B (2006) Combination effect of oncolytic adenovirotherapy and TRAIL gene therapy in syngeneic murine breast cancer models. Cancer Gene Ther 13:82–90

Halin C, Gafner V, Villani ME, Borsi L, Berndt A, Kosmehl H, Zardi L, Neri D (2003) Synergistic therapeutic effects of a tumor targeting antibody fragment, fused to interleukin 12 and to tumor necrosis factor alpha. Cancer Res 63:3202–3210

Hehlgans T, Pfeffer K (2005) The intriguing biology of the tumour necrosis factor/tumour necrosis factor receptor superfamily: players, rules and the games. Immunology 115:1–20

Herrmann T, Grosse-Hovest L, Otz T, Krammer PH, Rammensee HG, Jung G (2008) Construction of optimized bispecific antibodies for selective activation of the death receptor CD95. Cancer Res 68:1221–1227

Hiramoto K, Inui M, Kamei T, Okumura K, Nakase M, Tagawa T (2006) mHFE7A, a newly identified monoclonal antibody to Fas, induces apoptosis in human melanoma cells in vitro and delays the growth of melanoma xenotransplants. Oncol Rep 15:409–415

Holler N, Tardivel A, Kovacsovics-Bankowski M, Hertig S, Gaide O, Martinon F, Tinel A, Deperthes D, Calderara S, Schulthess T, Engel J, Schneider P, Tschopp J (2003) Two adjacent trimeric Fas ligands are required for Fas signaling and formation of a death-inducing signaling complex. Mol Cell Biol 23:1428–1440

Hyer ML, Voelkel-Johnson C, Rubinchik S, Dong J, Norris JS (2000) Intracellular Fas ligand expression causes Fas-mediated apoptosis in human prostate cancer cells resistant to monoclonal antibody-induced apoptosis. Mol Ther 2:348–358

Hylander BL, Pitoniak R, Penetrante RB, Gibbs JF, Oktay D, Cheng J, Repasky EA (2005) The anti-tumor effect of Apo2L/TRAIL on patient pancreatic adenocarcinomas grown as xenografts in SCID mice. J Transl Med 3:22

Ichikawa K, Yoshida-Kato H, Ohtsuki M, Ohsumi J, Yamaguchi J, Takahashi S, Tani Y, Watanabe M, Shiraishi A, Nishioka K, Yonehara S, Serizawa N (2000) A novel murine anti-human Fas mAb which mitigates lymphadenopathy without hepatotoxicity. Int Immunol 12:555–562

Igney FH, Krammer PH (2005) Tumor counterattack: fact or fiction? Cancer Immunol Immunother 54:1127–1136

Ito A, Shinkai M, Honda H, Kobayashi T (2001) Heat-inducible TNF-alpha gene therapy combined with hyperthermia using magnetic nanoparticles as a novel tumor-targeted therapy. Cancer Gene Ther 8:649–654

Jacob D, Davis J, Zhu H, Zhang L, Teraishi F, Wu S, Marini FC, III, Fang B (2004) Suppressing orthotopic pancreatic tumor growth with a fiber-modified adenovector expressing the TRAIL gene from the human telomerase reverse transcriptase promoter. Clin Cancer Res 10:3535–3541

Jacob D, Davis JJ, Zhang L, Zhu H, Teraishi F, Fang B (2005) Suppression of pancreatic tumor growth in the liver by systemic administration of the TRAIL gene driven by the hTERT promoter. Cancer Gene Ther 12:109–115

Jee YS, Lee SG, Lee JC, Kim MJ, Lee JJ, Kim DY, Park SW, Sung MW, Heo DS (2002) Reduced expression of coxsackievirus and adenovirus receptor (CAR) in tumor tissue compared to normal epithelium in head and neck squamous cell carcinoma patients. Anticancer Res 22:2629–2634

Jiang YY, Liu C, Hong MH, Zhu SJ, Pei YY (2007) Tumor cell targeting of transferrin-PEG-TNF-alpha conjugate via a receptor-mediated delivery system: design, synthesis, and biological evaluation. Bioconjug Chem 18:41–49

Katz MH, Spivack DE, Takimoto S, Fang B, Burton DW, Moossa AR, Hoffman RM, Bouvet M (2003) Gene therapy of pancreatic cancer with green fluorescent protein and tumor necrosis factor-related apoptosis-inducing ligand fusion gene expression driven by a human telomerase reverse transcriptase promoter. Ann Surg Oncol 10:762–772

Kelley SK, Ashkenazi A (2004) Targeting death receptors in cancer with Apo2L/TRAIL. Curr Opin Pharmacol 4:333–339

Kelley SK, Harris LA, Xie D, DeForge L, Totpal K, Bussiere J, Fox JA (2001) Preclinical studies to predict the disposition of Apo2L/tumor necrosis factor-related apoptosis-inducing ligand in humans: characterization of in vivo efficacy, pharmacokinetics, and safety. J Pharmacol Exp Ther 299:31–38

Kim CY, Jeong M, Mushiake H, Kim BM, Kim WB, Ko JP, Kim MH, Kim M, Kim TH, Robbins PD, Billiar TR, Seol DW (2006a) Cancer gene therapy using a novel secretable trimeric TRAIL. Gene Ther 13:330–338

Kim EY, Teh HS (2001) TNF type 2 receptor (p75) lowers the threshold of T cell activation. J Immunol 167:6812–6820

Kim EY, Teh HS (2004) Critical role of TNF receptor type-2 (p75) as a costimulator for IL-2 induction and T cell survival: a functional link to CD28. J Immunol 173:4500–4509

Kim EY, Priatel JJ, Teh SJ, Teh HS (2006b) TNF receptor type 2 (p75) functions as a costimulator for antigen-driven T cell responses in vivo. J Immunol 176:1026–1035

Kircheis R, Ostermann E, Wolschek MF, Lichtenberger C, Magin-Lachmann C, Wightman L, Kursa M, Wagner E (2002a) Tumor-targeted gene delivery of tumor necrosis factor-alpha induces tumor necrosis and tumor regression without systemic toxicity. Cancer Gene Ther 9:673–680

Kircheis R, Wightman L, Kursa M, Ostermann E, Wagner E (2002b) Tumor-targeted gene delivery: an attractive strategy to use highly active effector molecules in cancer treatment. Gene Ther 9:731–735

Koschny R, Ganten TM, Sykora J, Haas TL, Sprick MR, Kolb A, Stremmel W, Walczak H (2007a) TRAIL/bortezomib cotreatment is potentially hepatotoxic but induces cancer-specific apoptosis within a therapeutic window. Hepatology 45:649–658

Koschny R, Holland H, Sykora J, Haas TL, Sprick MR, Ganten TM, Krupp W, Bauer M, Ahnert P, Meixensberger J, Walczak H (2007b) Bortezomib sensitizes primary human astrocytoma cells of WHO grades I to IV for tumor necrosis factor-related apoptosis-inducing ligand-induced apoptosis. Clin Cancer Res 13:3403–3412

Koschny R, Walczak H, Ganten TM (2007c) The promise of TRAIL – potential and risks of a novel anticancer therapy. J Mol Med 85:923–935

Kreuz S, Siegmund D, Rumpf JJ, Samel D, Leverkus M, Janssen O, Hacker G, Dittrich-Breiholz O, Kracht M, Scheurich P, Wajant H (2004) NFkappaB activation by Fas is mediated through FADD, caspase-8, and RIP and is inhibited by FLIP. J Cell Biol 166:369–380

Krippner-Heidenreich A, Grunwald I, Zimmermann G, Kuhnle M, Gerspach J, Sterns T, Shnyder SD, Gill JH, Mannel DN, Pfizenmaier K, Scheurich P (2008) Single-chain TNF, a TNF derivative with enhanced stability and antitumoral activity. J Immunol 180:8176–8183

Kristensen CA, Nozue M, Boucher Y, Jain RK (1996) Reduction of interstitial fluid pressure after TNF-alpha treatment of three human melanoma xenografts. Br J Cancer 74:533–536

Kuijlen JM, de Haan BJ, Helfrich W, de Boer JF, Samplonius D, Mooij JJ, de Vos P (2006) The efficacy of alginate encapsulated CHO-K1 single chain-TRAIL producer cells in the treatment of brain tumors. J Neurooncol 78:31–39

Kuijlen JM, Mooij JJ,Helfrich W,den Dunnen WF (2007) A single chain (scFv425):sTRAIL fusion protein with specificity for the EGF receptor is effective in vitro but not in an in vivo brain tumor animal model. J Neurooncol DOI 10.1007/s11060-006-9322-4

Kursa M, Walker GF, Roessler V, Ogris M, Roedl W, Kircheis R, Wagner E (2003) Novel shielded transferrin-polyethylene glycol-polyethylenimine/DNA complexes for systemic tumor-targeted gene transfer. Bioconjug Chem 14:222–231

Larbouret C, Robert B, Linard C, Teulon I, Gourgou S, Bibeau F, Martineau P, Santoro L, Pouget JP, Pelegrin A, Azria D (2007) Radiocurability by targeting tumor necrosis factor-alpha using a bispecific antibody in carcinoembryonic antigen transgenic mice. Int J Radiat Oncol Biol Phys 69:1231–1237

Lejeune FJ, Lienard D, Matter M, Ruegg C (2006) Efficiency of recombinant human TNF in human cancer therapy. Cancer Immun 6:6

Libutti SK, Paciotti GF, Myer L, Haynes R, Gannon WE, Eugeni M, Seidel G, Shutack Y, Yuldasheva N, Tamarkin L (2007) Preliminary results of a phase I clinical trial of CYT-6091: A pegylated colloidal gold-TNF nanomedicine. J Clin Oncol (ASCO Meeting Abstracts) 25:3603

Lin T, Gu J, Zhang L, Huang X, Stephens LC, Curley SA, Fang B (2002a) Targeted expression of green fluorescent protein/tumor necrosis factor-related apoptosis-inducing ligand fusion protein from human telomerase reverse transcriptase promoter elicits antitumor activity without toxic effects on primary human hepatocytes. Cancer Res 62:3620–3625

Lin T, Huang X, Gu J, Zhang L, Roth JA, Xiong M, Curley SA, Yu Y, Hunt KK, Fang B (2002b) Long-term tumor-free survival from treatment with the GFP-TRAIL fusion gene expressed from the hTERT promoter in breast cancer cells. Oncogene 21:8020–8028

Liu Y, Cheung LH, Marks JW, Rosenblum MG (2004) Recombinant single-chain antibody fusion construct targeting human melanoma cells and containing tumor necrosis factor. Int J Cancer 108:549–557

LoRusso P, Hong D, Heath E, Kurzrock R, Wang D, Hsu M, Goyal L, Wiezorek J, Storgard C, Herbst R (2007) First-in-human study of AMG 655, a pro-apoptotic TRAIL receptor-2 agonist, in adult patients with advanced solid tumors. J Clin Oncol (ASCO Meeting Abstracts) 25:3534

MacGill RS, Davis TA, Macko J, Mauceri HJ, Weichselbaum RR, King CR (2007) Local gene delivery of tumor necrosis factor alpha can impact primary tumor growth and metastases through a host-mediated response. Clin Exp Metastasis 24:521–531
Marr RA, Addison CL, Snider D, Muller WJ, Gauldie J, Graham FL (1997) Tumour immunotherapy using an adenoviral vector expressing a membrane-bound mutant of murine TNF alpha. Gene Ther 4:1181–1188
Menon C, Ghartey A, Canter R, Feldman M, Fraker DL (2006) Tumor necrosis factor-alpha damages tumor blood vessel integrity by targeting VE-cadherin. Ann Surg 244:781–791
Meurette O, Fontaine A, Rebillard A, Le Moigne G, Lamy T, Lagadic-Gossmann D, Dimanche-Boitrel MT (2006) Cytotoxicity of TRAIL/anticancer drug combinations in human normal cells. Ann N Y Acad Sci 1090:209–216
Mezhir JJ, Smith KD, Posner MC, Senzer N, Yamini B, Kufe DW, Weichselbaum RR (2006) Ionizing radiation: a genetic switch for cancer therapy. Cancer Gene Ther 13:1–6
Mocellin S, Pilati P, Nitti D (2007) Towards the development of tumor necrosis factor (TNF) sensitizers: making TNF work against cancer. Curr Pharm Des 13:537–551
Mortara L, Balza E, Sassi F, Castellani P, Carnemolla B, De Lerma BA, Fossati S, Tosi G, Accolla RS, Borsi L (2007) Therapy-induced antitumor vaccination by targeting tumor necrosis factor alpha to tumor vessels in combination with melphalan. Eur J Immunol 37:3381–3392
Murugesan SR, Akiyama M, Einfeld DA, Wickham TJ, King CR (2007) Experimental treatment of ovarian cancers by adenovirus vectors combining receptor targeting and selective expression of tumor necrosis factor. Int J Oncol 31:813–822
Muruve DA, Nicolson AG, Manfro RC, Strom TB, Sukhatme VP, Libermann TA (1997) Adenovirus-mediated expression of Fas ligand induces hepatic apoptosis after Systemic administration and apoptosis of ex vivo-infected pancreatic islet allografts and isografts. Hum Gene Ther 8:955–963
Nawroth P, Handley D, Matsueda G, De Waal R, Gerlach H, Blohm D, Stern D (1988) Tumor necrosis factor/cachectin-induced intravascular fibrin formation in meth A fibrosarcomas. J Exp Med 168:637–647
Nishimura Y, Hirabayashi Y, Matsuzaki Y, Musette P, Ishii A, Nakauchi H, Inoue T, Yonehara S (1997) In vivo analysis of Fas antigen-mediated apoptosis: effects of agonistic anti-mouse Fas mAb on thymus, spleen and liver. Int Immunol 9:307–316
Ogasawara J, Watanabe-Fukunaga R, Adachi M, Matsuzawa A, Kasugai T, Kitamura Y, Itoh N, Suda T, Nagata S (1993) Lethal effect of the anti-Fas antibody in mice. Nature 364:806–809
Okada Y, Okada N, Mizuguchi H, Hayakawa T, Mayumi T, Mizuno N (2003) An investigation of adverse effects caused by the injection of high-dose TNFalpha-expressing adenovirus vector into established murine melanoma. Gene Ther 10:700–705
Paciotti GF, Myer L, Weinreich D, Goia D, Pavel N, McLaughlin RE, Tamarkin L (2004) Colloidal gold: a novel nanoparticle vector for tumor directed drug delivery. Drug Deliv 11:169–183
Park SM, Schickel R, Peter ME (2005) Nonapoptotic functions of FADD-binding death receptors and their signaling molecules. Curr Opin Cell Biol 17:610–616
Park SM, Rajapaksha TW, Zhang M, Sattar HA, Fichera A, Ashton-Rickardt PG, Peter ME (2008) CD95 signaling deficient mice with a wild-type hematopoietic system are prone to hepatic neoplasia. Apoptosis 13:41–51
Pasut G, Veronese FM (2007) Polymer–drug conjugation, recent achievements and general strategies. Prog Polym Sci 32:933–961
Pfeffer K (2003) Biological functions of tumor necrosis factor cytokines and their receptors. Cytokine Growth Factor Rev 14:185–191
Pillay V, Dass CR, Choong PF (2007) The urokinase plasminogen activator receptor as a gene therapy target for cancer. Trends Biotechnol 25:33–39
Plummer R, Attard G, Pacey S, Li L, Razak A, Perrett R, Barrett M, Judson I, Kaye S, Fox NL, Halpern W, Corey A, Calvert H, de Bono J (2007) Phase 1 and pharmacokinetic study of lexatumumab in patients with advanced cancers. Clin Cancer Res 13:6187–6194
Ranges GE, Bombara MP, Aiyer RA, Rice GG, Palladino MA, Jr (1989) Tumor necrosis factor-alpha as a proliferative signal for an IL-2-dependent T cell line: strict species specificity of action. J Immunol 142:1203–1208

Rasmussen H, Rasmussen C, Lempicki M, Durham R, Brough D, King CR, Weichselbaum R (2002) TNFerade Biologic: preclinical toxicology of a novel adenovector with a radiation-inducible promoter, carrying the human tumor necrosis factor alpha gene. Cancer Gene Ther 9:951–957
Robert B, Mach JP, Mani JC, Ychou M, Folli S, Artus JC, Pelegrin A (1996) Cytokine targeting in tumors using a bispecific antibody directed against carcinoembryonic antigen and tumor necrosis factor alpha. Cancer Res 56:4758–4765
Rubinchik S, Wang D, Yu H, Fan F, Luo M, Norris JS, Dong JY (2001) A complex adenovirus vector that delivers FASL-GFP with combined prostate-specific and tetracycline-regulated expression. Mol Ther 4:416–426
Sacchi A, Gasparri A, Curnis F, Bellone M, Corti A (2004) Crucial role for interferon gamma in the synergism between tumor vasculature-targeted tumor necrosis factor alpha (NGR-TNF) and doxorubicin. Cancer Res 64:7150–7155
Samel D, Muller D, Gerspach J, Assohou-Luty C, Sass G, Tiegs G, Pfizenmaier K, Wajant H (2003) Generation of a FasL-based proapoptotic fusion protein devoid of systemic toxicity due to cell-surface antigen-restricted Activation. J Biol Chem 278:32077–32082
Schenk-Braat EA, van Mierlo MM, Wagemaker G, Bangma CH, Kaptein LC (2007) An inventory of shedding data from clinical gene therapy trials. J Gene Med 9:910–921
Schneider P, Holler N, Bodmer JL, Hahne M, Frei K, Fontana A, Tschopp J (1998) Conversion of membrane-bound Fas(CD95) ligand to its soluble form is associated with downregulation of its proapoptotic activity and loss of liver toxicity. J Exp Med 187:1205–1213
Seki N, Brooks AD, Carter CR, Back TC, Parsoneault EM, Smyth MJ, Wiltrout RH, Sayers TJ (2002) Tumor-specific CTL kill murine renal cancer cells using both perforin and Fas ligand-mediated lysis in vitro, but cause tumor regression in vivo in the absence of perforin. J Immunol 168:3484–3492
Seth P (2005) Vector-mediated cancer gene therapy: an overview. Cancer Biol Ther 4:512–517
Seynhaeve AL, Hoving S, Schipper D, Vermeulen CE, Wiel-Ambagtsheer G, van Tiel ST, Eggermont AM, Ten Hagen TL (2007) Tumor necrosis factor alpha mediates homogeneous distribution of liposomes in murine melanoma that contributes to a better tumor response. Cancer Res 67:9455–9462
Shankar S, Chen X, Srivastava RK (2005) Effects of sequential treatments with chemotherapeutic drugs followed by TRAIL on prostate cancer in vitro and in vivo. Prostate 62:165–186
Shankar S, Ganapathy S, Chen Q, Srivastava RK (2008) Curcumin sensitizes TRAIL-resistant xenografts: molecular mechanisms of apoptosis, metastasis and angiogenesis. Mol Cancer 7:16
Shanker A, Brooks AD, Tristan CA, Wine JW, Elliott PJ, Yagita H, Takeda K, Smyth MJ, Murphy WJ, Sayers TJ (2008) Treating metastatic solid tumors with bortezomib and a tumor necrosis factor-related apoptosis-inducing ligand receptor agonist antibody. J Natl Cancer Inst 100:649–662
Shay JW, Bacchetti S (1997) A survey of telomerase activity in human cancer. Eur J Cancer 33:787–791
Shibata H, Yoshioka Y, Ikemizu S, Kobayashi K, Yamamoto Y, Mukai Y, Okamoto T, Taniai M, Kawamura M, Abe Y, Nakagawa S, Hayakawa T, Nagata S, Yamagata Y, Mayumi T, Kamada H, Tsutsumi Y (2004) Functionalization of tumor necrosis factor-alpha using phage display technique and PEGylation improves its antitumor therapeutic window. Clin Cancer Res 10:8293–8300
Shiraishi T, Suzuyama K, Okamoto H, Mineta T, Tabuchi K, Nakayama K, Shimizu Y, Tohma J, Ogihara T, Naba H, Mochizuki H, Nagata S (2004) Increased cytotoxicity of soluble Fas ligand by fusing isoleucine zipper motif. Biochem Biophys Res Commun 322:197–202
Smith MR, Jin F, Joshi I (2007) Bortezomib sensitizes non-Hodgkin's lymphoma cells to apoptosis induced by antibodies to tumor necrosis factor related apoptosis-inducing ligand (TRAIL) receptors TRAIL-R1 and TRAIL-R2. Clin Cancer Res 13:5528s–5534s
Stefanidakis M, Koivunen E (2006) Cell-surface association between matrix metalloproteinases and integrins: role of the complexes in leukocyte migration and cancer progression. Blood 108:1441–1450
Stieglmaier J, Bremer E, Kellner C, Liebig TM, ten Cate B, Peipp M, Schulze-Koops H, Pfeiffer M, Buhring HJ, Greil J, Oduncu F, Emmerich B, Fey GH, Helfrich W (2008) Selective induction

of apoptosis in leukemic B-lymphoid cells by a CD19-specific TRAIL fusion protein. Cancer Immunol Immunother 57:233–246

Tarrus M, van der Sloot AM, Temming K, Lacombe M, Opdam F, Quax WJ, Molema G, Poelstra K, Kok RJ (2008) RGD-avidin-biotin pretargeting to alpha v beta 3 integrin enhances the proapoptotic activity of TNF alpha related apoptosis inducing ligand (TRAIL). Apoptosis 13:225–235

Todaro M, Lombardo Y, Francipane MG, Alea MP, Cammareri P, Iovino F, Di Stefano AB, Di Bernardo C, Agrusa A, Condorelli G, Walczak H, Stassi G (2008) Apoptosis resistance in epithelial tumors is mediated by tumor-cell-derived interleukin-4. Cell Death Differ 15:762–772

Tolcher AW, Mita M, Meropol NJ, von Mehren M, Patnaik A, Padavic K, Hill M, Mays T, McCoy T, Fox NL, Halpern W, Corey A, Cohen RB (2007) Phase I pharmacokinetic and biologic correlative study of mapatumumab, a fully human monoclonal antibody with agonist activity to tumor necrosis factor-related apoptosis-inducing ligand receptor-1. J Clin Oncol 25:1390–1395

Tsurushima H, Yuan X, Dillehay LE, Leong KW (2007) Radioresponsive tumor necrosis factor-related apoptosis-inducing ligand (TRAIL) gene therapy for malignant brain tumors. Cancer Gene Ther 14:706–716

Tsutsumi Y, Kihira T, Tsunoda S, Kubo K, Miyake M, Kanamori T, Nakagawa S, Mayumi T (1994) Intravenous administration of polyethylene glycol-modified tumor necrosis factor-alpha completely regressed solid tumor in Meth-A murine sarcoma model. Jpn J Cancer Res 85:1185–1188

Tsutsumi Y, Kihira T, Tsunoda S, Kanamori T, Nakagawa S, Mayumi T (1995) Molecular design of hybrid tumour necrosis factor alpha with polyethylene glycol increases its anti-tumour potency. Br J Cancer 71:963–968

Uno T, Takeda K, Kojima Y, Yoshizawa H, Akiba H, Mittler RS, Gejyo F, Okumura K, Yagita H, Smyth MJ (2006) Eradication of established tumors in mice by a combination antibody-based therapy. Nat Med 12:693–698

van Mierlo GJ, Scherer HU, Hameetman M, Morgan ME, Flierman R, Huizinga TW, Toes RE (2008) Cutting edge: TNFR-shedding by CD4+CD25+ regulatory T cells inhibits the induction of inflammatory mediators. J Immunol 180:2747–2751

Van Molle W, Van Roy M, Van Bogaert T, Dejager L, Van Lint P, Vanlaere I, Sekikawa K, Kollias G, Libert C (2007) Protection of zinc against tumor necrosis factor induced lethal inflammation depends on heat shock protein 70 and allows safe antitumor therapy. Cancer Res 67:7301–7307

Van Roy M, Van Lint P, Van L, I, Wielockx B, Wilson C, Lopez-Otin C, Shapiro S, Libert C (2007a) Involvement of specific matrix metalloproteinases during tumor necrosis factor/IFNgamma-based cancer therapy in mice. Mol Cancer Ther 6:2563–2571

Van Roy M, Wielockx B, Baker A, Libert C (2007b) The use of tissue inhibitors of matrix metalloproteinases to increase the efficacy of a tumor necrosis factor/interferon gamma antitumor therapy. Cancer Gene Ther 14:372–379

VanOosten RL, Griffith TS (2007) Activation of tumor-specific CD8+ T Cells after intratumoral Ad5-TRAIL/CpG oligodeoxynucleotide combination therapy. Cancer Res 67:11980–11990

Visaria R, Bischof JC, Loren M, Williams B, Ebbini E, Paciotti G, Griffin R (2007) Nanotherapeutics for enhancing thermal therapy of cancer. Int J Hyperthermia 23:501–511

Visaria RK, Griffin RJ, Williams BW, Ebbini ES, Paciotti GF, Song CW, Bischof JC (2006) Enhancement of tumor thermal therapy using gold nanoparticle-assisted tumor necrosis factor-alpha delivery. Mol Cancer Ther 5:1014–1020

Volkmann X, Fischer U, Bahr MJ, Ott M, Lehner F, MacFarlane M, Cohen GM, Manns MP, Schulze-Osthoff K, Bantel H (2007) Increased hepatotoxicity of tumor necrosis factor-related apoptosis-inducing ligand in diseased human liver. Hepatology 46:1498–1508

Wada A, Tada Y, Kawamura K, Takiguchi Y, Tatsumi K, Kuriyama T, Takenouchi T, Wang J, Tagawa M (2007) The effects of FasL on inflammation and tumor survival are dependent on its expression levels. Cancer Gene Ther 14:262–267

Wadler S, Yu B, Tan JY, Kaleya R, Rozenblit A, Makower D, Edelman M, Lane M, Hyjek E, Horwitz M (2003) Persistent replication of the modified chimeric adenovirus ONYX-015 in both tumor and stromal cells from a patient with gall bladder carcinoma implants. Clin Cancer Res 9:33–43

Wajant H (2004) TRAIL and NFkappaB signaling – a complex relationship. Vitam Horm 67:101–132

Wajant H (2006) CD95L/FasL and TRAIL in tumour surveillance and cancer therapy. Cancer Treat Res 130:141–165

Wajant H, Moosmayer D, Wuest T, Bartke T, Gerlach E, Schonherr U, Peters N, Scheurich P, Pfizenmaier K (2001) Differential activation of TRAIL-R1 and -2 by soluble and membrane TRAIL allows selective surface antigen-directed activation of TRAIL-R2 by a soluble TRAIL derivative. Oncogene 20:4101–4106

Wajant H, Pfizenmaier K, Scheurich P (2003) Tumor necrosis factor signaling. Cell Death Differ 10:45–65

Wajant H, Gerspach J, Pfizenmaier K (2005) Tumor therapeutics by design: targeting and activation of death receptors. Cytokine Growth Factor Rev 16:55–76

Walther W, Stein U, Fichtner I, Alexander M, Shoemaker RH, Schlag PM (2000) Mdr1 promoter-driven tumor necrosis factor-alpha expression for a chemotherapy-controllable combined in vivo gene therapy and chemotherapy of tumors. Cancer Gene Ther 7:893–900

Walther W, Arlt F, Fichtner I, Aumann J, Stein U, Schlag PM (2007) Heat-inducible in vivo gene therapy of colon carcinoma by human mdr1 promoter-regulated tumor necrosis factor-alpha expression. Mol Cancer Ther 6:236–243

Wang Y, Huang F, Cai H, Zhong S, Liu X, Tan WS (2008) Potent antitumor effect of TRAIL mediated by a novel adeno-associated viral vector targeting to telomerase activity for human hepatocellular carcinoma. J Gene Med 10:518–526

Watermann I, Gerspach J, Lehne M, Seufert J, Schneider B, Pfizenmaier K, Wajant H (2007) Activation of CD95L fusion protein prodrugs by tumor-associated proteases. Cell Death Differ 14:765–774

Whiteside TL (2007) The role of death receptor ligands in shaping tumor microenvironment. Immunol Invest 36:25–46

Wielockx B, Lannoy K, Shapiro SD, Itoh T, Itohara S, Vandekerckhove J, Libert C (2001) Inhibition of matrix metalloproteinases blocks lethal hepatitis and apoptosis induced by tumor necrosis factor and allows safe antitumor therapy. Nat Med 7:1202–1208

Wirth T, Kuhnel F, Kubicka S (2005) Telomerase-dependent gene therapy. Curr Mol Med 5:243–251

Wu X, He Y, Falo LD, Jr, Hui KM, Huang L (2001) Regression of human mammary adenocarcinoma by systemic administration of a recombinant gene encoding the hFlex-TRAIL fusion protein. Mol Ther 3:368–374

Xu Y, Szalai AJ, Zhou T, Zinn KR, Chaudhuri TR, Li X, Koopman WJ, Kimberly RP (2003) Fc gamma Rs modulate cytotoxicity of anti-Fas antibodies: implications for agonistic antibody-based therapeutics. J Immunol 171:562–568

Yagita H, Takeda K, Hayakawa Y, Smyth MJ, Okumura K (2004) TRAIL and its receptors as targets for cancer therapy. Cancer Sci 95:777–783

Yamamoto Y, Tsutsumi Y, Yoshioka Y, Nishibata T, Kobayashi K, Okamoto T, Mukai Y, Shimizu T, Nakagawa S, Nagata S, Mayumi T (2003) Site-specific PEGylation of a lysine-deficient TNF-alpha with full bioactivity. Nat Biotechnol 21:546–552

Yamini B, Yu X, Pytel P, Galanopoulos N, Rawlani V, Veerapong J, Bickenbach K, Weichselbaum RR (2007) Adenovirally delivered tumor necrosis factor-alpha improves the antiglioma efficacy of concomitant radiation and temozolomide therapy. Clin Cancer Res 13:6217–6223

Yee L, Fanale M, Dimick K, Calvert S, Robins C, Ing J, Ling J, Novotny W, Ashkenazi A, Burris H (2007) A phase IB safety and pharmacokinetic (PK) study of recombinant human Apo2L/TRAIL in combination with rituximab in patients with low-grade non-Hodgkin lymphoma. J Clin Oncol (ASCO Meeting Abstracts) 25:8078

Zerafa N, Westwood JA, Cretney E, Mitchell S, Waring P, Iezzi M, Smyth MJ (2005) Cutting edge: TRAIL deficiency accelerates hematological malignancies. J Immunol 175:5586–5590

Zhang X, Cheung RM, Komaki R, Fang B, Chang JY (2005) Radiotherapy sensitization by tumor-specific TRAIL gene targeting improves survival of mice bearing human non-small cell lung cancer. Clin Cancer Res 11:6657–6668

Zhao X, Mohaupt M, Jiang J, Liu S, Li B, Qin Z (2007) Tumor necrosis factor receptor 2-mediated tumor suppression is nitric oxide dependent and involves angiostasis. Cancer Res 67:4443–4450

Index

D

E

FSC
www.fsc.org
MIX
Papier aus verantwortungsvollen Quellen
Paper from responsible sources
FSC® C105338

If you have any concerns about our products,
you can contact us on
ProductSafety@springernature.com

In case Publisher is established outside the EU,
the EU authorized representative is:
Springer Nature Customer Service Center GmbH
Europaplatz 3, 69115 Heidelberg, Germany

Printed by Libri Plureos GmbH
in Hamburg, Germany